INDUSTRIAL ECOLOGY AND SUSTAINABILITY

INDUSTRIAL ECOLOGY AND SUSTAINABILITY

Thomas E Graedel
Yale University, USA

Matthew J Eckelman
Northeastern University, USA

World Scientific

NEW JERSEY • LONDON • SINGAPORE • BEIJING • SHANGHAI • HONG KONG • TAIPEI • CHENNAI • TOKYO

Published by

World Scientific Publishing Co. Pte. Ltd.
5 Toh Tuck Link, Singapore 596224
USA office: 27 Warren Street, Suite 401-402, Hackensack, NJ 07601
UK office: 57 Shelton Street, Covent Garden, London WC2H 9HE

Library of Congress Cataloging-in-Publication Data
Names: Graedel, T. E., author. | Eckelman, Matthew J., author.
Title: Industrial ecology and sustainability / Thomas E Graedel, Yale University, USA, and
Matthew J Eckelman, Northeastern University, USA.
Description: New Jersey : World Scientific, [2024] | Includes bibliographical references and index.
Identifiers: LCCN 2023023156 | ISBN 9789811277603 (hardcover) |
ISBN 9789811277610 (ebook) | ISBN 9789811277627 (ebook other)
Subjects: LCSH: Industrial ecology. | Sustainable development.
Classification: LCC TS161 .G7424 2024 | DDC 338.9/27--dc23/eng/20230731
LC record available at https://lccn.loc.gov/2023023156

British Library Cataloguing-in-Publication Data
A catalogue record for this book is available from the British Library.

For any available supplementary material, please visit
https://www.worldscientific.com/worldscibooks/10.1142/13447#t=suppl

Desk Editors: Aanand Jayaraman/Amanda Yun

Typeset by Stallion Press
Email: enquiries@stallionpress.com

To Steven, Iris, and Hazel
The Next Generation

Preface

Whether we know it or not, whether we like it or not, technology strongly interacts with almost every facet of our lives. The food we eat, the energy we use, every product we purchase, our vehicles, clothes, and electronics, every building we live and work inside, all of the physical infrastructure that connects us and provides us with essential services, nearly every aspect of modern civilization relies on designs and techniques that humanity has developed and passed down through the generations. Technology in turn has also come to interact with almost every facet of the natural world. The industrial system built up to supply us with technology requires materials, energy, and water resources that are extracted from nature. In turn, the use of these resources often leads to wastes and emissions back to nature. The central theme of this book, and the grand challenge that underpins the fields of industrial ecology and sustainable engineering, is how to design technology and our industrial system so that it can continue to benefit humanity while simultaneously safeguarding or improving the health of natural systems.

Technology today is undergoing a rapid, unprecedented, and accelerating period of evolutionary growth. The overall effect is to undermine many of the approaches underlying current engineering disciplines and policy frameworks. The implications for industrial ecologists and sustainability scientists and engineers are profound as societies attempt to harness these new technologies to address critical global environmental challenges, often without a full understanding of the long-term consequences.

As industrial ecology has developed, ideas that were embryonic a decade ago have matured, and new ways of approaching technology–environment interactions have emerged. Accordingly, we spend significant parts of this book on topics largely or completely unaddressed in earlier textbooks. In particular, the following transitions have struck us as particularly notable: a progression of the field from completely qualitative to increasingly quantitative, an enlarged vision that moves from a focus on products to a consideration of systems, a transition from making use of a biological ecology metaphor to the use of the tools of modern ecology, and a progression from a concern for environmental effects to the challenges of long-term sustainability, explicitly including social dimensions. In the process, industrial ecology and sustainable engineering are becoming increasingly interdisciplinary (as evidenced by the fact that our far-from-exhaustive Further Reading lists cite more than 30 different scholarly journals) and ever more broadly relevant.

We anticipate three audiences for the book. The first is comprised of graduate students in industrial ecology, industrial engineering, civil engineering, and materials science. These fields are advancing rapidly, so previous textbooks are now rather dramatically out of date. In addition, the linkage of environmental, material, and energy sciences is increasingly interconnected so that students and scholars in those areas need an up-to-date treatise on fields with which they may not be familiar.

The second audience is undergraduate students in majors such as environmental studies, sustainable development, and similar fields. To our knowledge, no current textbook provides a comprehensive approach to industrial ecology at the advanced undergraduate level.

A third audience is comprised of scientists and engineers in industries dealing with product design, product manufacture, corporate sustainability departments, and related areas. Such specialists often have no formal training in the topics treated in this book but are expected to make knowledgeable decisions regarding how the activities of their employers interface with increasingly important areas such as materials sustainability, product life cycle assessment (LCA), energy analysis, the circular economy, and related fields.

A feature of the book is the inclusion of worked examples in the text. Industrial ecology utilizes life cycle assessment (LCA), material flow analysis (MFA), matrix analytics, in-use stock derivation, and other mathematical and analytic tools. Most students will be acquainted with one or two of these, but few will have experience with them all. To develop familiarity with the tools, appropriate chapters include worked examples to illustrate their use.

A textbook approach that we believe has real value is the emphasis on the usefulness of a wide variety of visual displays of information, and the discussions of their benefits. As is increasingly the case in science and engineering, and certainly in the areas of industrial ecology and sustainable engineering, the use of computer graphics to illustrate the properties of large data assemblages enables much quicker and more understandable features of data sets. The uses of each graphical type are discussed upon their initial appearance, and where possible, the student exercises exploit the use of computer graphics as a pedagogical tool.

We have attempted to minimize the prerequisites necessary for this book, but we assume reasonable familiarity with introductory college-level biology and chemistry and some knowledge of elementary statistics.

Each chapter opens with three or four concepts central to the topic to be discussed and illustrated. Interested students can pursue any chapter topic in additional depth by making use of the Further Reading list. Exercises help the students test their understanding of the material.

Three appendices provide helpful background information: Units of Measurement, Glossary of Relevant Terms, and Diagrams and Displays Used in Industrial Ecology.

As aids to instructors, we make available worked answers to the student exercises, sets of slides for each chapter, suggested examination questions, and alternative syllabi for students at different levels and of different levels and topics of academic experience.

We are grateful to many people for their assistance during the preparation of this book, especially our many professional colleagues whose work we have attempted to summarize and convey. Our students over the years have both inspired us and provided reactions to topics, teaching approaches, and tools they found helpful. Several anonymous external reviews were also very helpful in determining the appropriate contents and presentations that we have included.

This book is an outgrowth of two previous books involving the senior author of this one: *Industrial Ecology* by Graedel, T. E. and Allenby, B. R. (1995, 2003) and *Industrial Ecology and Sustainable Engineering* by Graedel, T. E. and Allenby, B. R. (2010). We thank Professor Allenby for his insightful contributions to those works and for his work more generally in furthering the knowledge and application of environmental and social issues related to industrial ecology.

Thomas E. Graedel
New Haven, CT, January 30, 2022

Matthew J. Eckelman,
Boston, MA, January 30, 2022

About the Authors

Thomas E. Graedel joined Yale University in 1997 after 27 years at AT&T Bell Laboratories. He is currently a Professor Emeritus and Senior Research Scientist at Yale. One of the founders of the field of industrial ecology, he co-authored the first textbook in that specialty and has published extensively and lectured widely on industrial ecology's implementation and implications. His characterizations of the cycles of industrially used metals have explored the aspects of resource availability, potential environmental impacts, opportunities for recycling and reuse, materials criticality, and resources policy. He was the inaugural President of the International Society for Industrial Ecology from 2002–2004 and winner of the ISIE Society Prize for excellence in industrial ecology research in 2007. He has served three terms on the United Nations International Resource Panel and was elected to the US National Academy of Engineering in 2002.

Matthew J. Eckelman is an Associate Professor of civil and environmental engineering at Northeastern University and an Adjunct Associate Professor at the Yale School of Public Health. His research laboratory builds process-based emissions models and life cycle sustainability assessment tools for a wide range of sectors and environmental technologies. He has served as a Board Member and Treasurer of the International Society for Industrial Ecology and received the Laudise Medal for research in industrial ecology in 2013. For the past 15 years, he has been the CTO of the green chemistry and engineering firm Sustainability A to Z, LLC. He is a member of the Lancet Countdown on Health and Climate Change and the National Academy of Medicine Decarbonization Action Collaborative. Dr Eckelman worked previously for the Massachusetts Executive Office of Environmental Affairs and received a PhD in Chemical and Environmental Engineering from Yale University in 2010.

Contents

Setting the Stage

Chapter 1

Humanity, Technology, and Sustainability

Chapter Concepts

- Common-pool resources are threatened by individual self-serving actions unless some form of community action intervenes to sustain the resources
- Technological innovations are proposed to be the enablers of the waves of economic development and decline
- The "master equation" of industrial ecology proposes that evolving environmental impact is the product of growth in global population, affluence, and technology

1.1 Our Technological World

Planet Earth has never been static, even well before the arrival of humans. Glacial action, weathering, erosion, sediment transport, and other natural processes have moved large amounts of soil and sediment for eons. Since human populations grew and began to change the Earth's surface for their own purposes, however, the relative importance of natural processes compared to those of human actions has dramatically changed. As Figure 1.1 demonstrates, human agricultural land-altering activity exceeded natural processes several millennia ago, but since the Industrial Revolution around the latter part of the 17th century, human action has accelerated rapidly. A dramatic milestone for the human footprint is shown in Figure 1.2: that as of 2020, the amount of anthropogenic (human-made) materials (concrete, bricks, gravel, asphalt, metals, plastics, etc.) had equaled or exceeded all of Earth's biomass and is continuing to increase rapidly. It has become clear that on almost every part of our planet, technologically-enabled human action has

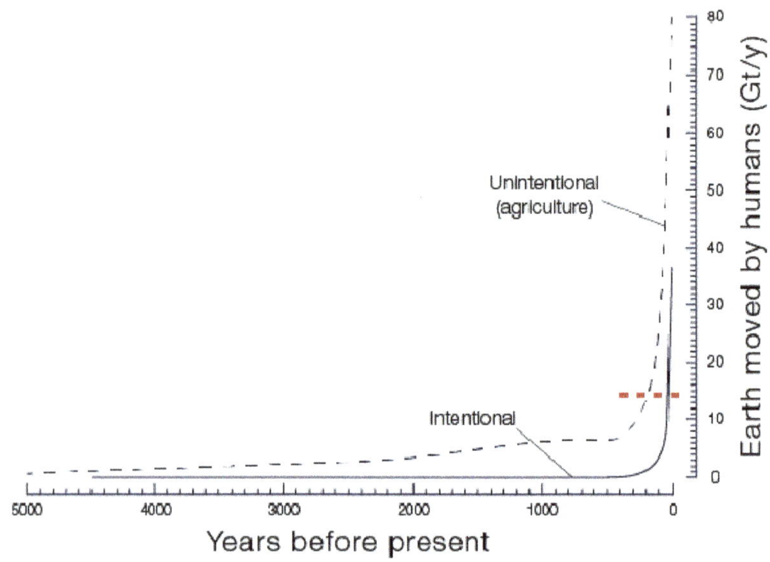

Figure 1.1. Estimates of total amount of soil and rock moved by humans since 5000 years before present (adapted from Hooke, 2000). The red line at lower right is the estimated rate of sedimentation by nature (Syvitski and Kettner, 2011).
Note: This is a *time series graph* that illustrates the values of a specific variable taken at regular time intervals.

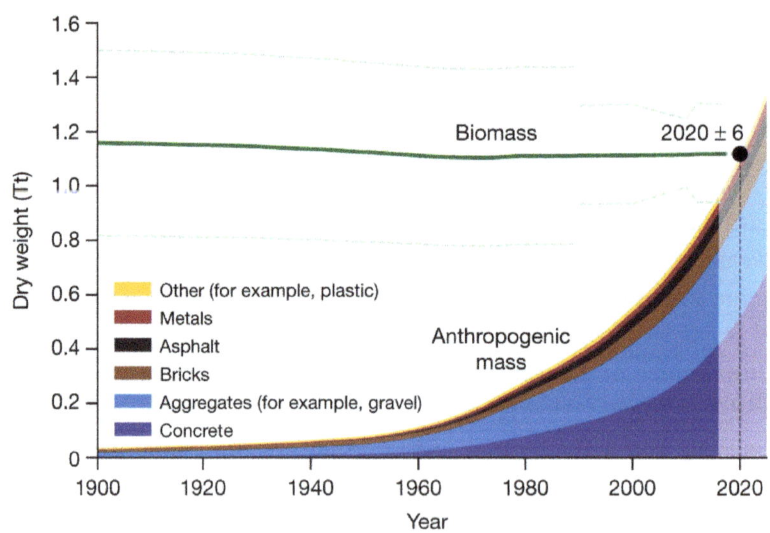

Figure 1.2. Biomass and anthropogenic mass estimates since 1900 (dry-mass basis); the dashed green lines represent one standard deviation in the biomass estimate (Elhacham *et al.*, 2020).
Note: This is an *area chart*.

become dominant over the forces of geology and ecology. Images like those of Figures 1.1 and 1.2 provide much of the impetus for referring to the current time period as the Anthropocene: the epoch of human dominance of Planet Earth. Industrial ecology, the science that quantitates the human use of resources and the related implications for the environment, is at the center of understanding how we have changed the face of the planet and what might be done going forward, at a civilizational scale.

Visual Display of Information in Industrial Ecology

Visual displays of information have become common tools in understanding the implications of data, and in presenting scientific and technological results to others. Industrial ecology is particularly likely to illustrate its processes and results using innovative diagrams and displays.

In this book, we will identify types of information displays the first time they appear. Figure 1.1 provides an example — it is a *time series graph* that illustrates the values of a specific variable taken at regular time intervals.

1.2 The Tragedy of the Commons

Garrett Hardin, a professor at the University of California, Santa Barbara, published an article in 1968 that has remained relevant, perhaps even more so with each passing year. Hardin's principal argument was that a society that permitted perfect freedom of action in activities that adversely influenced common properties was eventually doomed to failure. He cited a community pasture area as an example, used by any local herdsman who wishes to. Each herdsman, seeking to maximize his financial well-being, concludes independently that he should enlarge his herd. By so doing, he derives additional income but is only weakly influenced by the effects of overgrazing, at least in the short term. His colleagues do the same. At some point, however, depending on the size and lushness of the common pasture and the increasing population of animals, the overgrazing destroys the pasture and disaster strikes all.

A modern version of Hardin's "Tragedy of the Commons" was posed by Harvey Brooks of Harvard University, who pointed out that the convenience, privacy, and safety of travel by private automobile encourages each individual to drive to work, school, or stores. At low levels of traffic density, this is a perfectly logical approach to the demands of modern life. At some critical density, however, the road network commons are incapable of dealing with the traffic, and the smallest disruption (i.e., a stalled vehicle or a minor accident) dooms drivers to minutes or hours of idleness, the exact opposite of what they had in mind. Examples of the frequent collapse of road network commons systems are now legendary: traffic congestion in Los Angeles, Tokyo, Naples, Bangkok, or Mexico City.

The common pasture and the common road network are examples of societal systems that are basically local in extent and can be addressed by local societal action if desired. In some cases, the same is true of portions of the *environmental* commons: improper trash disposal or soot emissions from a combustion process are basically local problems, for example. Perturbations to water and air do not follow this pattern, however. The hydrosphere and the atmosphere are examples not of "local commons", but of "global commons", a system that can be altered by individuals the world over for their own gains, but if abused, can injure all. Much of society's functions are embodied in industrial activity (where the word "industrial" should be interpreted broadly to mean any human action involving the transformation of materials or energy), and it is the relationships between industry, the environment, and society, especially the global commons, that are the subject of this book.

When we objectively view the recent past, and 200 years is recent even in terms of human cultural evolution, and certainly in terms of our biological evolution, one fact becomes clear: the Industrial Revolution as we now know it is not sustainable over time. We cannot keep using materials and resources the way we do now, especially as is the case in high-income countries. But what is the alternative?

1.3 Technological Evolution

It is undeniable that modern technology has provided enormous benefits to the world's people: a longer life span, increased mobility, decreased manual labor, and widespread literacy, to name a few. Nonetheless, there are growing concerns about the relationships between industrial activity and Earth's environment, nowhere better captured than in the pathbreaking report, *Our Common Future*, produced by the World Commission on Environment and Development in 1987. The concerns raised in that report gather credence as we place some of the impacts in perspective. Since 1700, the volume of goods traded internationally has increased some 800 times. In the last 100 years, the world's industrial production has increased more than 100-fold in physical terms. In the early 1900s, production of synthetic organic chemicals was minimal; today, it is over 225 petagrams (billion metric tons) per year in the U.S. alone (see the Appendix on Units of Measurement in Industrial Ecology for the conversion between the metric and imperial unit systems). Since 1900, the rate of global consumption of fossil fuels has increased by a factor of 50. What is important is not just the numbers themselves, but their magnitude and the relatively short historical time they represent.

Together with these obvious pressures on the Earth system, several underlying trends deserve attention. The first is the diminution of regional and global capacities to deal with anthropogenic emissions. For example, carbon dioxide production associated with human economic activity has grown dramatically, largely because of extremely rapid growth in energy consumption. This pattern is in keeping with the evolution of the human economy to a more complex state, the increasing growth in materials use and consumption, and an increased use of capital. The societal evolution has been accompanied by a shift in the form of energy consumed, which is increasingly electrical (secondary) as opposed to biomass or direct fossil fuel use (primary), the result being the now familiar exponential increase in atmospheric carbon dioxide that has occurred since the beginning of the Industrial Revolution. This trend is evidence that human activities are rapidly compromising the ability of the atmosphere to act as an effective sink for the by-products of our economic practices.

Human population growth is, of course, a major factor fueling this explosive industrial growth and expanded use and consumption of materials. Since 1700, human population has grown 10-fold. While this growth is generally recognized, it is less widely appreciated how closely human population growth patterns are tied to technological and cultural evolution. Indeed, the three great jumps in human population have accompanied the initial development of tool use, the agricultural revolution, and the Industrial Revolution. The Industrial Revolution actually consisted of both a technological revolution and a "neo-agricultural" revolution (the advent of modern agricultural practices), which created what appeared to be unlimited resources for population growth. Our current population levels, patterns of urbanization, economies, and cultures are now inextricably linked to how we use, process, dispose of, and recover or recycle natural and synthetic materials and energy, and the innumerable products made from them.

Technological evolution generally proceeds in one of two ways. Most of the time technological evolution is incremental, marked by small improvements or changes in existing products or systems that, taken together, improve the quality of life but do not significantly change economic, cultural, or natural systems. In some periods, however, so-called "transformative technologies" change the technological landscape so profoundly that change in the related systems is significant and often difficult. Indeed, economists have identified stages in economic development that can be associated with particular enabling technologies. For example, the introduction of the railroad, the automobile, and electricity changed not just economic and related technological systems, but also culture, national competitiveness, political systems, and most people's way of life at the individual level. It is indeed accurate to say that the railroad was a necessary, and enabling, technology for the rise of Britain as a world economic and political power. It also necessitated other technologies, such as national communication systems using Morse Code, and required accurate timekeeping, thus changing the way time was perceived and measured around the world. Among the other effects of extensive railroad infrastructure was helping to make the American Midwest a viable agricultural enterprise, feeding products to the American East Coast and from there to global markets (and leading to the development of Chicago in the process). That a single technology so changed the structure and use of vast areas of land and affected the economic well-being and personal lives of so many helps make clear the relationship between sustainability and technology.

Technological change is thus a major part of the context within which sustainable engineering and industrial ecology studies are conducted. At smaller scales, it may be adequate to simply explicate and revalidate assumptions about technology that underlie the methodology and particular project. At larger scales, however, technological change in itself must be part of industrial ecology studies; indeed, the transdisciplinary nature of industrial ecology is a critical framework for such studies.

In the early 20th century, Nikolai D. Kondratieff, a Russian economist, introduced the concept of long-range economic cycles, depicted in Figure 1.3(a). He described five of them as follows:

- The first cycle occurred from 1780 to 1830 and resulted principally from the invention of the steam engine and the growth of textile manufacturing.
- The second cycle lasted from 1830 to 1880 and was triggered by the genesis of the steel industry, coinciding with the invention of the Bessemer converter. Increasing rail transport supported mass transportation of both people and cargo, which led to rapid economic growth.
- The third cycle lasted from 1880 to 1930. It is notable for being the first wave to be enabled by the practical application of scientific knowledge. It occurred during the rise of electrical power, which was also the time period during which innovations in the chemical industry allowed for the mass production of commodities.
- The fourth cycle occurred from 1930 to 1970, fueled by the growth of the petrochemical industry. The petrochemical industry's growth also supported the growth of the auto market. The cycle ended when the Organization of Petroleum

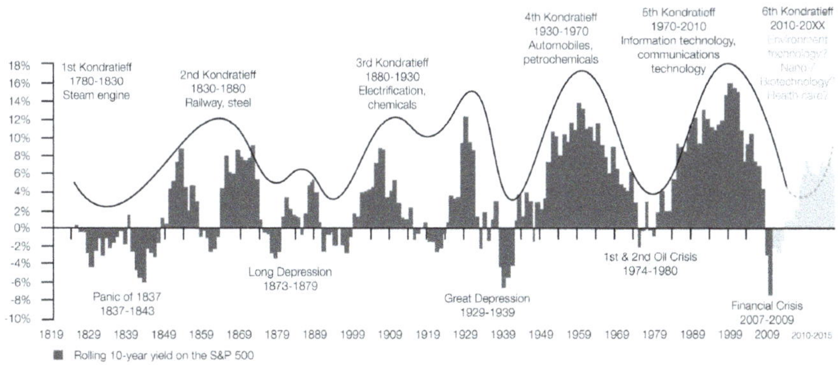

(a)

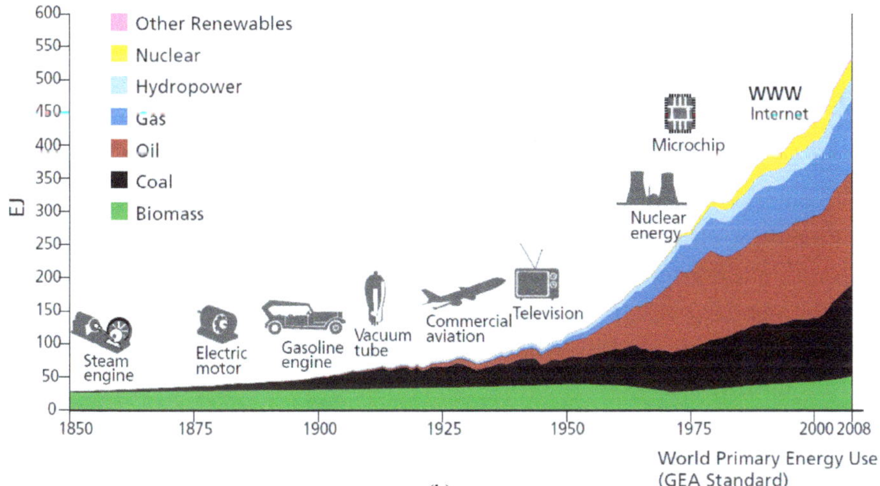

(b)

Figure 1.3. (a) Depiction of Kondratieff cycles, adapted from "The Next Wave", 2015. (b) Global energy consumption (in exajoules, or EJ) over roughly the period covered by the Kondratieff economic cycles (Global Energy Assessment, 2012).

Notes: (a) This is a bar chart with a speculative wave pattern added; (b) This is a line chart — a series of data points connected by line segments.

Exporting Companies (OPEC) increased the price of crude oil in the 1970s, triggering a recession.

- The fifth cycle started in the 1970s, and it was triggered by the advent of computer-based information technology. The industrial society began transitioning to an information society, which effectively turned the world into a global village as the information technology sector became the main driver of economic growth. This cycle is regarded as having begun to close around the end of the 20th century.

It has now been proposed that we are in a sixth Kondratieff wave that began around 2005, and that economic growth in this cycle will be fueled by advances in human-centered technology, perhaps with healthcare as a central focus.

An important consideration with regard to the economic cycles is their relationship to technology-enabled energy use, as shown in Figure 1.3(b). Over the 150 years of this figure, global energy consumption increased by approximately 20 times, with major transitions associated with technological breakthroughs. If there is to be a sustained sixth Kondratieff wave, it will likely be enabled by large amounts of energy, and that energy will presumably need to be renewable — an argument for energy sustainability that we discuss later in Chapter 22.

1.4 The Master Equation

A useful way to focus thinking on the most efficient response that society can make to environmental stresses is to examine the predominant factors involved in generating those stresses. Clearly, the stresses on many aspects of the Earth system are strongly influenced by the needs of the population that must be provided for, and by the standard of living that population desires. A famous expression of these driving forces is provided by what is sometimes referred to as the "master equation" of industrial ecology:

$$\text{Environmental Impact} = \text{Population} \times [\text{GDP / Person}] \times [(\text{Environmental Impact})/\{\text{unit of GDP}\}] \quad (1.1)$$

where GDP is a country's Gross Domestic Product, a measure of industrial and economic activity. This equation has traditionally been called the IPAT Equation, where I = impact, P = population, A = affluence (GDP/person), and T = technology (impact/unit of GDP). Let us examine the three terms in this equation and their probable change with time.

Earth's population has risen dramatically over the past several centuries. Thanks to the work of demographers and census-takers, the current population is well-known, and future populations in each region can be projected based on age structures, birth rates, and death rates. While the rate of population increase has slowed overall, some regions are still growing quickly; the global population now approaches eight billion and is anticipated to peak at nine billion or thereabouts during the 21st century.

The second term in Eq. (1.1), per capita gross domestic product, varies substantially among different countries and regions, responding to the forces of local and global economic conditions, the stage of historical and technological development, governmental factors, weather, and so forth. The general trend, however, is positive, reflecting the aspirations of humans for a better life. Although GDP and quality of life may not be fully connected, we can expect per capita GDP growth to continue, particularly in developing countries.

The third term in the master equation, the degree of environmental impact per unit of gross domestic product, is an expression of the degree to which technology is available to permit development without serious environmental consequences and the degree to which that available technology is deployed.

Although the master equation should be viewed as conceptual rather than mathematically rigorous, it can be used to suggest goals for technology and society. If our aim is to constrain the environmental impact of humanity to its present level (and one could make arguments that we need to do even better than that), we need to look at the probable trends in the three terms of the equation. The first, as discussed above, will likely increase by 20% to 40% before peaking in this century. The second term is thought likely to increase over the same time period by a factor of between three and five. Accordingly, to merely hold our environmental impact where it is today, the third term must decrease by something between 50% and 90%. This third term is the main focus of the fields of industrial ecology and sustainable engineering.

Of the trends for the three terms of the master equation, the one which has the greatest degree of support for its continuation is the second, the gradual improvement of the human standard of living, defined in the broadest of terms. The first term, population growth, is not primarily a technological issue but a social issue. Although countries and cultures approach the issue differently, population is clearly on a strong upward trend. The third term, the amount of environmental impact per unit of output, is primarily a technological term, though societal and economic issues provide strong constraints to changing the term rapidly and dramatically. It is this third term in the equation that appears to offer the greatest hope for a transition to sustainable development, especially in the next few decades.

1.5 Facing the Challenge

The 20th century was a period of enormous progress, progress that was achieved in part by ignoring the possible consequences. The conjunction of inadequately thought-out technological approaches with rapidly rising populations and an increasing culture of consumption is now producing stresses that are increasingly obvious and call out for change: a new perception of our industrialized society.

There are roles for many players in addressing the need to transform the technology–society–environment relationship. Social scientists need to understand consumption and how it may evolve and be modified. Environmental scientists and materials specialists need to understand the limits imposed by a planet with limited resources and limited assimilative capacity for industrial emissions. Technologists need to develop design and manufacturing approaches that are more environmentally sound. Industrialists need to understand all these frameworks for action and develop ways to integrate the concepts within today's corporate structures. Policymakers need to provide the proper mix of regulations and incentives to promote the long-term health of the planet rather than short-term fixes.

These are great challenges. They are the ones to which this book is addressed. We cannot treat all of them in detail, nor are all of them sufficiently developed to permit doing so even if we wished. Nonetheless, we can see many approaches that will take us in the right direction. It is time to get started.

Further Reading

Chertow, M. (2001). The IPAT equation and its variants: Changing views of technology and environmental impacts. *Journal of Industrial Ecology*, *4*(4), 13–29.

Elhacham, E., Ben-Uri, L., Grozovski, J., Bar-On, Y. M., and Milo, R. (2020). Global human-made mass exceeds all living biomass. *Nature*, *588*, 442–444.

Frischmann, B., Marciano, A., and Ramello, G. B. (2019). Tragedy of the commons after 50 years. *Journal of Economic Perspectives*, *23*(4), 221–228.

Global Energy Assessment. (2012). Laxenburg, Austria: International Institute for Applied Systems Analysis (IIASA). https://www.cambridge.org/us/academic/subjects/earth-and-environmental-science/environmental-policy-economics-and-law/global-energy-assessment-toward-sustainable-future?format=PB&isbn=9780521182935.

Hardin, G. (1968). The tragedy of the commons. *Science*, *162*, 1243–1248.

Hooke, R. L. (2000). On the history of humans as geomorphic agents. *Geology*, *28*(9), 843–846.

Janssen, M. A., Smith-Heiskers, S., Aggarwal, A. R., and Schoon, M. L. (2019). 'Tragedy of the commons' as conventional wisdom in sustainability education. *Environmental Education Research*, *25*(11), 1587–1604.

Syvitski, J. P. M. and Kettner, A. (2011). Sediment flux and the Anthropocene. *Philosophical Transactions of the Royal Society A*, *369*, 957–975.

The Next Wave. (2015). Retrieved from: https://deconstructingrisk.com/2015/10/25/the-next-wave/.

Chapter 2

The Linked Systems of Human Society and Industrial Ecology

Chapter Concepts

- The material footprints of modern societies accelerated rapidly in the mid-20th century, and show no signs of levelling off
- Consumption is ultimately the product of individual decisions, and sustainable levels of consumption at the societal level can only be achieved by influencing individual choice
- Attempts to measure social progress in different countries suggest that material stocks and energy use are not directly related to perceived quality of life, thereby suggesting that a less material-oriented world might become one that is accepted and valued by most people worldwide

2.1 Historical Linkages of Humans and Resources

2.1.1 *Social ecology*

Incorporating social and cultural relationships with issues of materials, energy, and environment is the focus of the scholarly practice termed *socioeconomic metabolism* (SEM). SEM is based on the concept that social systems reproduce themselves not only culturally but also biophysically (i.e., through agricultural practices, livestock ownership, built infrastructure, and so on). The embodiments of societies large and small can be quantified by SEM at least in part in terms of the flows of materials and energy, and each sociometabolic regime (community, region, or nation) possesses a unique metabolic profile. The higher the metabolic rate, the more resources per inhabitant are needed, the more outflows of wastes and emissions are produced, and the higher the environmental impact. SEM enlarges traditional definitions of metabolism to treat human communities (often, but not necessarily, countries) and their material and energy flows as constituting a social metabolism. Thus, just as the metabolic analysis of a bird can be expanded to include the nest it builds, so can that of a society address aqueducts and railroads.

Throughout early human existence, a hunter/gatherer lifestyle (i.e., "living off the land") was the workable approach to life and society. Somewhat later, agrarian lifestyles became the norm. Those who live in industrialized countries that are technologically advanced may not always realize that a hunter/gatherer or agrarian lifestyle remains the case for much of the Earth's people, and remote regions still retain societal groups who continue to live as hunter/gatherers. It is anticipated, however, that the next few decades will witness an unprecedented level of hunter/gatherer to agrarian to industrial transition. What might be the implications of this process?

Some clues are given in Figure 2.1, in which the energy and material properties of hunter/gatherer, agrarian, and industrial lifestyles are contrasted. It is striking that the material use of industrial societies is some five times that of

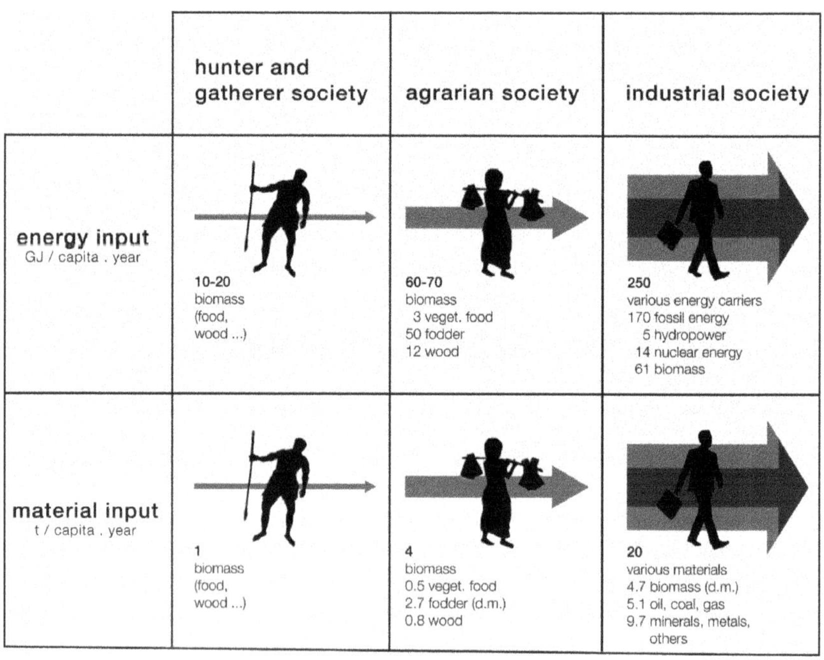

Figure 2.1. Typical resource consumption by the three traditional sociometabolic regimes (adapted from Fischer-Kowalski and Haberl, 1998).

agrarians and 20 times that of hunter/gatherers. Similar multipliers apply to energy use.

The source of energy also underwent a marked transition over time — from an almost total dependence on biomass in the agrarian regions to a strong emphasis on fossil fuels in the industrial era to the advent of renewable and transportable energy-on-demand in the near future. Many opportunities arise as a consequence of this evolution. One is that fossil fuel energy permits mechanized farming, which frees much of the labor force for other employment. Another is that the increased population density enables public transportation, better schooling, and improved healthcare. However, liabilities come along with the transition as well: a changed social structure and the spatial concentration of pollution sources, to name only two.

Although the sociometabolic regimes of Figure 2.1 have been widely accepted for many years, it has been argued that hunter/gatherers made (and consistently make) frequent use of fire to manage their proximate ecosystems. The biomass energy in those fires has not traditionally been taken into account in sociometabolic metabolism studies. However, Rambo (2015) argues that such energy should be incorporated into those analyses "because it helps to explain the otherwise anomalous finding that the quality of life of hunters and gatherers was generally much better than that of peasant farmers in traditional agrarian states." Whatever the resolution of this scholarly debate, traditional societies have harnessed non-food energy to at least some degree, and that energy use has expanded very sharply in recent decades.

The metabolic structure of societies can be studied not only at degrees of consumption of resources but also at various spatial levels. At the local level, studies address villages and small islands; such studies tend to focus on societies in the midst of the agrarian–industrial transition. At the country level, societal use of resources and energy can be compared among countries differing in their locations along the transitional path. Finally, in cities, the comprehensive metabolism of modern industrial societies can be examined in detail.

2.1.2 *The pyramid of needs*

All humans have innate predispositions that favor specific actions. The classic interpretation called Maslow's "hierarchy of needs" is that human needs form a sequence: first, physiological needs such as food and shelter; next, safety; next, societal relationships; and finally, self-actualization (Figure 2.2). Potential desires thus form a hierarchy in which some desires generally have precedence over others, with societal needs generally being favored at times when immediate needs have been satisfied.

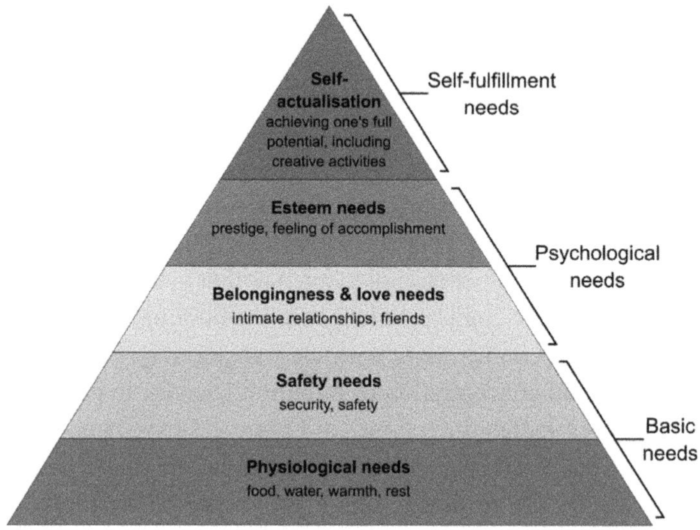

Figure 2.2. Maslow's hierarchy of needs.

From the perspective of industrial ecology, it is interesting to examine the hierarchy of needs from the standpoint of the physical resources needed to satisfy them. Physiological needs require food and water, which in all but hunter/gatherer communities can be met only by systems that acquire those resources and then transport them to the consumers. Self-protection is enabled through defensive structures of various types, i.e., fortresses, homes, etc. The more sociological desires (i.e., affiliation, status, mate acquisition, etc.) favor more communal resources such as mechanisms of communication, transport, communal service systems, and the like.

Modern humans use far more resources per capita than was the case historically. Earth's human population is increasing, so it would be reasonable to expect that global resource use would be increasing as well. And, because many features of resource use are related to population and income, one would expect specific metrics such as fertilizer consumption and energy use to be increasing. Additionally, if environmental impacts are a consequence of such actions, those impacts probably have also increased. But by how much?

This question was addressed by a group of researchers under the auspices of the International Geosphere-Biosphere Program (IGBP). The results for a dozen globally aggregated socioeconomic indicators are shown in Figure 2.3. The data, extending where possible back to 1750 at the beginning of the Industrial Revolution, communicate their message in a quick glance: Not only are all the indicators increasing over time but starting in about 1950, they also began to rise

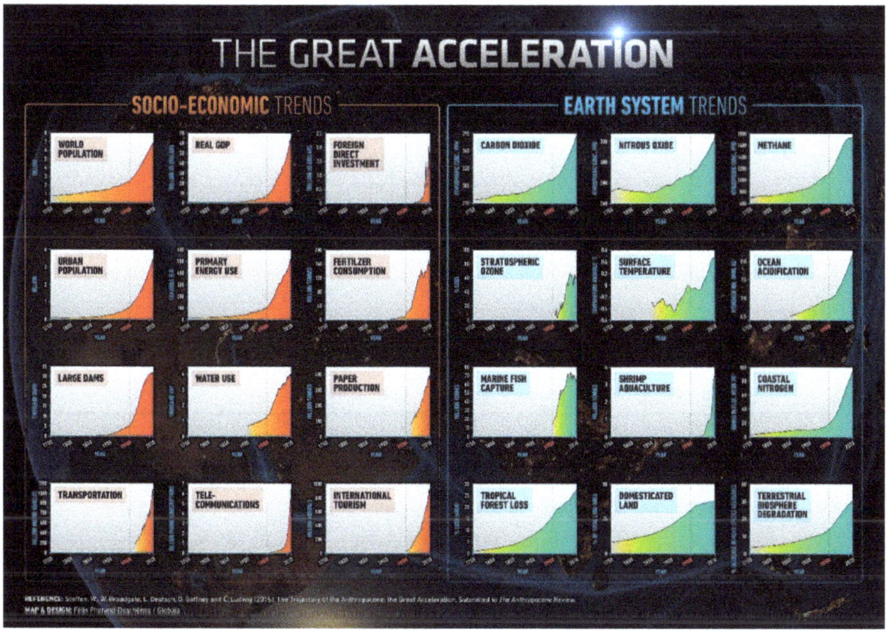

Figure 2.3. Socioeconomic and earth system trends, 1750–2010 (Steffen *et al.*, 2015).

quite rapidly. The sharp mid-20th century transition is mirrored in the trends for a variety of earth system indicators.

The impact of this work on socioeconomic and earth system trends was so dramatic that it inspired a new label for the post-1950 period: *The Great Acceleration*. The graphs of Figure 2.3 visually demonstrate what a shorter-term material flow analysis or environmental assessment cannot — that the impact of humans on the planet has become so amazingly large as to instigate panicky thoughts about the prospects for our planet and society. In this context, it is useful to recall the IPAT equation presented in Chapter 1, where A = affluence (GDP/person) and T = technology (impact/unit of GDP).

2.2 Evolutionary Approaches to Human Wants and Needs

The picture painted earlier regarding human desires and their satisfaction is a highly problematic one so far as the environment and long-term sustainability are concerned. As will be discussed in more detail in Chapter 5, the natural world has evolved over time to be cyclic in resource use, or nearly so, with populations of species constrained by available habitats and resources. The analogous human picture is problematic instead of cyclic, as can be seen by considering the

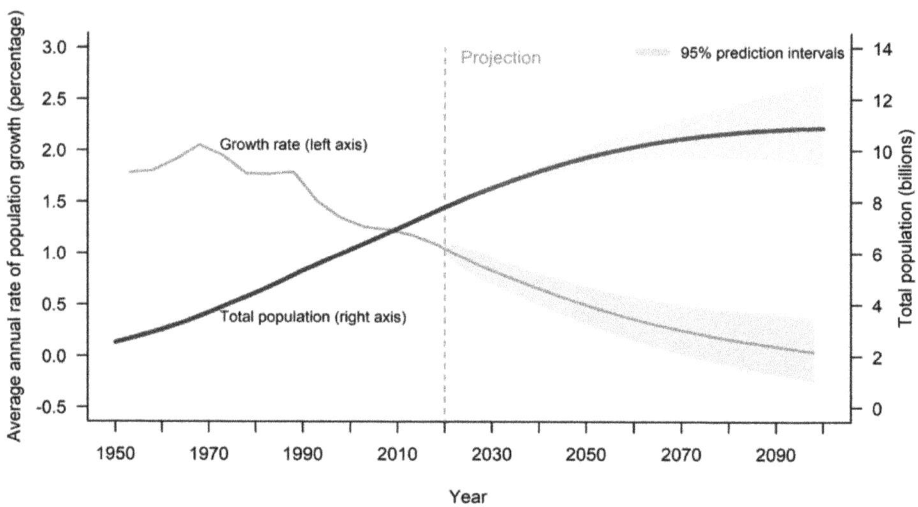

Figure 2.4. Global population trends (UNDP, 2019).

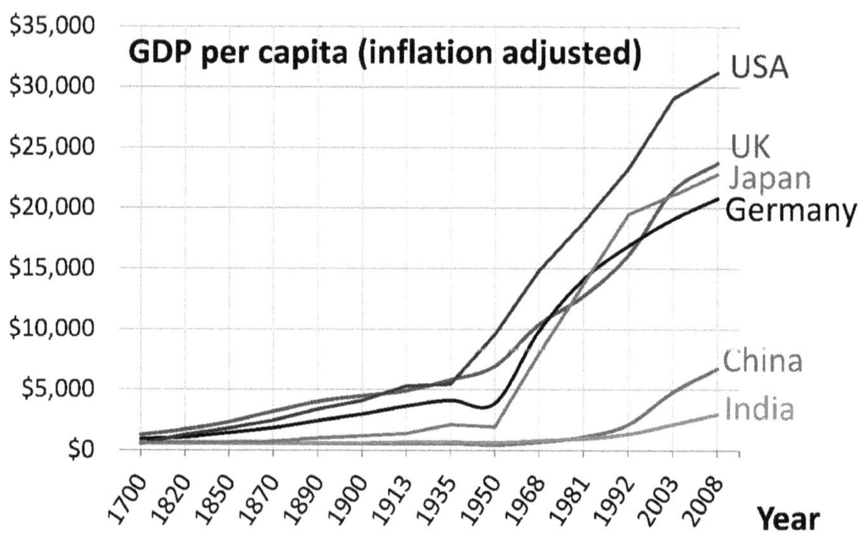

Figure 2.5. The evolution of gross domestic product per capita for the period 1700–2008 (Wikipedia).

Source: https://en.wikipedia.org/wiki/File:1700_AD_through_2008_AD_per_capita_GDP_of_China_Germany_India_Japan_UK_USA_per_Angus_Maddison.png.

IPAT equation. Figure 2.4 shows that the human population is predicted to grow for the rest of this century, stabilizing by perhaps 2100: an encouraging if long-term sign. However, Figure 2.5 exemplifies that gross domestic product per capita, a reliable measure of affluence, continues to increase (with moderate

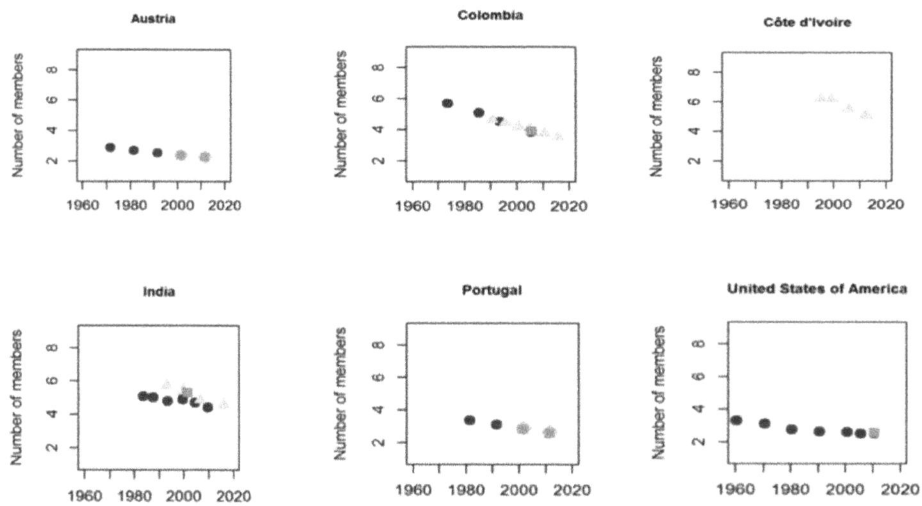

Figure 2.6. Trends in average household size in selected countries, 1970–2010 (UNDP, 2019).

economic fluctuations) all over the world. A related sign of affluence as well as social behavior is illustrated by Figure 2.6: all over the world, the average number of members in families or family-like groups is steadily decreasing. Nonetheless, each small group needs to be housed and so the number of housing units are still increasing, along with their associated building materials and appliances.

If "I" (the environmental impact term of the IPAT equation) is likely to be stressed by increasing affluence ("A") in the coming decades and by population ("P") expansion for many years to come, there is a big job ahead for technology ("T") — to serve the needs and desires of billions of people while sharply decreasing the related environmental implications ("I"). Transitioning from technology's use of fossil fuels is surely part of what is needed, but much more dramatic initiatives will be required to respond to the challenge, and those initiatives will need to be sociological as well as technological because the former drives the latter. We will explore this topic in more detail in Chapter 35.

Further Reading

Fischer-Kowalski, M., and H. Haberl, H. (1998). Sustainable development: Socioeconomic metabolism and the colonization of nature. *International Social Science Journal, 158*, 573–587.

Kenrick, D. T., Griskevicius, V., Neuberg, S. L., and Schaller, M. (2010). Renovating the pyramid of needs: Contemporary extensions built upon ancient foundations. *Perspectives on Psychological Science, 5* (3), 292–314.

Maslow's Hierarchy of Needs (2020). Retrieved from: https://en.wikipedia.org/wiki/Maslow%27s_hierarchy_of_needs.

Rambo, A. T. (2015). A burning issue: Rethinking the transition from hunter-gatherer to industrial sociometabolic regimes. *Journal of Industrial Ecology, 19*, 82–92.

Steffen, W., Broadgate, W., Deutsch, L., Gaffney, O., and Ludwig, C. (2015). The trajectory of the Anthropocene: The great acceleration. *The Anthropocene Review, 2*, 81–98.

United Nations Department of Economic and Social Affairs, Population Division. (2019). *Patterns and trends in household size and composition: Evidence from a United Nations dataset.* (ST/ESA/SER A/433). Retrieved from: https://www.un.org/en/development/desa/population/publications/pdf/ageing/household_size_and_composition_technical_report.pdf. https://www.un.org/en/development/desa/population/publications/pdf/popfacts/PopFacts_2019-6.pdf.

Chapter 3

Carrying Capacity and Grand Challenges

Chapter Concepts

- For millennia, humans have been employing nature's bounty with little or no appreciation of the fact that the gifts provided by nature have limits
- Industrial ecologists measure the pressures on nature by three approaches: (1) sustainable yield assessments, (2) carrying capacity evaluations, and (3) planetary boundary exceedances
- The United Nations measures human progress by the Sustainable Development Goals and the degree to which those goals are being met. Many of the goals are closely related to industrial ecology concerns and methodologies

3.1 Human and Natural Systems are Intertwined

Imagine for yourself the most remote location on the planet, one so far from population centers or shipping routes that it seems free of human impact. Now go to that location, set up your analytical instruments, and take samples of the air and water. You will readily find molecules synthesized only by modern industry, never by nature: the footprints of humanity. Or, go to a location as intensely human as you can imagine, one showing no indications (other than human beings) of "nature": a skyscraper in a major city, an automobile assembly line, or a museum of modern art. Observe with an ecologist's eye, or take samples of a few surfaces and examine them with a microscope. You will readily find plants, bacteria, insects, and perhaps small mammals and birds: the footprints of nature. Human and natural systems are intertwined at a global scale; we cannot understand the behavior of one without accounting for the influence of the other.

It may seem obvious to you reading this book that the actions of humans affect the planet and vice versa, but the ethos of human development has often been to treat them separately, especially since the Industrial Revolution. Nature was often seen in the past as a limitless store of natural resources that exists for our benefit, to do with as we wish. Throughout history, we have built vast technological systems to harness nature's services, to provide food, water, energy, and materials, and to absorb our wastes, without much consideration of the consequences of how human actions affect natural systems. We have simply assumed that the rivers will keep flowing, plants will keep growing, and that if we run out of something locally, we can go find what we need somewhere else. In this mindset, natural systems can be tamed completely through human ingenuity and engineering prowess.

But obviously, the planet is not limitless in its land, resources, or ability to cope with exploitation. Modern environmentalism was sparked by a series of major events when natural systems became so degraded that they essentially shut down and their effects on humans became all too clear. For example, the proliferation of gasoline-powered tractors and inappropriate farming practices in the American plains coupled with severe drought led to the Dust Bowl in the 1930s. In some regions, 75% of all the topsoil was lost, ruining agriculture and causing the largest internal migration in American history. In Cleveland, Ohio, a major manufacturing hub, the local Cuyahoga River became so laden with industrial pollution that it famously caught fire more than a dozen times. Across the world, the effects of air pollution on human health were sadly demonstrated during acute events such as the London "Great Smog" in the 1950s that killed thousands of people. These events (and many others in history) clearly demonstrated that natural systems are finite and can fail.

As we learned in Chapter 2, natural systems are typically complex. How they will respond to human perturbations is not always predictable and is often nonlinear. A wetland ecosystem may be able to naturally filter pollutants and maintain good water quality up to a point, but it may be overwhelmed and stop functioning once the pollutant loading becomes too great. Complex systems also have the ability to adapt to changing circumstances. This "adaptive capacity" enables many natural systems to heal themselves, given enough time. For hikers enjoying a walk in the seemingly pristine woods of New England, it is hard to fathom that nearly all of that land was clear-cut in the 1800s. The forests that have regrown there are very different in composition compared to what was there before, but they are still forests, providing habitat for wildlife, supplying oxygen and clean water, and providing quiet space for reflection.

Humans are a part of nature and, like any organism, need to utilize resources from the environment to survive. But how are we to know when we are using too many resources, when the natural systems we rely upon are in danger of losing their adaptive capacity, and when our actions pose a danger to ourselves?

3.2 Sustainable Yield, Carrying Capacity, and Planetary Boundaries

One of the central challenges posed by sustainability is how to "live within our means" in a planetary sense, to ensure that we can "[meet] the needs of the present without compromising the ability of future generations to meet their own needs," in the words of the seminal 1987 report *Our Common Future*, also known as the Brundtland Report. These words convey imperative general goals but must be operationalized and quantified in order to be useful.

"Sustainable yield" is an important concept and metric that is used for natural systems that provide us with physical resources, such as aquifers that bring water or forests that can provide timber. Aquifers can be replenished by groundwater infiltration, and trees can obviously grow back in forests following cutting, so some uses of these resources will not cause these natural systems to fail. Sustainable yield is defined as the amount of resources that can be taken while still maintaining the normal functioning of an ecosystem. If the sustainable yield is exceeded, then the natural system will need time to recover, during which harvesting or extraction may need to be halted or greatly diminished.

Determining a sustainable yield is necessary to solving the Tragedy of the Commons dilemma presented in Chapter 1. As long as the extraction of resources can be controlled, thus allowing some extraction to occur up to the maximum sustainable yields, both current and future generations can benefit from the continued healthy functioning of natural systems. Many fisheries around the world that had been in dire conditions due to collective overfishing have recovered as a result of instituting quotas built around sustainable yield.

Clearly, oceans do more than supply fish, and forests do more than supply trees. Humans and all organisms need to utilize a broad range of resources from their environments. An ecosystem can only support so many individuals before the extraction of resources exceeds the sustainable yield. This maximum population is called the "carrying capacity" of the system. Carrying capacity can be estimated for small, local ecosystems all the way up to planetary scales. Scientists have long worked to estimate the carrying capacity of Earth. The eminent Harvard biologist E.O. Wilson puts the number at 9–10 billion humans, but estimates range widely, in part because the sustainable yield of so many systems is unknown, as are the dynamics among them.

Just because we are currently below Wilson's estimated carrying capacity for Earth does not mean that everything is okay. Some systems are highly degraded and headed for catastrophe, while others are less disturbed. A final concept to introduce here is that of "planetary boundaries". Similar to the idea of sustainable yield, these are thresholds past which natural systems on Earth can no longer maintain themselves and become degraded, sometimes irreversibly. While yields focus on the extraction of resources, planetary boundaries can also consider Earth's capacity for absorbing pollution and other impacts from human activities.

The concept of planetary boundaries was introduced in a seminal article in 2009 by a large group of earth and environmental scientists headed by Johan Rockström. In it, they defined nine planetary systems and estimated the planetary boundary for each, as well as whether or not the boundary was being exceeded at that time. Figure 3.1 reproduces the main figure from the article. Their estimates were semi-quantitative but show clearly that, in their judgment, human activities are causing much more harm to some systems than others. Climate crisis is one of

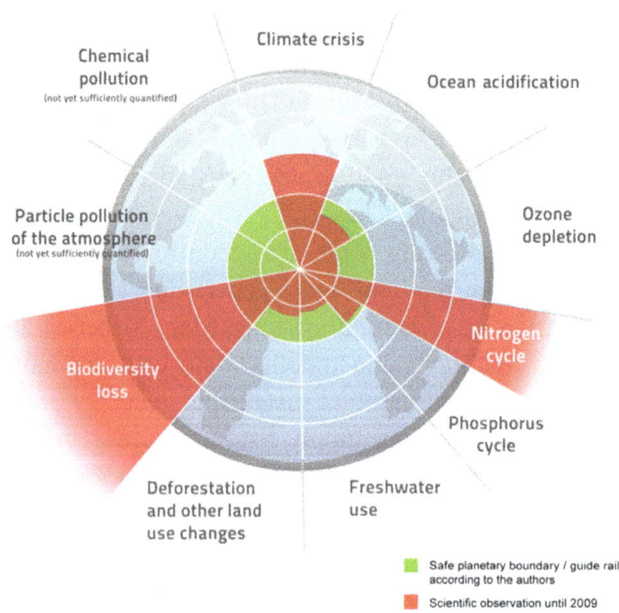

Figure 3.1. Planetary boundaries (Rockström *et al.*, 2009).

the categories that is exceeding planetary boundaries, with dire consequences predicted, if we fail to mitigate emissions or adapt to the effects of climate change. But the climate crisis is not the only crisis. The article asserts that humanity's effects on the nitrogen cycle and biodiversity loss are far beyond planetary boundaries and thus constitute large-scale emergencies that deserve our attention.

The concepts of sustainable yield, carrying capacity, and planetary boundaries help set the stage for many of the methods and analyses employed in industrial ecology. Being able to quantify these thresholds is central to assessment tools such as criticality assessment, environmental footprinting, or life cycle impact assessment (LCIA), as we will see in future chapters.

3.3 Grand Challenges for Sustainability

Advancements in technology and human industry, broadly speaking, have been hugely beneficial to us as a species. It is not simple to demonstrate this statement, but a widely recognized effort in that direction is by Spanish scholar Leandro Prados de la Escosura. He forms his index seen in Figure 3.2 around several measures of human development: life expectancy, health, education, and standard of living. While the rankings of individual countries vary widely, all have shown sharp improvement in the past several decades.

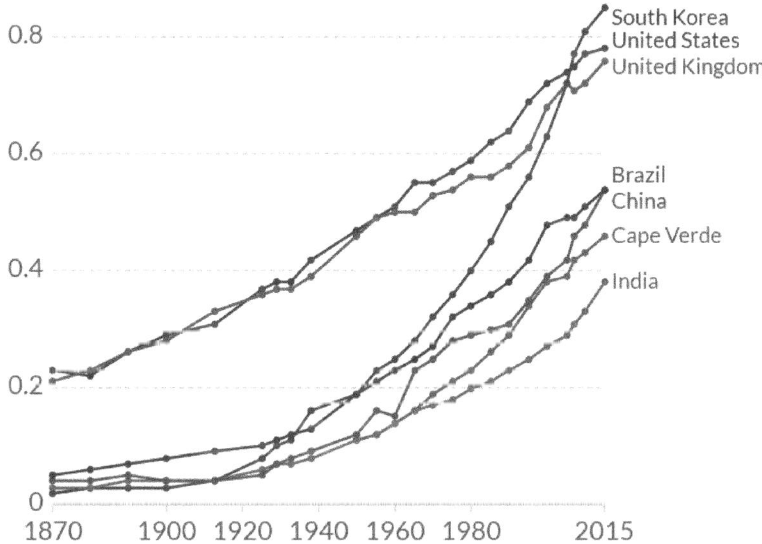

Figure 3.2. The historical index of human development for several countries, 1870–2015 (Prados de la Escosura, 2018).
Source: https://ourworldindata.org/grapher/human-development-index-escosura.

A central theme has been that advances in technology have allowed us to extract resources more efficiently, spreading material wealth but at potentially great cost to natural systems. It is clear that we face many challenges ahead, and that our current models of production and consumption are not sustainable. So, how should we focus our collective attention?

Grand challenges are long-term, systems-level problems that require global action and cooperation to solve. Lists of grand challenges are released by different institutions periodically. In 2019, the National Academy of Engineering (NAE) in the United States (U.S.) released a report titled *Environmental Engineering for the 21st Century: Addressing Grand Challenges*, which laid out what the science and technology establishment viewed as the areas needing the most attention and having the most promise to improve the environmental sustainability of our civilization. They were as follows:

- sustainably supply food, water, and energy;
- curb climate change and adapt to its impacts;
- design a future without pollution and waste;
- create efficient, healthy, and resilient cities;
- foster informed decisions and actions.

Solving each of these challenges necessitates understanding the complex relationships between human and natural systems and using this understanding to design our farms, cities, and our entire human industry to be as resource efficient, safe, and equitable as possible. This is the vision of industrial ecology.

In this book, you will learn about industrial ecology principles, methods, and tools that are used to tackle these grand challenges. Life cycle assessment (LCA) and network modeling reveal how food–energy–water systems are interdependent and provide metrics for tracking whether production is in fact "sustainable". Input–output analysis is used to track how carbon emissions occur throughout the global economy, while scenario modeling allows us to consider how technology, development, and population trends may affect future emissions. Closing materials loops is a central principle of industrial ecology. Material flow analysis provides a quantitative picture of our physical economy, while design strategies and new circular economy business models lay out frameworks for delivering innovative solutions. There are many inspiring case studies from around the world that demonstrate successful resource cycling strategies and can serve as models for the future, from islands to industrial parks to innovative urban designs.

Grand challenges are, by their nature, complex and involve many trade-offs and unintended consequences. You will learn throughout this book that the same topic can be analyzed in multiple ways. The oft-used metaphor is that industrial ecology provides different tools in your toolbox, which can be used to generate results and provide interpretation to designers and policymakers who are responsible for making decisions and taking action. Sustainability is inherently multi-disciplinary and multi-factorial; it is not just about carbon emissions, or toxicity, or equity, or efficiency. So, sustainability frameworks like industrial ecology need robust methods for interpreting multiple metrics and types of results. As the NAE list suggests, developing effective communication and decision methods is itself a grand challenge! Across the tools and topics covered here, you will see how industrial ecologists have grappled with complexity and multi-dimensionality through both quantitative and visual means.

While the NAE grand challenges focus on engineered systems, other groups have taken broader lenses. The most influential list of Grand Challenges globally is the 17 Sustainable Development Goals (SDGs) (Figure 3.3). Some SDGs are very closely tied to industrial ecology concerns, such as SDG 9 — Industry, Innovation and Infrastructure and SDG 12 — Responsible Consumption and Production. But in truth, industrial ecology has an important role to play in designing systems and assessing progress toward all of the SDGs, which each involve interlinked human and natural systems, and so benefit from an integrated systems perspective. Industrial ecology's support of the SDGs has been reflected in numerous reports, academic publications, and professional meetings, and new

Figure 3.3. UN Sustainable Development Goals (United Nations, 2015).

work continues to emphasize the SDGs as a central theme in industrial ecology research.

In the final chapter of this book, we will revisit the SDGs as we attempt to link the visions of sustainable development and living within planetary boundaries to assess industrial ecology's view of the relationships between humans and the planet on which we live.

Further Reading

Brundtland, G. H. (1987). *Chairperson, World Commission on Environment and Development, Our Common Future*. Oxford: Oxford University Press.

National Academies of Sciences, Engineering, and Medicine. (2019). *Environmental Engineering for the 21st Century: Addressing Grand Challenges*. Washington, DC: National Academies Press.

Roser, M., Historical index of human development (2019). Retrieved from: https://ourworldindata.org/human-development-index.

Rockström, J., *et al.* (2009). A safe operating space for humanity. *Nature*, *461*, 472–475.

United Nations (2015). Do you know all 17 SDGs? Retrieved from: https://sdgs.un.org/goals.

Wilson, E. O. (2002). *The Future of Life*. New York: Alfred A. Knopf.

Chapter 4

Industrial Ecology and Sustainable Engineering Concepts

Chapter Concepts

- Industrial ecology and sustainable engineering reject the concept of waste in favor of reuse, as is the case in nature
- Sustainable engineering can be regarded as the operational arm of industrial ecology. Together, they provide a template for the environmental and societally sustainable redesign of the modern world
- Six comprehensive goals and 10 operational tools constitute the basic framework of the practice of industrial ecology and sustainable engineering

4.1 From Contemporaneous Thinking to Forward Thinking

Since the Industrial Revolution, the activities of firms large and small have largely defined the interactions between humanity and the environment, and have significantly shaped the structure and dynamics of our society. These interactions have, however, traditionally of major significance for corporate decision-makers. Technology's influence on nature, and especially the potential magnitude of that influence across the full spectrum of economic activity, has been under-appreciated in the business world.

No individual firm exists in a vacuum. Every activity is linked to thousands of processes and transactions and to their environmental and social impacts. A large firm manufacturing products will have tens of thousands of suppliers located all around the world, potentially changing on a daily basis. The firm may manufacture and offer for sale thousands of individual products to a myriad of customers, each of which has their own needs and cultural preferences. Each customer, in turn, may treat the product very differently and live in areas with very different environmental characteristics; these are considerations of importance when the use and maintenance of the product may be a source of potential environmental impact (e.g., used oil from vehicles). The services and cultural patterns that the technology enables will also differ significantly in different communities and societies. When finally discarded, the product may end up in almost any country, in a landfill, an incinerator, dumped beside a road, or in a river that supplies drinking water to local populations.

In such complex circumstances, with social and environmental impacts at many scales, how has industry approached its relationships with the outside world? Satisfying the needs of its customers has always been pretty well done, at least in market economies. Industry has, however, been less adept at identifying some of the consequences, especially the long-term consequences of the ways in which it goes about satisfying needs. Since the beginning of the 1970s, when modern environmentalism began to arise, analysis of environmental consequences has increased; examples of a few of these consequences have been collected by Dr. James Wei of Princeton University (adapted and displayed as Table 4.1). The table indicates some of the environmental difficulties created for society in a world in which industrial operations have been perceived as essentially unrelated to the wider world, as suggested in the left side of Figure 4.1.

It is important to note that the relationships in Table 4.1 were not the planned result of disdain for the external world by industry. Several of the solutions were,

Table 4.1. Relating Current Environmental Problems to Industrial Responses to Yesterday's Needs.

Yesterday's need	**Yesterday's solution**	**Today's problem**
Non-toxic, non-flammable refrigerants	Chlorofluorocarbons	Ozone hole
Automobile engine knock	Tetraethyl lead	Widespread lead poisoning and contamination
Locusts, mosquitoes, malaria	DDT	Adverse effects on birds and mammals
Expanded food production	Synthetic fertilizer	Eutrophication
Food safety and preservation	Plastic packaging	Macroplastic and microplastic contamination

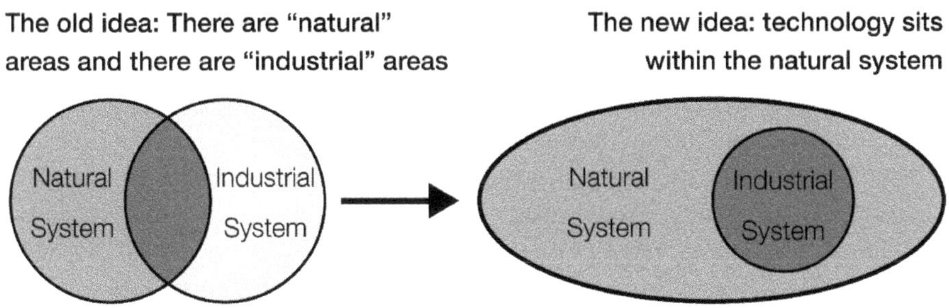

Figure 4.1. The transformation from natural and industrial systems as essentially independent entities to the realization that technology is embedded within the natural system.

in fact, great improvements over the practices they replaced, and their eventual consequences could not have been forecast with any precision. What was missing, however, was any structured attempt to relate the techniques for satisfying customer needs to possible environmental consequences; similar efforts understand any systemic long-term social implications of industrial activity lag even further behind. For example, we can speak of Design for Environment (DfE) with some clarity, but the development of Design for Sustainability (DfS) is in the early stages. While making such attempts does not ensure that no deleterious impacts will result from industrial activity, proactive consideration has the potential to avoid or minimize the most egregious of the impacts, and to contribute toward incremental improvements in impacts that are now occurring or can be well forecast. How are such attempts best made?

The broad approach to understanding the industry–environment–sustainability interactions described in this book is termed "industrial ecology (IE)". The overall concept is, in part, technological. As applied in manufacturing, industrial ecology involves the design of industrial processes, products, and services from the perspectives of product competitiveness, environmental concerns, and society. Industrial ecology is also, in part, sociological. In that regard, it recognizes that human culture, individual choice, and societal institutions play major roles in defining the interactions between our technological society and the environment. It recognizes as well that modern technological and societal systems are fully connected with and embedded within the natural world, as indicated on the right side of Figure 4.1.

As we shall see, industrial ecology and sustainable engineering focus in many different ways on topics that ultimately are connected to and enabled by physical resources. Because of this focus, Gjalt Huppes of Leiden University proposes a succinct definition of the field:

Industrial ecology is the science of the physical functioning of society.

This compact definition can be expanded to a more comprehensive form:

Industrial ecology is the means by which humanity can deliberately approach and maintain sustainability, given continued economic, cultural, and technological evolution. The concept requires that an industrial system be viewed not in isolation from its surrounding systems, but in concert with them. It is a systems view in which one seeks to optimize the total materials cycle from virgin material, to finished material, to component, to product, to obsolete product, and to ultimate disposal, and to carefully evaluate the impacts of the industrial system on the environment.

In this latter definition, the emphasis on *deliberate* differentiates the industrial ecology path from unplanned, precipitous, and perhaps quite costly and disruptive alternatives. By the same token, the definition indicates that industrial ecology practices have, or should have, the potential to support a sustainable world with a satisfactory quality of life for all. A bit later in this book, we will present a third definition of industrial ecology that uses a biological analogy to describe the perspective from which industrial ecology views an industrial system.

All of these definitions provide insight into a field that has become increasingly sophisticated both conceptually and methodologically and one that has become increasingly relevant in our modern materials-oriented world.

Practitioners of industrial ecology and sustainable engineering interpret the word *industry* very broadly. It is intended to represent the sum total of human activity, encompassing mining, manufacturing, agriculture, construction, energy generation and utilization, transportation, product use by customers and service providers, infrastructure systems, service networks, and waste disposal. IE is not limited to the domain within the factory walls but extends to all the impacts on the planet resulting from the presence and actions of human beings. IE thus encompasses society's use of resources of all kinds.

In considering manufactured products, for example, IE may focus on the study of individual products and their environmental impacts at different stages in their life cycles, but a complementary focus is the study of a facility where products are made. In such a facility, raw materials, processed materials, and perhaps finished components produced by others are the input streams, along with energy. The emergent streams are the manufactured product itself, as well as residues to land, water, and air, plus energy residues in the form of emissions, heat, and noise. The IE approach to such a facility treats the budgets and cycles of the input and output streams, and it seeks to devise ways in which smaller portions of the residues are lost and more are retained and recycled within the facility itself or into the facilities of others. Key concepts include the conservation of mass (all material must be accounted for), the conservation of energy (all energy must be accounted for), and the technological arrow of time — the realization that as society becomes more technologically advanced, it builds on its past technological base and so cannot sustain or improve itself without strong reliance on technology.

One of the most important concepts of industrial ecology is that it rejects the concept of waste. Dictionaries define waste as useless or worthless material. In nature, however, nothing is eternally discarded; almost all materials are reused in various ways, often with great efficiency. Natural systems have evolved these patterns because acquiring materials from their reservoirs is costly in terms of energy and resources, and thus something to be avoided whenever possible. In our industrial world, discarding materials wrested from the Earth system at great cost is also generally unwise. Hence, materials and products that are obsolete should be termed *residues* rather than *wastes*. It should be recognized that wastes are merely residues that our economy has not yet learned to use efficiently. We will sometimes use the term "wastes" in this book where the context refers to material that is or has been discarded, but we encourage the use instead of the term "residues", or perhaps the even less pejorative *experienced resources*, thereby calling attention to the engineering potential and societal value contained in obsolete products of all sizes and types. In doing so, we acknowledge that the law of entropy prohibits complete reuse without loss, but vision is more useful than scientific rigor in establishing the experienced resource perspective.

A full consideration of industrial ecology includes the entire scope of economic activity: mining, agriculture, forestry, manufacturing, service sectors, consumer behavior, and so forth. It is obviously impossible to cover the full scope of industrial ecology in one book, particularly given the reality that most services, no matter how abstruse, must rely on physical platforms and consume energy. Accordingly, we largely limit the discussion herein to manufacturing activities, although some of the final chapters explore the subject in more general terms.

4.2 The Greening of Engineering

Engineering has traditionally been regarded as the specialty that employs scientific principles to achieve practical ends. Because engineers are problem-solvers in the context of their cultures, resources and disposal sites were historically regarded as limitless. As a result, engineering designs made profligate use of resources and unintentionally caused a great deal of environmental damage. Moreover, considerations of the social implications of their professional actions tended to be limited to issues that directly affected the use of products or designs. These approaches are now clearly recognized as outdated, and modern engineers acknowledge the need to do better. But how?

The first step in this increasingly vigorous transformation of the engineering profession is to practice *green engineering*. There is a variety of definitions of this concept, which applies to all the engineering disciplines, but the essence is

> Green engineering is the design, commercialization, and use of engineering solutions, viewed from the perspective of human and environmental health.

Green chemistry is the parallel specialty that focuses on the chemicals industry. Green engineering and green chemistry practices focus on minimizing pollution and risk as a consequence of product manufacture and product use, that is, of being more environmentally responsible than had been the case previously.

The second step in this transformation is to move beyond green engineering (including green chemistry) to *sustainable engineering*. The distinction is that the first step moves in the direction of more responsible technology, but does not ask "How can social and environmental considerations be fully integrated into the engineering profession?" Nor does green engineering ask "How far do we need to go?" This latter question is, in many ways, the province of industrial ecologists, who study the interactions between technology and the wider world. Sustainable engineering draws upon this information to inform itself about limits and goals and then places them within the traditional framework for accomplishing practical ends. Thus, sustainable engineering can legitimately be regarded as an operational arm of industrial ecology, and the essence of industrial ecology and sustainable engineering can be briefly stated as follows:

> Industrial ecology and sustainable engineering provide a template for the environmentally and societally sustainable redesign of the modern world.

A template is particularly useful only if those who used it understand and address the characteristics of the system they are attempting to redesign. It is useful, therefore, to present some of the characteristics of modern technology:

- *technology is uncertain* (the best solution is never obvious, and experimentation is vital);
- *technology is progressive* (change occurs by evolution and transformation);
- *technology is analytical* (it measures actions and new ideas);

- *technology is cumulative* (it builds on previous knowledge and existing capabilities);
- *technology is systemic* (interdependence of technologies is required for progress);
- *technology is embedded* (technology sits within natural systems);
- *technology is accelerating* (the waves of technological transitions are ever shorter).

Any attempt to refashion technology must be responsive to these characteristics and adapt as they change, else the attempt is doomed to failure.

In 2002, William McDonough and Michael Braungart published a book titled *Cradle to Cradle* in which they proposed that a sensible goal for society (and thus for sustainable engineering) was not to be "less bad" but rather to be good, advocating a philosophy centered on restorative architecture and planning, use of benign materials and reductions in material use, a focus on reuse, and a commitment to recycling. These are important aspirational goals but say little about how those goals might be achieved. That part of the task belongs to industrial ecologists and sustainability engineers.

Perhaps the best contemporary example of sustainable engineering is that of Interface, Inc. of Lagrange, Georgia, US, a maker of floor coverings. A recent carpet designed by that corporation and marketed for use in offices, airports, and other high-traffic areas does not only use less energy during its production than the carpets of its competitors, it is actually carbon negative, that is, its materials store more carbon than is consumed in manufacture. The process and its philosophy are described by Gertner (2021). This approach is technology at its best — helping to sustain the planet rather than damaging it to a smaller degree than its competitors.

4.3 Linking Industrial Activity with Environmental and Social Sciences

The contrast between traditional environmental approaches to industrial activity and those suggested by industrial ecology can be demonstrated by considering several time scales and types of activity, as shown in Table 4.2. The first topic, remediation, deals with such things as removing toxic chemicals from soil. It concerns past mistakes, is very costly, and adds nothing to the productivity of industry. The second topic, treatment, storage, and disposal, deals with the proper handling of residual streams from today's industrial operations. The costs are embedded in

Table 4.2. Aspects of Industry–environment Interactions.

Activity	Time	Focus	Endpoint	Corporate view
Remediation	Past	Local site	Reduce human risk	Overhead
Treatment, disposal	Present	Local site	Reduce human risk	Overhead
Industrial ecology	Future	Global	Sustainability	Strategic

the price of doing business, but accomplishing this task contributes little or nothing to corporate success except to prevent criminal actions and lawsuits. Neither of these activities is industrial ecology. In contrast, industrial ecology deals with practices that look to the future, and seeks to guide industry to cost-effective methods of operation that will less its impacts on the environment and will optimize the entire manufacturing process for the general good (and, we believe, for the financial good of the corporation). Corporate executives are familiar with the liabilities of past and present industry-environment interactions. A challenge to the industrial ecologist is to demonstrate that viewing this interaction from the perspective of the future is a corporate asset, not a liability.

Chapter 1 began by discussing the "Tragedy of the Commons", in which a large number of individual actions, clearly beneficial over the short term to those making the decisions, eventually overwhelm a common resource and produce tragedy for all. Originally formulated to describe such local commons resources as public grazing lands, the concept was extended to global resources with discoveries such as the Antarctic ozone hole. Industrial processes and products interact with many different commons regimes, and design engineers should interpret the concept of the commons in a very broad way. Commons resources certainly include local venues such as city air, local watersheds, and natural habitats. Regional resources are also included, groundwater and precipitation (and their possible chemical alteration) being examples. The global commons calls other regimes to mind: the deep oceans (can oil spills or ocean dumping significantly degrade this resource, or is humanity's influence modest?), the Antarctic continent (50 years of international scientific activity with minimal environmental controls have left a dubious legacy), and, of course, the atmosphere (and its ozone and climate).

In practice, every product and process interact with at least one commons regime and probably with several, fragile and robust, monitored and ignored, close and distant. Analyses of industrial designs may produce very different results if the same product or process is to be used under the sea, in the Arctic, or on a spacecraft rather than in a factory. Just as industrial ecology extends to all parts of the life cycle, it extends as well to all commons regimes it may affect.

4.4 The Challenge of Quantification and Rigor

Science and engineering are quantitative sciences, drawing much of their value from their exactness. Industrial ecology and sustainable engineering, however, spring from a few concepts, and rather simplistic ones at that. The following are some examples:

- *Mimic nature*: Nature does many things well, such as minimizing energy use. It also sometimes does things rather badly, such as depositing so many bird droppings on small oceanic islands that the islands become completely hostile environments. Nature's engineering is often a good model to follow, but not always, and the quantification of the consequences is probably needed to indicate which path to follow.
- *The "Circular Economy"*: The idea of never throwing anything away is a good starting point, but with complex modern technology, it is often very expensive in terms of energy and other resources to recover everything. Deciding when to be circular and when not to be is an analytical issue, not a conceptual one.
- *"Reduce, reuse, recycle"*: Again, a good general rule, but sometimes reusing an old, inefficient product can result in more harm than good. Is following this rule "committing a little less sin rather than solving the problem", as William McDonough puts it?
- *The "environmental footprint"*: It sounds really bad that Amsterdam's footprint is more than 10 times its physical area, but is it? We only know by understanding the assumptions built into the methodology, as well as the consequences of that footprint, in some detail.

Concepts are important and played a big role in the inauguration of the field that is described and discussed in this book. As the field moves into its more mature phase, however, it is becoming increasingly quantitative and rigorous. We will present and discuss the foundational concepts in the chapters that follow, but we will present and discuss even more the ways in which those concepts are now being quantified through formalized methodologies and tools.

4.5 The Goals and Tools of Industrial Ecology and Sustainable Engineering

As in any field, there are key questions in industrial ecology and sustainable engineering. Unlike biological ecology, we are interested not only in the functioning of the technological system *per se* but also in the industrial ecosystem's interactions with and the implications for the natural and social systems of the planet. We specifically concentrate on a single species (humans), our relationship with the environment, and the impacts of technological operations and choices on social systems. From this broad framework, it is possible to specify a set of key questions that this book explores:

(1) How do modern technological cycles operate?
  (1.1) How are industrial sectors linked?
  (1.2) What are the environmental and social opportunities and threats related to specific technologies or products?
  (1.3) How are technological products and processes designed, and how might those approaches be usefully modified?
  (1.4) Can cycles from extraction to final disposal be established for the technological materials used by our modern society?
  (1.5) How do technological cycles interact with culture and society, and what are the implications inherent in these "second-order" effects of technology?
(2) How do the resource-related aspects of human cultural systems operate?
  (2.1) How do corporations manage their interactions with the environment and society, and how might corporate environmental management evolve?
  (2.2) How can the culture/consumption influence on materials cycles be modulated?
  (2.3) How can engineers appreciate their relationships with environment and society?
  (2.4) How might industrial ecology systems be better understood?
(3) What are the limits to the interactions of technology with the world within which it operates?
  (3.1) What limits are imposed by non-renewable, non-fossil resource availability?
  (3.2) What limits are imposed by the availability of energy?
  (3.3) What limits are imposed by the availability of water?
  (3.4) What limits are imposed by environmental and/or sustainability concerns?
  (3.5) What limits are imposed by institutional, social, and cultural systems?
(4) What is the future of the technology-environment-society relationship?
  (4.1) What scenarios for development over the next several decades form plausible pictures of the future of technology and its relationship to the environment and social systems?

(4.2) Should systems degraded by technological activity, local to global, be restored, and if so how?

These are broad questions that get to the nature of the basic physical relationships between humans and our environment. A complementary approach is to consider what goals industrial ecology should address, especially given the urgent need for more sustainable systems, and what industrial ecology tools can be used to address those goals. Although opinions could well differ regarding this list of goals, here is what will be addressed in this book:

- *Goal 1*: Characterize and quantify material flows and stocks, in order to understand our changing demand for resources, how they might best be supplied, and the potential for circularity.
- *Goal 2*: Characterize and quantify the environmental and health impacts of modern technologies and their uses, in order to redesign them for lower impacts and positive outcomes.
- *Goal 3*: Describe how physical flows move through the economy and among firms, in order to enable effective supply chain management and utilization of resources.
- *Goal 4*: Understand the evolving relationship between humans and technology and the role of human behavior and psychology in effective management of technological systems.

To address these goals and work toward a more sustainable future, industrial ecology and sustainable engineering employ 10 primary tools that you will learn about in the following chapters:

Tool 1	In-use stock modeling and circularity	Chapters 7, 9, and 37
Tool 2	Material flow analysis (MFA) and urban metabolism	Chapters 8 and 33
Tool 3	Environmental input–output assessment	Chapter 10
Tool 4	Life cycle assessment (LCA)	Chapters 11 and 12
Tool 5	Technoeconomic analysis (TEA)	Chapter 13
Tool 6	Network analysis	Chapter 14
Tool 7	Material criticality	Chapter 23
Tool 8	Sustainable engineering design frameworks	Chapter 24
Tool 9	Industrial symbiosis (IS)	Chapter 30
Tool 10	Scenario modeling	Chapter 34

Each one of these tools could (and does!) take up an entire course or textbook. Here we present the basics, explaining the basic structure of each approach, and showing examples of how industrial ecologists have utilized them to shed light on society's most pressing and large-scale environmental challenges.

Further Reading

Allen, D. T. and Shonnard, D. R. (2002). *Green Engineering: Environmentally Conscious Design of Chemical Processes.* Upper Saddle River, NJ: Prentice Hall.

Ayres, R. U. and Ayres, L. W. (2001). *A Handbook of Industrial Ecology.* Cheltenham, UK: Edward Elgar.

Frosch, R. A. (1992). Industrial ecology: A philosophical introduction. *Proceedings of the National Academy of Sciences of the United States of America, 89*, 100–103.

Gertner, J. (2021, June 27). Better living through CO_2. *New York Times Magazine*, 27–33, 61.

Graedel, T. E. and Howard-Grenville, J. A. (2005). *Greening the Industrial Facility.* New York, USA: Springer.

McDonough, W. and Braungart, M. (2010). *Cradle to Cradle: Remaking the Way we Make Things.* New York: North Point Press.

Chapter 5

Organisms — Biological and Industrial

Chapter Concepts

- Biological and industrial organisms share some (but not all) common characteristics
- The types of biological and industrial systems provide insight into the degrees of reuse of resources
- Some analytical methods developed by biologists can be adapted to industrial ecology

5.1 Biological Organisms

The elementary unit of study in biological ecology (BE) is the organism, which is defined as "an entity internally organized to maintain vital activities". Organisms share several characteristics, broadly defined, and it is instructive to list and comment on some of them.

1. *A biological organism is capable of independent activity*: Although biological organisms vary greatly in the degree of independence, each can take actions on its own behalf.
2. *A biological organism utilizes energy and material resources*: Biological organisms expend energy to transform materials into new forms suitable for use. They also release waste heat and material residues. Excess energy is released by biological organisms into the surroundings, as are materials residues (feces, urine, expelled breath, etc.).

3. *A biological organism is capable of reproduction*: Biological organisms are all able to reproduce their own kind, though the lifetimes and number of offspring vary enormously.
4. *A biological organism responds to external stimuli*: Biological organisms react to such factors as temperature, humidity, resource availability, potential reproductive partners, and so on.
5. *All multicellular organisms originate as one cell and move through stages of growth*: This characteristic is commonly recognized in every living being from moths to humans.
6. *A biological organism has a finite lifetime*: Unlike some physical systems such as igneous and metamorphic rocks, which for most purposes can be regarded as having existence of unconstrained length, biological organisms generally have variable but limited lifetimes.

5.2 Metabolisms of Biological Organisms

A biological organism takes in a variety of resources (plants, smaller animals, etc.) which it uses to enable a variety of bodily functions (breathing, motion, etc.) and to build muscles, bones, and other parts of its body, and then emits or excretes a variety of waste products. The first step in characterizing an organism's metabolism is identifying and quantifying these inputs and outputs. The next step is to identify and understand the *pathways*, or internal chemical transformation sequences, that occur. In principle, every transformation can be described as a reaction or a series of reactions. As with all chemical reaction sets, however, complexity is inherent in the process. For example, an intermediate product, termed a *metabolite*, may often be formed by several different pathways. The flow of material down these pathways is generally controlled by *enzymes*, which are molecules that act as catalysts. Byproducts are common, with a reaction often producing both the desired metabolic product (generally an intermediate product to be built upon) and a less desired product (to be removed by further reaction or discard).

What materials are available to nature for use in the construction and maintenance of organisms? The starting point for beginning to answer this question is the periodic table of the elements, shown in Figure 5.1. In this version, the astrophysical processes deemed to be responsible for the creation of each of the elements (mostly exploding massive stars and perhaps merging neutron stars) are indicated. As astronomer Carl Sagan said several decades ago, "We're made of star stuff".

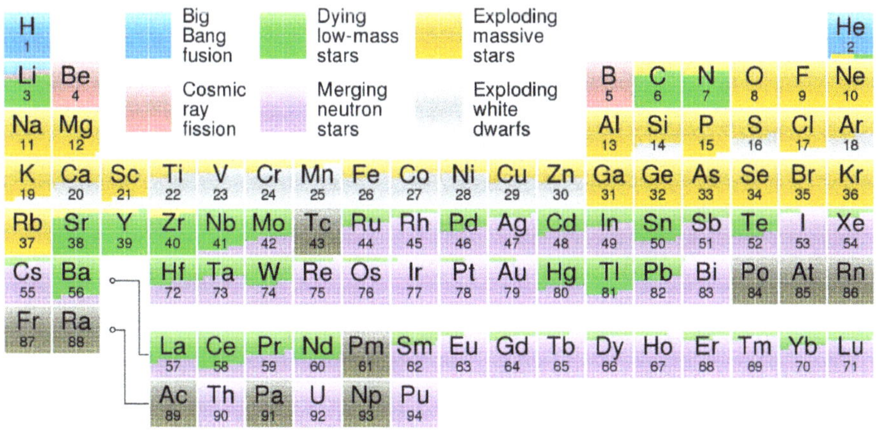

Figure 5.1. The periodic table of the elements, showing the astrophysical processes deduced to be responsible for element creation (Johnson, 2017).

Every secondary school chemistry course deals with the periodic table, the locations of the elements reflecting the number of protons and electrons in the atoms. The horizontal rows of the table are termed *periods*, and the vertical columns *families*. For the purposes of this book, the elements of interest are those utilized by natural organisms and by modern industry; both are discussed later in this chapter. Some elements see little or no use in either regard, however: the noble gases of group 18 (except helium) and the radioactive elements, largely at the right end of period 6 and beyond.

The usefulness for the life of various combinations of the elements has doubtless been explored by organisms on Earth throughout history, but contemporary biological organisms today utilize only 27 elements to function. Moreover, only five of these elements constitute 99.6% of total organismal mass: hydrogen, oxygen, carbon, nitrogen, and calcium; these are generally elements that are quite abundant and have high reactivity. Other elements are utilized only for specific purposes, such as iron in red blood cells.

Nature utilizes its choice of elements to perform the functions that make life possible. The operational features of that metabolism can be illustrated by the classic tricarboxylic acid cycle shown in Figure 5.2. The formulae of the entities that make up the cycle are shown, together with enzymes such as nicotinamide adenine dinucleotide (NADH) that catalyze the reactions. Many (but not all) of the reactions can be reversible depending on factors such as enzyme concentrations and temperature.

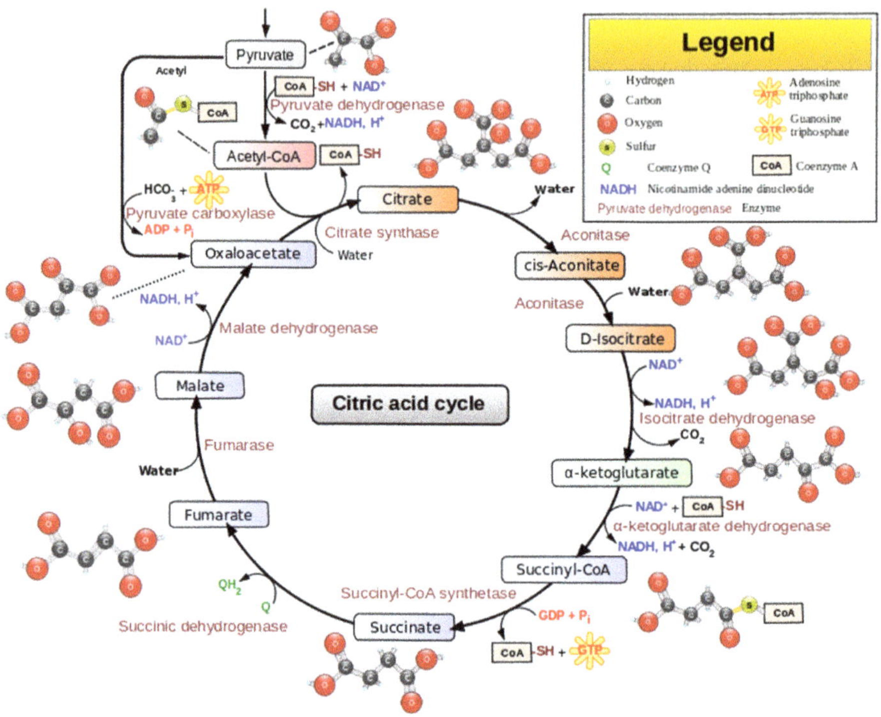

Figure 5.2. A metabolic example from biology: The tricarboxylic acid (Krebs) cycle. (U.S. National Institutes of Health).

The first step in constructing the tricarboxylic acid cycle was the realization that a Resource I gives rise to a Product O. With better analytical tools, intermediate metabolites could be identified. More detailed analysis eventually produced the system shown in the upper left of Figure 5.2. This system is certainly complex, but it is only a small part of the total metabolism of a typical organism. The metabolic network of the bacterium *Escherichia coli*, for example, is comprised of several hundred metabolites, perhaps twice as many reactions, and more than hundred pathways of interest. As a result, in the words of bioinformatics specialist Peter Karp,

> The theoretical understanding of a system such as the biochemical network of *E. coli* is too large for a single scientist to grasp … As scientific theories reach a certain complexity, it becomes essential to encode those theories in a symbolic form within a computer database.

Why might one wish to understand the complete set of metabolites, reactions, and pathways? One reason might simply be because that understanding is

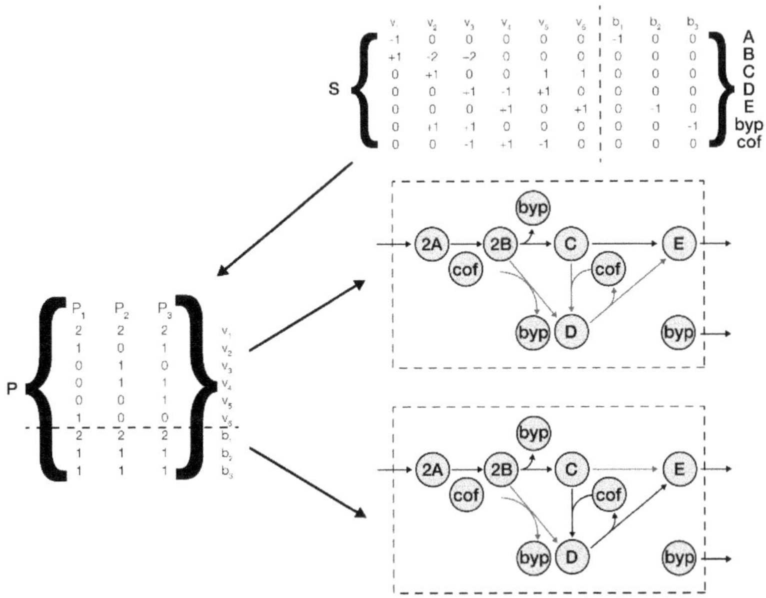

Figure 5.3. The progressive analysis of metabolic pathways in biology, as further discussed in the text (adapted from Papin *et al.*, 2003).

intellectually satisfying. A more purposeful reason, however, is to understand biological systems at the level of detail that will enable a more comprehensive understanding of nature than had previously been possible or to enable a more directed design of pharmaceuticals for specific diseases. Such goals generally possess the complexity described by Karp above: a system too large for a single mind to grasp. This realization has spawned the development of computer approaches termed "systems biology".

The systems biology approach (termed "biological informatics") begins by representing the networks with a matrix 'S', where the rows correspond to the metabolites and the columns to the reactions (Figure 5.3, upper right). The matrix elements are the stoichiometric coefficients of the reactions. If all metabolites are identified, and all fluxes determined, the result then specifies matrix "P", the potential alternative pathways for the transformation of resources to products (Figure 5.3, lower left). Two of these pathways are shown in Figure 5.3, lower right, in which the dark lines indicate the deduced reactions and the faint lines inactive reactions; in any particular circumstance, the pathway taken may be a function of temperature, cofactor concentration, or some other system parameter. With the availability of computer analysis, very large networks can be analyzed in this way, and the existence and properties of unknown metabolites and unnamed reactions can be predicted.

The pathway analysis approach is data-intensive, as befits such a complex system. The result of utilizing this information from biochemistry, physiology, and genomics is the elucidation of the processes within an organism, thereby leading to increased understanding and even the possible design of engineered biological systems.

5.3 Industrial Organisms

Industrial ecology (IE) consciously incorporates the word *ecology*, a term originated with reference to biological systems. On the face of it, combining a word associated with the natural world with one associated with its exact opposite seems inappropriate, or at least injudicious. However, the idea of conceptualizing human social systems (including industrial systems) from an organismic point of view is at least a century old. The distinction between previous thinking along this line and the evolving IE approach is that the former concentrated on behavior and social structure, whereas the latter focuses especially on physical and chemical parameters: resource flows, energy use, and the like. The vision of industry in all its facets engaged in the cycling of resources rather than in one-time resource use was presented in 1989 by Robert Frosch and Nicholas Gallopoulos of the General Motors Research Laboratories in Michigan. Their paper is regarded by many as the first publication in the field of industrial ecology.

Just as organisms possess sets of elements from which they construct molecules and products (bones, muscle, etc.), corporations have periodic tables from which they manufacture products. For example, General Electric Healthcare, a single business unit of GE, utilized *69* different elements in building its diagnostic imaging machines and related contrast agents in 2015. A similar review that year for Nokia demonstrated that their periodic table for mobile phones involved *65* different elements. Similar results could be derived for many high-technology companies. These illustrations demonstrate that with the exception of the radioactive elements, modern technology utilizes essentially the entire periodic table for one purpose or another.

One of the most complete contemporary industrial versions of physical metabolism, that of the Toyota Motor Company, is shown in Figure 5.4. The following are the major inputs and outputs are identified — water, energy, carbon dioxide, and so forth. Flows subject to regulations of one sort or another are quantified. Transformations are not indicated, however, and individual materials are not identified. The result is an analysis that parallels the metabolic studies that biologists were developing some three or four decades earlier.

Just as for biological organisms, human or otherwise, it is possible and useful to analyze the metabolism of an industrial organism (i.e., a factory or corporation)

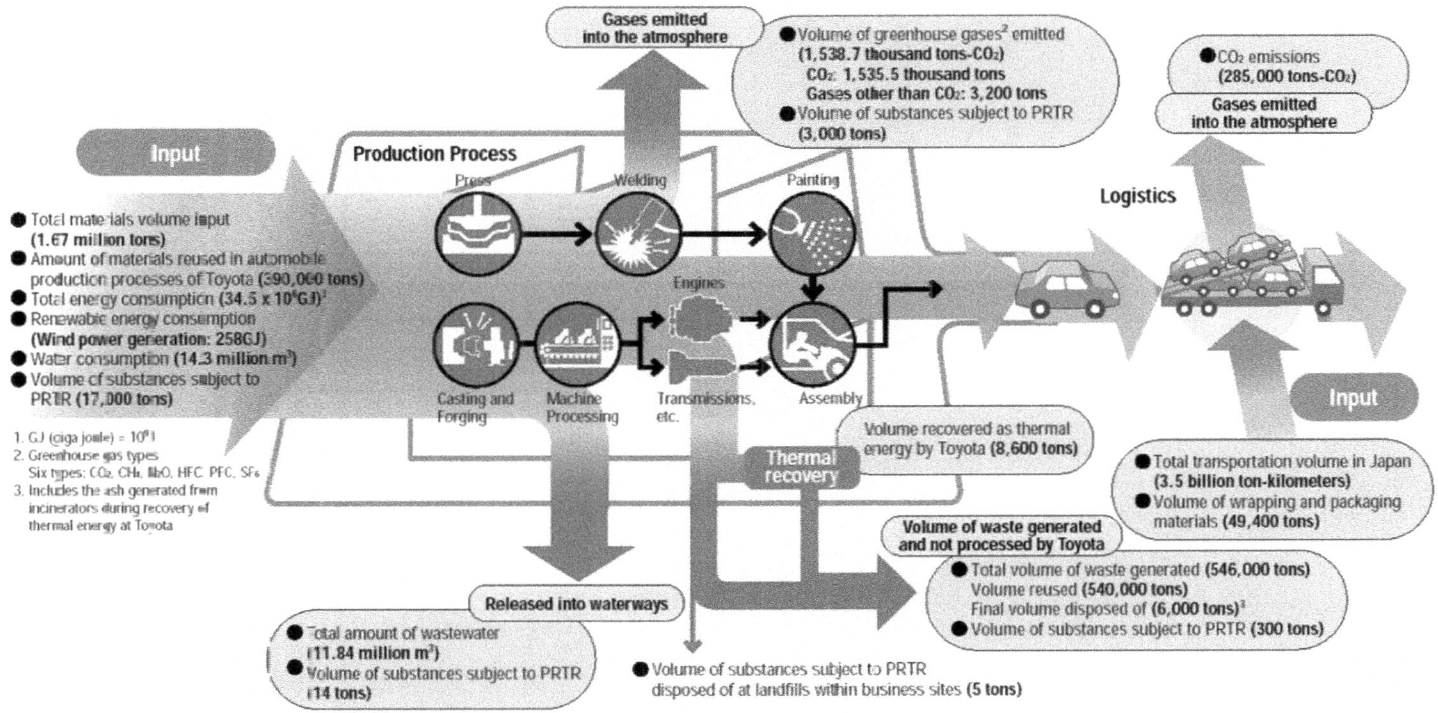

Figure 5.4 The industrial metabolism for Toyota's manufacturing processes in 2005. Only some of the materials and flows are indicated, but clearly there is much underlying detail from which this figure was drafted (OECD, 2008).

in much more detail than is indicated in the Toyota diagram. To do so, we need to revise somewhat the definitions of metabolic terminology:

Units of Measurement

All of science and most other activities worldwide employ the International System of Units (termed "SI Units" from the French abbreviation), which was established in Paris by 17 countries in 1835. You may encounter, however, systems of units other than the SI. The most common is the Imperial Units System, defined by the British Empire in 1824 and officially used as of 2021 in only three countries in the world: Myanmar, Nigeria, and the United States. The Imperial System has forced young students for almost two centuries to convert inches to feet and feet to yards.

Perhaps even worse is a second variation: systems that mix units from the two systems or use units in such a way that it is unclear which system is being used. A classic example of this type is of units related to amount of weight. In the SI system, the official unit is the gram, but some utilize the tonne, defined as one million grams. The customary Imperial unit of weight is the ton, which should be straightforward enough but it is defined in two distinct forms: the short ton (2000 pounds) and the long ton (2240 pounds).

Some people or some documents overlook this difficulty or assume the reader will know what is meant. The consequence of all this is that if the weight of something is expressed in tons, the reader cannot be sure whether the reference is to short tons, long tons, or metric tons. A related difficulty is that within the metric system, a series of prefixes are specified so that no confusion need occur, but some users prefer to avoid the prefixes and use unauthorized symbols such as hundred cubic meters or thousand tons. To avoid such unnecessary confusion, SI units and nothing else should be employed when expressing numerical values. We will do so in this book.

- *Industrial metabolite* — an intermediate product in the transformation of resources into final products. Industrial metabolites can also be termed parts, subassemblies, etc.
- *Industrial enzyme* — an industrial process or piece of equipment that results in a transformation (also termed reactor, milling machine, lathe, etc.).
- *Industrial pathway* — the sequence of transformations that convert resources into final products.

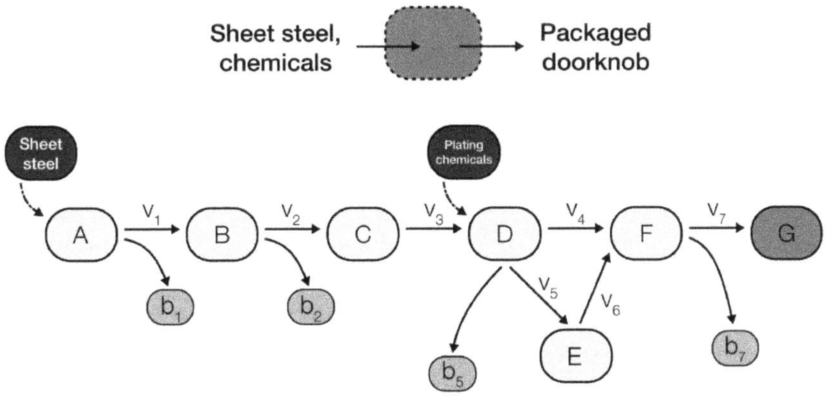

Figure 5.5. Industrial metabolic reaction pathways, from simple to complex: the doorknob cycle; (A) sheet steel in receiving bin; (B) thin coil of sheet steel; (C) disk of sheet steel; (D) shaped knob; (E) plated knob; (F) assembled doorknob; (G) packaged doorknob; (v_1) slitter; (v_2) punch press; (v_3) drawing press; (v_4) assembly station; (v_5) plating bath; (v_6) assembly station; (v_7) packaging station.

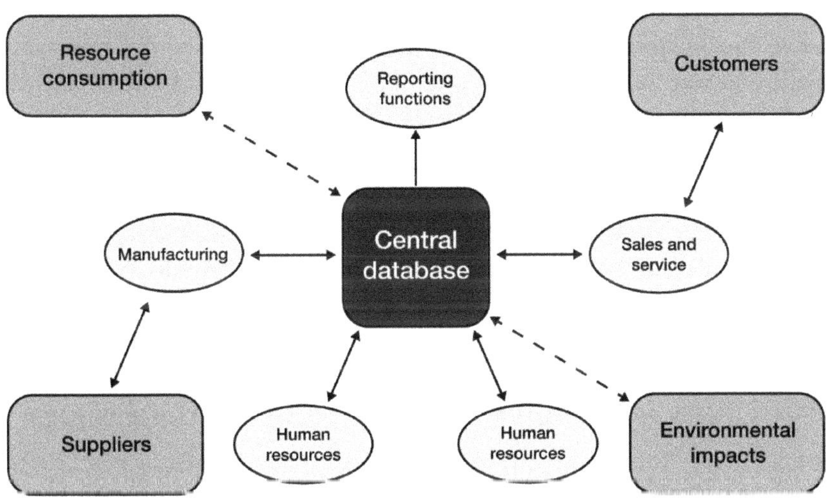

Figure 5.6. The scope of an environmentally extended enterprise resource planning (ERP) system.

An industrial pathway shares many characteristics with a biochemical pathway, as can be seen in the simplified diagram of Figure 5.5 (top). In overview, at the top of the figure steel, parts, and chemicals enter the facility, and a packaged product emerges. To construct the diagram for a given material (and there could be concurrent diagrams for other materials), one needs to know which process transforms or transports the material and which final products contain the material. To specify the fluxes, one needs to know the mass fluxes of the material inputs, the material concentrations in the inputs, and the details of the transformation process. This information then generates the industrial metabolic diagram (Figure 5.5, bottom) which indicates the metabolites, enzymes, and byproducts. It is straightforward to form the "S" matrix for this facility from the diagram. The ultimate choice between routes depends in large part on the fluxes of the pathways and the degree and types of byproduct generation. In a more realistic example of a factory manufacturing several final products, each containing scores of parts and requiring 10 or 20 reactions or transformations, computer approaches become essential if optimization is to be achieved.

It turns out that a form of metabolic analysis in industry has existed for about a quarter of a century under the name *material requirements planning* (MRP). MRPs are software programs that link production schedules with "bills of materials" files (sources of materials information on every part and component), inventory status files, and materials requirements files. A conceptual picture of a modern Enterprise Resource Planning (ERP) system including environmental extensions is shown in Figure 5.6. This approach has enabled businesses to more efficiently manage manufacturing so as to "build to order" rather than "build to stock". In essence, MRPs are the industrial equivalent of modern biological informatics systems. Factors other than resource transformation options enter into decisions about the industrial metabolism, of course: financial systems, labor, management, etc. Analytical (software) approaches that encompass this entire suite of considerations are termed ERP tools, and are in extensive use in industry.

5.4 The Utility of Metabolic Analysis in Industrial Ecology

There are clearly several important distinctions between biological metabolism analysis and industrial metabolism analysis:

- In principle, each industrial metabolite is known, as is each industrial enzyme, because they are specified by the designer rather than having to be deduced.
- The biological approach assumes a steady state for all reactions, thus simplifying the mathematics at the cost of simplifying the problem. Industrial approaches often look at disruptions caused by a perturbation.

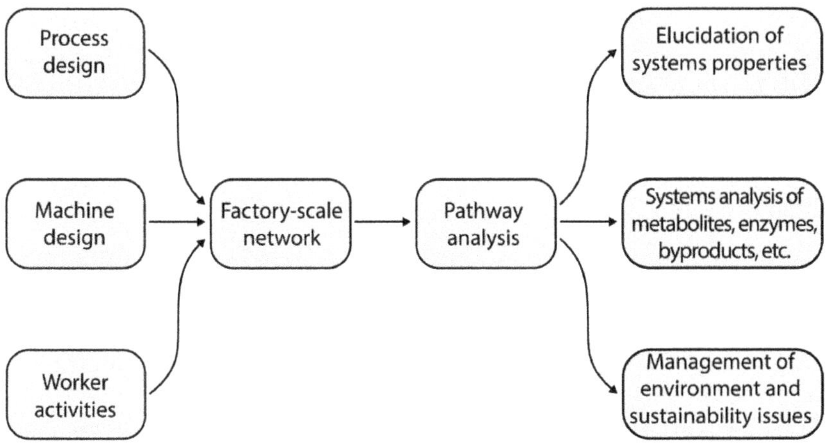

Figure 5.7. The linkage of industrial metabolic data, pathways, and applications (adapted from Papin *et al.*, 2003).

- MRP is used not to deduce how a given organism functions, as is generally the case in biology, but to explore the desirability of alternative ways of functioning (although we acknowledge that drug design involves exploring perturbations of biological pathways).

Moreover, the motivation for metabolic analysis in biology and industrial ecology is clearly different. The biological focus is on how substances are changed, what the products of the change may be, and what regulatory mechanisms direct the system. The industrial ecology focus is more toward mass flows in the system, and understanding how a disruption to one part of the system ripples through the system as a whole. In both cases, however, increased knowledge is at the center of the activity, and physical laws constrain the possibilities.

The industrial metabolic analyses described in this chapter are capable of addressing the concerns of the environment or of sustainability as well as manufacturing and efficiency. Industrial ecologists will thus incorporate flux analysis and follow byproduct generation. In this way, they integrate the tools of biology and industrial management to serve larger societal goals.

The information flow in metabolic industrial analysis is sketched in Figure 5.7, where the metabolic information is seen to feed the industrial networks and thus to enable pathway analysis. The elucidation of systems properties then permits the active management of those aspects of industrial or societal activity that link to environmental or sustainability concerns.

The discussion in this chapter is centered on individual organisms — the squirrel or the factory. The approach can be scaled up, however: some biologists study the metabolism of an ecosystem, and some industrial ecologists study the metabolism of a city. We will see aspects of metabolic analyses at different scales later in this book, in discussions of industrial ecosystems, material flow analysis, and urban ecology.

Further Reading

Frosch, R. A. and Gallopoulos, N. E. (1989). Strategies for manufacturing. *Scientific American, 261* (3), 144–1552.

Graedel, T. E. (1996). On the concept of industrial ecology. *Annual Reviews of Energy and the Environment*, *21*, 69–98.

Hess, G. (2010). The ecosystem: Model or metaphor? *Journal of Industrial Ecology*, *14*, 270–285.

Johnson, J. (2017). The Ohio State University, Retrieved from: https://commons.wikimedia.org/wiki/File:Nucleosynthesis_periodic_table.svg.

Jones, C. G., Lawton, J. H., and Shachak, M. (1994). Organisms as ecosystem engineers. *Oikos*, *69*, 373–386.

Karp, P. D. (2001). Pathway databases: A case study in computational symbolic theories. *Science*, *293*(5537), 2040–2044.

Kusiak, A. (1990). *Intelligent Manufacturing Systems*. Englewood Cliffs, NJ: Prentice Hall, 341.

Organization for Economic Co-operation and Development (2008). *Measuring Material Flows and Productivity, Volume 1. The OECD Guide*. Retrieved from: https://www.oecd.org/environment/indicators-modelling-outlooks/MFA-Guide.pdf.

Papin, J. A., *et al.* (2003). Metabolic pathways in the post-genome era. *Trends in Biochemical Sciences*, *28*, 250–258.

Chapter 6

Ecosystems — Biological and Industrial

Chapter Concepts

- Systems are fundamental in the study of biological and industrial ecology and can be characterized as simple or complex
- Systems analysis studies complex systems holistically rather than focusing on individual components, and addresses the driving factors in technology and society that manifest themselves in the properties of a system
- The study of biological metabolism provides a conceptual model for understanding the complex materials network (or "industrial metabolism") of resource extraction, materials processing, product manufacture, and eventual material loss at end of useful product life
- Metabolic analysis in industrial ecology can be applied at a variety of spatial or analytical scales, from an industrial facility that transforms and utilizes a variety of resources to a country that hosts resource extraction facilities and is dependent on ocean shipping for most of its final products that it utilizes

6.1 The Systems Concept

Perhaps the most important operational feature of the ecology approach is its ability to focus not only on the product or process itself, but also on related systems and their behavior. Identifying the appropriate system level and properly relating it to its technological context can be a critical step in any successful industrial ecology assessment. Accordingly, it is useful to introduce and discuss the general concept of systems and how they can be studied in industrial ecology.

A system may be pictured as a group of interacting, interdependent parts linked together by exchanges of energy, matter, and/or information and subject to a common plan or serving a common purpose. There are two general types of systems: simple and complex. The terms are used here not in the sense of classical physics (e.g., the "simple" harmonic oscillator) but of the dictionary definitions: "complex" meaning "consisting of interconnected parts so as to make the whole difficult to understand" and "simple" is thus "not combined or in compound form". Some of the implications of this distinction are:

1. Simple systems tend to behave in a linear fashion: the output of the system is linearly related to the input. That is, a system with stressors A and B will have an impact Y given by

$$Y = A + B \tag{6.1}$$

This is a reductive view of the world, explained by linear cause and effect relationships. The complementary approach to studying systems is to realize that complex systems require deeper and broader analysis. Behaviors of such systems can emerge from interactions not readily apparent. For example, we cannot predict the characteristics of an anthill by summing up the observed behavior of individual ants because we have omitted interactive effects. In a complex system with stressors A and B, impact Y may be dictated by some function of interactive AB terms, the magnitude and behavior of which may be far from obvious:

$$Y = A + B + f(AB) \tag{6.2}$$

2. Simple systems can generally be evaluated in terms of cause and effect: an action is easily traceable through the system to its predictable effect. With complex systems, on the other hand, feedback loops and other response properties often make the linkage between cause and effect difficult to establish.
3. Complex systems, unlike simple systems, are characterized by significant time and space discontinuities. A problem for many people in comprehending the issues surrounding global climate change is that there are large time lags and spatial discontinuities between the forcing function (driving automobiles or using electricity, for example), and the resulting shifts in regional and global climate patterns.
4. Simple systems tend to be characterized by a stable and known equilibrium point to which they return in a predictable fashion if perturbed. Many complex systems, on the other hand, operate far from equilibrium, in a state of constant adaptation to changing conditions. Complex systems often evolve, simple systems generally stay more or less as they are.

Another attribute of systems that helps define and classify them is whether or not they are purposeful, that is, guided by a definite aim. Consider, for example, private transportation. The automobile is a planned system, predictable when operated according to its design, and developed with a purpose in mind, i.e., transportation. The traffic flow on an interstate highway (a "higher level" in the transportation system) reflects independent human agents with a common purpose, i.e., reaching their destinations, but with the unpredictable changes in speed and traffic jams that emerge from the system as a whole as a result of interactions among individual decisions.

Nature has levels as well, but contrasts with the purposeful actions of transportation. Nature can be predictive, in the sense that an experienced naturalist can anticipate with great certainty the actions of bees in a beehive, but the bees have no other plans than survival. Nature can also be unpredictable, as in the behavior of an ecosystem with many different species, each trying to find its own niche. In many cases, natural systems will tend towards homeostasis or a dynamic stability; while this may appear to be purposeful, homeostasis is a characteristic of the system that emerges over time rather than the result of an overall plan.

Imagine for yourself the most remote location on the planet, one so far from population centers or shipping routes that it seems free of human impact. Now go to that location, set up your analytical instruments, and take samples of the air and water. You will readily find molecules synthesized only by modern industry, never by nature: the footprints of humanity. Or, go to a location as intensely human as you can imagine, one showing no indications of "nature" (other than human beings): a skyscraper in a major city, an automobile assembly line, or a museum of modern art. Observe with an ecologist's eye, or take samples of a few surfaces and examine them with a microscope. You will readily find plants, bacteria, insects, and perhaps small mammals and birds: the footprints of nature. Although the two systems are intertwined to a least a modest degree, we will examine them separately in this chapter while realizing their interconnectivity.

Fundamentally, industrial systems are purposeful. To the degree that they restrict their scope to strictly technological products and processes, they are also predictive. Once independent human actors are added, however, complexity sets in. The results cannot be studied solely by natural science approaches, but inextricably involve specialists from the social sciences. This is another example of the fact that industrial ecology is inherently multidisciplinary and, given the coupling in such systems between the social and environmental, an appropriate framework for sustainable engineering.

Systems thinking originated in biology with the realization that many characteristics of organisms and their behavior were emergent, interactive, and variable.

Orgill *et al.* (2019) have defined some characteristics of systems thinking, which include the following:

- considering the system as a whole, not simply as a collection of parts;
- focusing on the variables that cause rather than correlate with system behavior;
- understanding the organization and interrelationships among the parts in the system;
- taking account of driving factors (chemical, economic, social, etc.) that lead to the system properties;
- deriving the relationships between a system and its environment.

Systems thinking reveals many otherwise obscure characteristics of the systems of nature and humanity, and the results from one area of specialty have the potential to inform other areas.

6.2 Metabolism of Biological Ecosystems

Groups of organisms, whether biological or industrial, utilize resources of various kinds to perform their functions. The study of these processes is metabolic analysis, where *metabolism* is defined as

> The aggregate of all physical and chemical processes taking place within a group or organisms.

The study of metabolisms is centered on processes taking place within a system of organisms — it is a dynamic form of analysis.

An ecosystem is a biological community (i.e., a group of organisms) present in a specific locale, together with the locale's physical and chemical abiotic attributes. In this connection, it is instructive to think of the materials cycles associated with a postulated primitive biological system such as must have existed early in Earth's history. At that time, the potentially usable resources were so large and the amount of life so small that the existence of life forms had essentially no impact on available resources. This process was one in which the flow of material from one stage to the next was essentially independent of any other flows. Schematically, what we might term a "Type I" ecosystem takes the form of Figure 6.1(a).

As the early life forms multiplied, external constraints on the sources and sinks of the Type I system began to develop. These constraints led in turn to the development of resource cycling as an alternative to linear materials flows. Feedback and cycling loops for resources were developed because scarcity drove

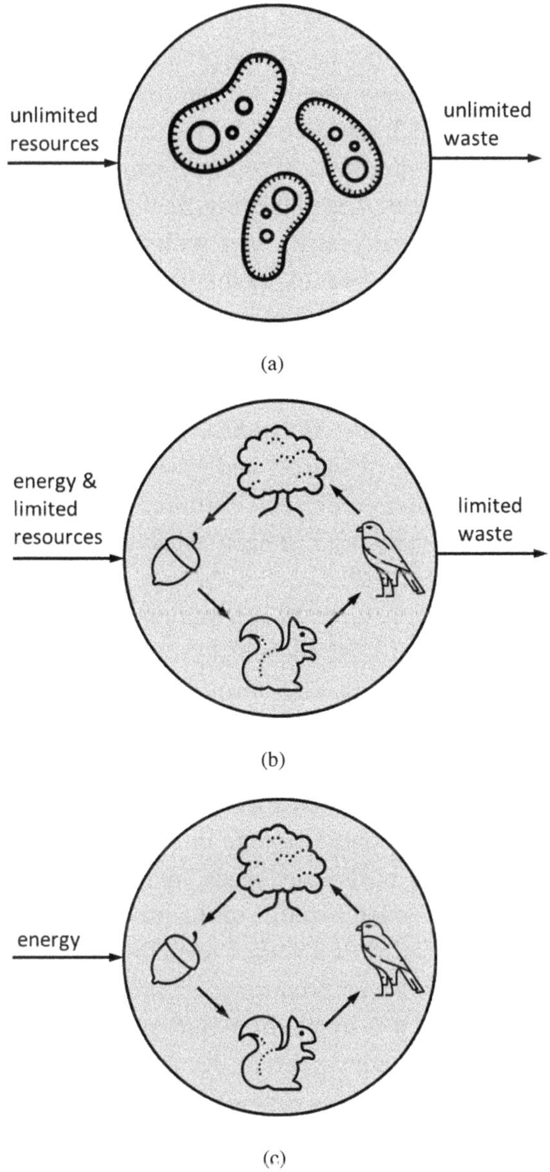

Figure 6.1. (a) Linear materials flows in Type I biological ecology. (b) Quasi-cyclic materials flows in Type II biological ecology. (c) Cyclic materials flows in Type III biological ecology.

the process of change. In such systems, the flows of material within the proximal domain could have been quite large, but the flows into and out of that domain (i.e., from resources and to waste) evidently were quite small. Schematically, such a Type II system might be expressed as shown in Figure 6.1(b). This transition from Type I to Type II, or from one Type II system to another, is termed *succession*.

The above discussion of Type I and Type II systems refers to the planet as a whole and regards the ecosystem types as sequential. Individual ecosystems, however, may be of either type in any epoch. A Type I system, known ecologically as an open system, is one in which resource flows into and out of the system are large compared with the flows within it. A Type II system, which can be regarded as a closed system, is the reverse. Complicating the issue, especially on the small scale, is the fact that an ecosystem can be open with respect to one resource (water, for example) and closed with respect to another (nitrogen, for example).

A Type II system is much more efficient than a Type I system, but on a planetary scale, it clearly is not sustainable over the long term because the flows are all in one direction; that is, the system is "running down". To avoid that trajectory and to be ultimately sustainable, the global biological ecosystem has evolved over the long term to the point where resources and waste are undefined, because waste to one component of the system represents resources to another. Such a Type III system, in which complete cyclicity has been achieved (except for solar energy), may be pictured as in Figure 6.1(c).

As suggested above, recycle loops have temporal and spatial scales as inherent properties. The ideal temporal scale for a resource recycling loop in a biological ecosystem is short, for two reasons. First, material viewed by a specific organism as a resource may degrade if left unused for a long period and thus become less useful to that organism. Second, resources not utilized promptly must be retained in some sort of storage facility, which must first be found or constructed, then defended. Still, examples of resource storage in biological systems are relatively common: squirrels store nuts, birds store seeds, and so forth; the organisms trade off the costs of storage against the benefits to themselves and their offspring.

The ideal spatial scale for most resource recycling loops in a biological ecosystem is small. One reason is that procuring resources from far away has a high energy cost. A second is that it is much more difficult to monitor and thus ensure the availability of spatially distant resources. As with temporal scales, counterexamples come readily to mind, such as eagles that hunt over wide areas. However, the eagles would probably have evolved to hunt close to home if such a strategy had provided adequate resources.

From a human perspective, the ideal use of the materials and resources available for industrial processes (broadly defined to include agriculture, urban infrastructure, etc.) might be one similar to the cyclic biological model. Historically, however, human resource use has mimicked the Type I unconstrained resource diagram. Such a mode of operation is essentially unplanned and imposes significant economic costs as a result. The implementation of IE is intended to accomplish the evolution of technological systems from Type I to Type II, and perhaps

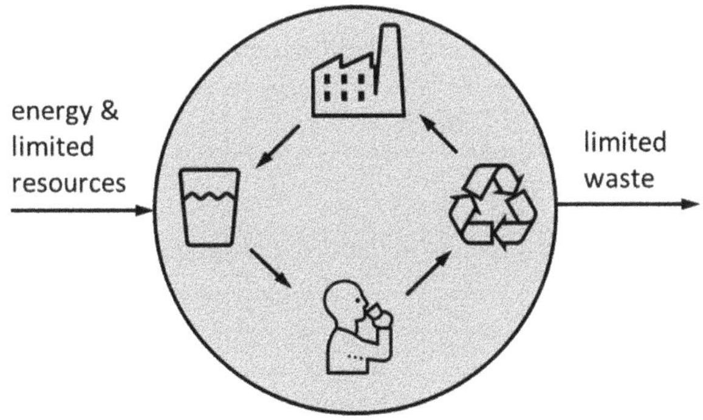

Figure 6.2. A Type II materials flow model of industrial ecology.

ultimately to Type III, by optimizing in ensemble all the flows and transformations involved, as shown in Figure 6.2.

6.3 Network Analysis of Ecosystems

The existence of multiple types of organisms (biological or industrial) in an ecosystem implies that interactions occur among the organisms. The study of such interactions is termed *network analysis*. In a network, a collection of individual compartments (nodes) are joined by connections (edges). Mass and energy flow along the edges, into and from the nodes. When the directivity of resource flow is specified, the result is a directed network. When the magnitudes of the flows are specified, the result is a weighted network. Constructing a network forms the relationships among the nodes and indicates what information (if any) is missing.

In analyzing a simple ecological network, the first task is to determine the nodes of the system, the second is to identify the directions of the flows, and the third is to quantify the flows along the edges. In Figure 6.3(a), inputs and outputs flow from and to the ecosystem, the different stocks exchange flows, and dissipation removes materials and energy from all stocks.

The generic network diagram is extended to represent a generic natural ecosystem in Figure 6.3(b). The diagram for an actual system would replace "detritivores" with actual animal prey such as rabbits, and "carnivores" with such species as foxes and hawks; the researchers would then attempt to determine the actual magnitudes of the interactions.

Figure 6.3 (c) is an attempt to reimagine Figure 6.3 (b) as an industrial ecosystem. It is worth spending a bit of time with the two figures to assess whether the latter provides any useful insight. If not, how might Figure 6.3 (c) be revised to be more useful as a model for industry?

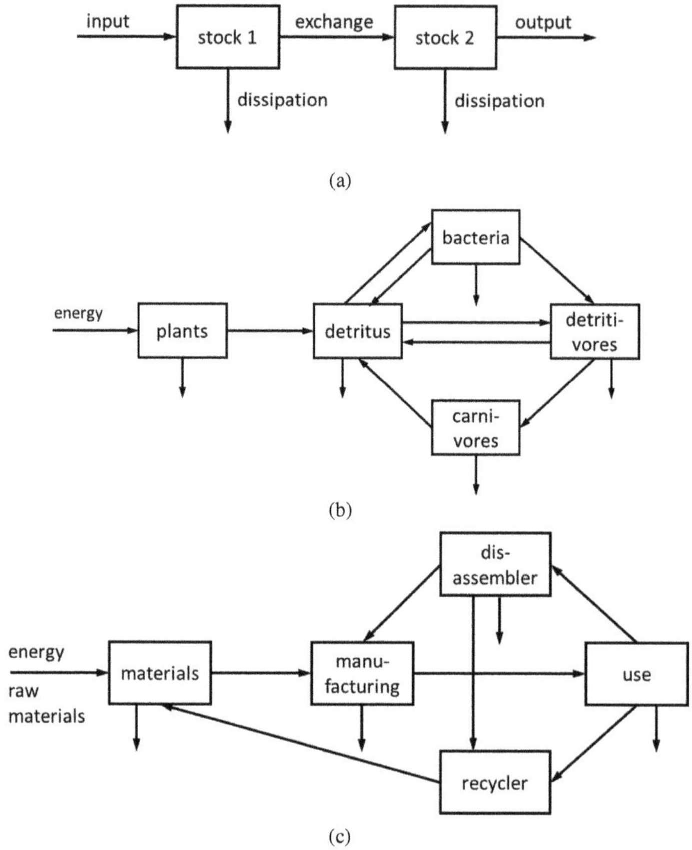

Figure 6.3 (a) Stock and flow designations in ecological networks. (b) A simple food chain. (c) A simple industrial network.

In a complex biological ecosystem, it is often a challenging task to answer the fundamental question "who is eating whom?" because such ecosystems can be very complex. An example, of the marine ecosystem off the coast of southwestern South Africa, is shown in Figure 6.4. Here, there are many diverse organisms at many different depths in the sea, and a few in the air above the sea. No doubt many students and fishermen were involved, and a great deal of time was expended in catching the organisms, conducting autopsies of the food in their stomachs, and generating the directed network shown here. Nonetheless, even casual examination of the diagram reveals a few obvious conclusions, such as that benthic filter feeders are always prey, snoek fish are mostly predators but are eaten by fur seals, and fur seals are predators only. Were it desired for all the interactions to be

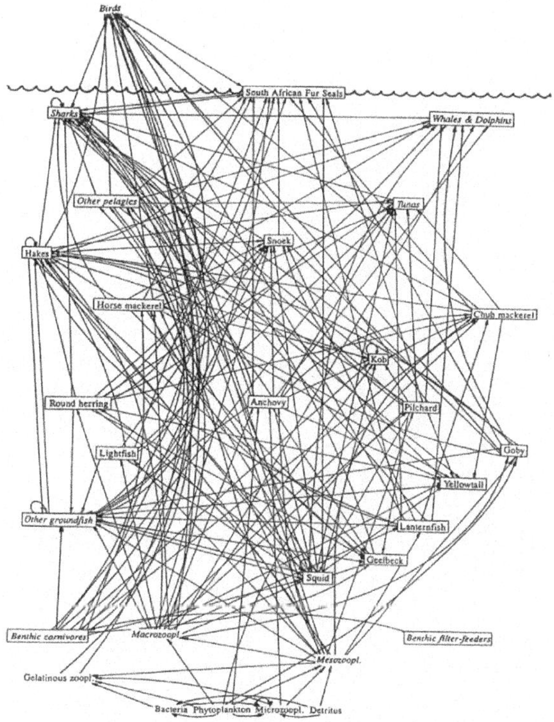

Figure 6.4. The Benguela marine ecosystem off the coast of southwestern South Africa (Yodzis, 2000).

quantified as well as their directions determined, an even more extensive research effort would be needed.

Industrial ecosystems share many similarities with biological systems, but with an important difference: in principle, the nodes, edges, and directivities of flows in industrial ecosystems are specified by their designers and only the flow magnitudes need to be determined. However, where diverse industrial ecosystems are linked (say to water treatment facilities in eco-industrial parks), measurement may be the quickest and most accurate way to determine the flow magnitudes.

6.4 Transformations in Natural Ecosystems

Organisms can act as engineers in the natural world; they may not design flow directions and magnitudes, but they certainly are capable of influencing them. Jones *et al.* (1994) identified several classes of materials transformation by organisms. The first case (Figure 6.5, Case 1) — the direct provisioning of resources by

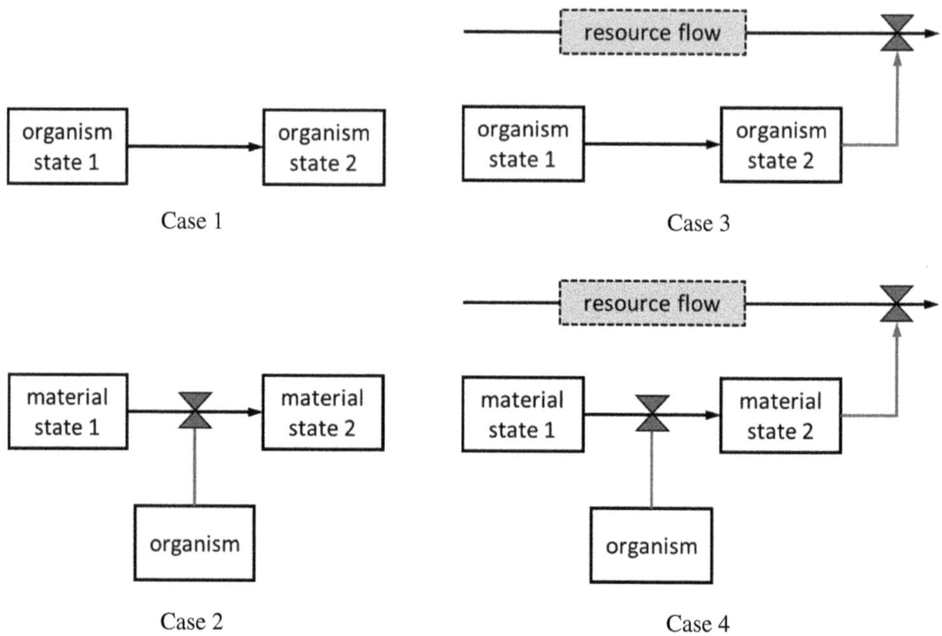

Figure 6.5. Conceptual models of engineering by organisms and industrial engineers. Terms are defined and examples given in Table 6.1. The "bow tie" symbol indicates points of material or resource transformation (adapted from Jones *et al.*, 1994).

one species to another, as in an osprey eating a fish — is not ecosystem engineering, because although materials are utilized (i.e., the fish's parts are transformed), no environmental change occurs. However, the simplest case of true ecosystem engineering (Case 2) occurs when an organism transforms materials into resources useful to other organisms. "Transform" is perhaps too strong a word here; "shape" or "rearrange" might be better, because the process need not involve modifying resource flows. A biological Case 2 example is a bird's nest in the above-water portion of a muskrat lodge, constructed by the muskrat without the slightest thought of providing avian living space.

In Cases 3 and 4, ecosystem engineering results in an alteration in the flows of resources. The Case 3 flow alteration arises from the tissues of the engineer, as in the alteration of solar energy to the upper ocean produced by planktonic absorption of solar radiation. In Case 4, the flow alteration is a consequence of the transformation of environmental materials. Case 4 is perhaps the easiest of all ecosystem engineering classifications to appreciate; it is exemplified by the beaver building a dam that results in changes in streamflow, nutrient transport, and wetlands habitat.

6.5 Transformations in Industrial Ecosystems

As with biological systems, modern industrial ecosystems are complex and not readily comprehended. An important first step, after one decides on a system to be studied, is to construct a provisional diagram of the system — what are the various actors or nodes and how do flows move among them, akin to Figure 6.3(c). Industrial ecosystems have similarities to biological networks, but there are differences:

1. The industrial food chain labels the biological "decomposer" as the "recycler". Unlike the former, who furnishes reusable minerals to primary producers, recyclers can often return resources one or two trophic levels higher, to primary or secondary consumers. In practice, of course, the system is seldom this prescribed: many recyclers send intermediate materials to the same smelters used for primary production and are thus more like biological decomposers.
2. The industrial ecology food chain has an additional actor, the disassembler, whose goal is to retain resources at high trophic levels, passing as little as possible on to the recycler.
3. Little in the way of resources is lost in the overall biological food chain, but much is lost in the industrial food chain. As a consequence, the industrial ecosystem must acquire a substantial portion of its resources from outside the immediate system, ecosystem must acquire only a small amount externally.

In practice, and especially at larger scales of regional, national, or international economies, the multiplicity of chemical reactions, the injections, transformations, incorporation into diverse entities, discards, and loss of what we might term industrial metabolites and industrial enzymes form technologically complex networks too large to readily comprehend.

The ecosystem concept of engineering thus mimics processes in industrial ecosystems and can be studied in similar ways. In industrial ecology, Case 1 activities are also those done to maintain organism function but not to manufacture products for others and include examples such as fixing the roof of a factory or repainting its interior. Examples of Case 2 are much less common today than they once were: Nomadic tribes moved rocks, trees, and brush to provide shelter; neolithic residents of Britain used tree trunks to make roads 6,000 years ago; and so forth. Even now, in certain parts of the world, one can find cobblestone and slate walkways and streets being constructed; if the materials are used in their natural state, the activities constitute Case 2 industrial engineering.

A common example of industrial Case 3 engineering is the alteration of the water budget of an urban area by buildings and their associated road and parking lot infrastructure. Most urban areas were originally settled because they possessed abundant natural water supplies, but in many modern cities, water must generally be transported to the city from outside sources, often quite remote ones.

Case 4 is, of course, exactly the historical concept of industrial engineering. A manufacturer consumes plastic and metal and produces a portable radio, for example, and during its life, the radio uses an occasional battery made by another supplier. Or, an automobile plant builds a gasoline-powered vehicle from metal, glass, plastic, and rubber, and the vehicle goes on to consume gasoline, oil, anti-freeze, and various replacement parts.

Demeester *et al.* (2004) compared the manufacturing systems of a biological cell and an industrial production facility, finding that biological system can readily be shown to be the more efficient and the more ecosystem-friendly. Moreover, the authors derived four sustainability principles from biology that they recommend for consideration in industrial systems:

1. *Customize local production*: Products are custom-built locally, with a very short delivery time and minimal inventory.
2. *Utilize universal components*: To a large degree, products are made from standard universal components or common raw materials that are bought in local markets.
3. *Employ just-in-time tools*: Desired production capacity is met with small-volume production machine tools that are easy to configure for universal application; these tools can easily be added, dismantled, or rearranged.
4. *Plan for local closed recycling loops*: All parts, generic and customized, are part of an efficient closed-loop cycle in which they are used again in local markets (as parts or raw materials), without having to be moved over long distances.

As will be seen in later chapters in this book, today's industrial approaches appear to be moving toward these four principles. Frosch and Gallopoulos (1989) stated in their seminal industrial ecology paper that "the analogy (between biological and industrial engineering)… is not perfect, but much could be gained if the industrial system were to mimic the best features of the biological analogue".

In fact, it is fair to say that the industrial system *already* mimics features of the biological system, for good and bad. Many similarities are suggested simply by the names of the specialties of the two systems (Table 6.1).

Table 6.1. Common Terms for Similar Systems in the Biosphere and the Technosphere.

Biosphere	Technosphere
Environment	Society
Organism	Company, city, country
Natural product	Industrial product
Anabolism	Manufacturing
Ecological niche	Market niche
Food web	Industrial ecosystem
Ecosystem	Eco-industrial park
Resource cycling	Remanufacturing and recycling
Natural selection	Competition
Adaptation	Innovation
Succession	Technological transformation

Note: Adapted and revised from Wikipedia, "Industrial Ecology", https://en.wikipedia.org/wiki/Industrial_ecology.

6.6 Tools of Ecosystems Analysis

This chapter has demonstrated that the concepts and tools of ecosystems ecology can be related to our perspective of industry and to industry's relationships to the natural world. The notion of scale is a central idea, enabling us to consider industrial ecosystems as biological ecosystems acting collectively. This perspective leads naturally to the idea of cycling and reuse of resources. In this regard, quantitative measures of efficiency in resource use provide useful comparative benchmarking. Also important is the concept of the food chain, which provides a framework for and an analytical approach to the flows of resources in a technological society. The notion of life cycles (of animals and industrial products) is also important, as is the realization that the notions of ecosystem succession in biology are mirrored by the effects produced in industry by invention and implementation. All of these biological ecology concepts help to inform industrial ecology discussions and analyses.

Over time, ecosystems analysts have developed a set of characteristics for assessing the material performance of ecosystems. These include the *standing stock* (how much is there?), the *mass balance* (are the material contents of the ecosystem stable?), the *material flux* or *material exchange rate* (how fast are the material contents changing?), the *residence time* (how long does material stay within the ecosystem?), and the *negative and positive feedbacks* (are resources recycled within the ecosystem, and how?). These analytical tools of biological ecosystems focus on a few central attributes that describe biological systems' processes and properties. Two complementary approaches, holarchic ecosystem structure and the benefits of symbiosis, add additional information to the understanding of biological ecosystems. Very similar approaches play central roles in understanding the functioning of industrial ecosystems, and discussions of them form a major portion of this book. Table 6.2 lists the somewhat different nomenclature employed in biological and industrial ecosystem analyses and indicates the book chapters in which they are discussed. Although ecosystem ecologist approaches have played a significant role in inspiring industrial ecology analytical methods, concepts and ideas from industrial ecology are beginning to reach out in turn to biological ecology as the influences of human actions on the biological world become ever more evident.

It is fair to point out that questions have sometimes been raised concerning the validity of the BE/IE analogy. Perhaps the most thoughtful critic has been Robert Ayres, who makes the following points:

- There is no primary producer in the economic system analogous to photosynthesizers.
- In the biosphere, there are no products as such.
- In the biosphere, there are no markets.
- Evolution in nature is driven by random variations, and in industry, by directed invention.

Table 6.2. Analytical Attributes of Biological and Industrial Ecosystems.

Biological ecosystems	Industrial ecosystems	Discussion chapter
Standing stock	In-use stock	Chapter 7
Material flow rate	Static material flow analysis	Chapter 8
Mass balance	Dynamic material flow analysis	Chapter 8
Residence time	Dynamic material flow analysis	Chapter 8
Feedbacks	Recycling	Chapter 9
Network characteristics	Systems analysis	Chapter 14
Energy generation and consumption	Energy analysis	Chapter 22
Holarchic structure	Holarchic structure	Chapter 29
Symbiosis	Industrial symbiosis	Chapter 34

Do Ayres's objections render the usefulness of the analogy problematic? In our view, they do not. No analogy implies perfect symmetry, only correspondence in some respects. If correspondence provides food for thought and potential analytical approaches drawn from what appears to be a complementary area of study, then the analogy will have had value.

In summary, the science of biological ecology appears to have much benefit for industrial ecology. The wide acceptance of the term industrial ecology has proven to be more than a convenient label; it is a reflection of the value that has resulted from explorations of the linkages between two apparently unrelated fields of study.

Further Reading

Ayres, R. U. (2004). On the life cycle metaphor: Where ecology and economics diverge. *Ecological Economics, 48*, 425–438.

Demeester, L., Eichler, K., and Loch, C. H. (2004). Organic production systems: What the biological cell can teach us about manufacturing. *Manufacturing and Service Operations Management*, *6*(2), 115–132.

Frosch, R. A. and Gallopoulos, N. E. (1989). Strategies for manufacturing. *Scientific American, 261*(3), 144–1552.

Hastings, A. *et al.* (2007). Ecosystem engineering in space and time. *Ecology Letters*, *10*, 153–164.

Jones, C. G., Lawton, J. H., and Shachuk, M. (1994). Organisms as ecosystem engineers. *Oikos*, *69*(3), 373–386.

Kay, J. J., Graham, L. A., and Ulanowicz, R. E. (1989). A detailed guide to network analysis. In F. Wulf *et al.* (Eds.), *Network Analysis in Modern Ecology*. Berlin, Germany: Springer-Verlag.

Orgill, M., York, S., and MacKellar, J. (2019). Introduction to systems thinking for the chemistry education community. *Journal of Chemical Education*, *96*, 2720–2729.

Ryen, E. G., Gaustad, G., Babbitt, C. W., and Babbitt, G. (2018). Ecological foraging models as inspiration for optimized recycling systems in the circular economy. *Resources, Conservation, and Recycling*, *135*, 48–57.

Schramski, J. R., Dell, A. I., Grady, J. M., Sibly, R. M., and Brown, J. H. (2015). Metabolic theory predicts whole-ecosystem properties. *Proceedings of the National Academy of Sciences of the United States of America*, *122*(8), 2617–2622.

Yodzis, P. (2000). Diffuse effects in food webs. *Ecology*, *81*(1), 261–266.

The Methodology of Industrial Ecology

Chapter 7

Providing Services to Society: In-Use Stocks

> **Chapter Concepts**
>
> - The benefits provided by materials in use in society explain the effort expended to acquire and utilize the materials
> - Because no routine data collection exists for material stocks, estimates of in-use stocks are potentially problematic
> - In-use stocks are predominantly functions of population, especially people with high incomes and/or those in large urban areas

7.1 Standing Stocks

The first measure of ecosystem function for an ecologist is often the *standing stock*: the amount (by number, mass, or some other convenient metric) of a particular species or species group in a particular area. This is generally done by sample collection and species counting but can also involve semi-quantitative estimates. Figure 7.1(a) illustrates the results of research on zooplankton in the Mediterranean Sea, which is important in determining mineral cycling, for example. In contrast to this very detailed analysis, Figure 7.1(b) shows an estimate of the total biomass of different taxa on the entire Earth, with increased detail shown for groupings of animal species. The results demonstrate that plants (mostly living on land) dominate global biomass, animals are a small fraction of the total, and humans are a small portion of the animal fraction. Such results provide the basis for more extensive studies, such as the rate of change of different biomass groupings over time (see Chapter 8).

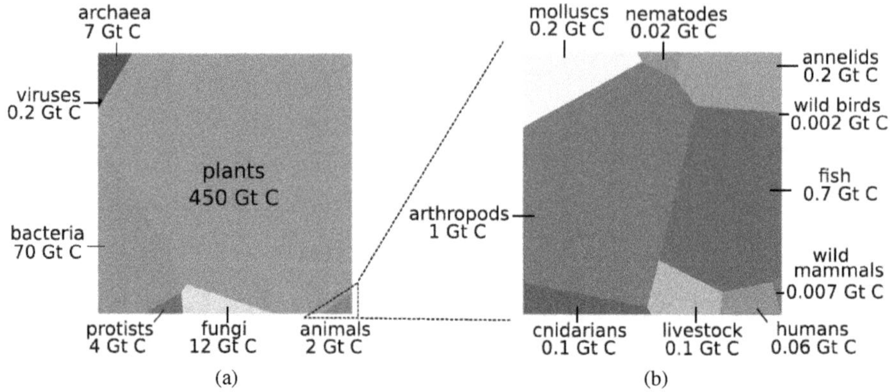

Figure 7.1. (a) Absolute biomass of the global distribution of biomass by taxa; (b) absolute global biomass of different animal taxa (Bar-On *et al.*, 2018). These are *tree charts*.

In an analogous fashion, similar approaches can be used to quantify the standing stock of material that has been mobilized and fashioned by humans — this is generally termed the *in-use stock*. In this chapter, we illustrate several results from this branch of industrial ecology and discuss the information that an in-use stock analysis provides.

7.2 The Value of Material Stocks

In previous chapters, we have shown how resource extraction, processing, manufacture, use, and discard create multiple interlocking flows of materials. We have also indicated that those flows, operating at different time scales, generate in-use stocks of materials. In this chapter, we take a more comprehensive look at those stocks, the stimuli that ultimately create those stocks, and the environmental implications of stock existence and use.

Humans do not mine and process materials because mining and processing is their ultimate purpose. Rather, they become miners and processors because humans value the benefits that they derive from materials when they become stocks in use. This reliance on stocks is easy to demonstrate, if probably self-evident. In Figure 7.2, we reproduce the results of a study that plots (1) per-capita in-use stocks (actually the in-use stocks of iron, aluminum, and concrete, generally the three largest mass flows) in most of the world's countries versus (2) the Human Development Index (HDI), an index generated by the United Nations Development Programme (UNDP) based on three indices: health (measured by life expectancy at birth), education (determined by schooling achieved and expected), and living standards (measured by income per capita). It is immediately clear that those with a higher HDI rank possess a much greater amount of in-use stock than those whose HDIs are lower; that is, as people become wealthier and better educated, they acquire more stuff. For those people, housing is larger and the associated amenities are more numerous, private transport is common, and public services such as roads, bridges, electronic systems, and the like are routinely available.

It is also clear from Figure 7.2 that those with HDI levels above about 0.8 can have a very wide range of in-use stock — that for Singapore is around twice that of the U.S. and Germany, for example. This may be due to geographical differences (as Singapore is an island, for example) or spatial patterns (which influences the per capita level of public transportation). In any case, it is easy to appreciate that those populations that are wealthier and better educated employ much larger quantities of stocks in the service of better qualities of life than those less well off.

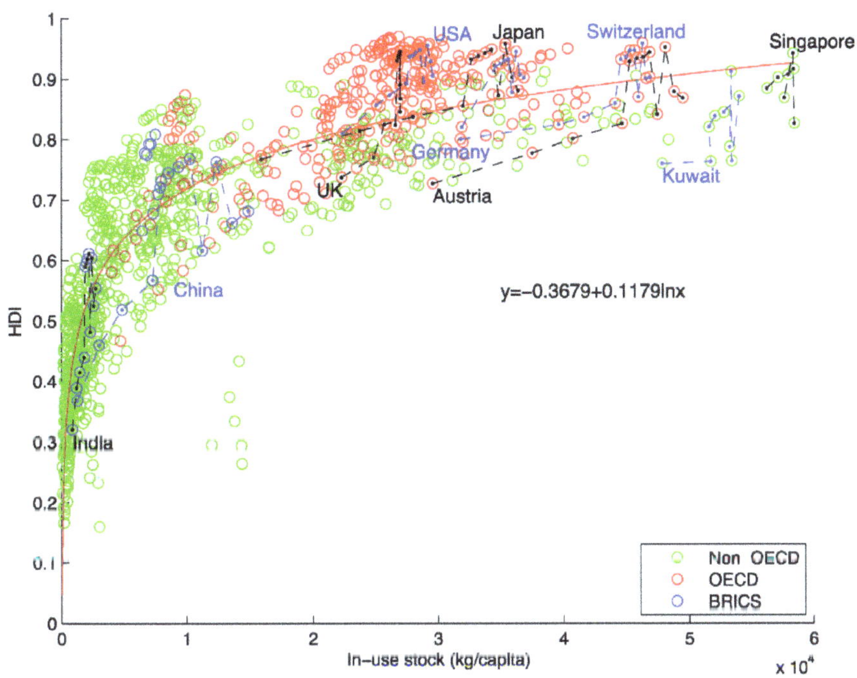

Figure 7.2. Human development index (HDI) as a function of per capita in–use stock (Lin *et al.*, 2017).

7.3 Creating Dynamic Material Flow Analyses

In principle, it is straightforward to determine the level of in-use stocks. There are two possible methods of doing so. One is termed the "bottom-up" method, and is achieved as follows:

- List the principal reservoirs of stocks of materials (buildings, transport, communication facilities, etc.).
- Determine in some way the typical stocks of materials of interest in those reservoirs (by consulting engineering documents, public records, etc.).
- Quantify the number of each of the typical stocks in the region of interest (by observation, tax records, etc.).

A challenge with this approach is that it is inherently incomplete (all reservoirs cannot be surveyed) and substantial on-the-ground effort is required. This approach can be illustrated by the bottom-up analysis of in-use stocks of nickel in New Haven, Connecticut, that is well described by the authors (Rostowski *et al.*, 2007, see References). The authors first determined the major societal sectors within which nickel was likely to be used. They identified six: (1) buildings and infrastructure, (2) transportation, (3) industrial machinery, (4) household appliances and electronics, (5) metal goods, and other end uses. For each of these, the researchers then needed to identify in detail all significant nickel repositories. Their list for one of the societal sectors appears in Table 7.1.

Finally, the research group needed to quantify each final repository, a task relatively easy for hospital equipment (i.e., few interviews needed, and data or measurements straightforward) but more challenging for categories such as the average nickel content in cookware per household. A further consideration was how to count automobiles, trains, and aircraft, some of which resided in New Haven most of the time, while some did not. The results for the different material sectors are given in Table 7.2.

The publication is listed in the References section, but the discussion noted here is sufficient to demonstrate the complexity and inherent inaccuracies of bottom-up approaches to in-use stock analysis.

The second method of stock determination, the "top-down" approach, is accomplished by generating a sequence of material flow analyses (see Chapter 8) for materials of interest. This approach requires estimations for a sequence of time intervals (usually yearly), in which input and output flows are quantified. Top-down approaches tend to be possible where measurements of appropriate flows are routinely available, a situation that favors countries with good import–export

Table 7.1. Weight of nickel in Metal Goods and Other End Uses in New Haven.

End use	Ni in use (Mg)	Ni in use per capita (g)
Domestic equipment	66.7	539
Tableware	1.3	11
Cookware	59.5	481
Tools	3.3	27
Household batteries	0.6	5
Disposable razors	0.1	0
Gardening tools	1.9	15
Commercial equipment	15.0	121
Kegs	10.0	81
Kettles	0.1	1
Tables/stations/bars	2.3	19
Tableware	0.1	0
Cash register drawers	2.4	20
Restaurant industry	15.9	129
Fasteners	9.3	75
Jewelry	0.4	3
Coins	5.0	41
Hospital equipment	1.6	13
Total	113.9	921

Note: *Extracted as an illustration from Rostkowski *et al.* (2007).

Table 7.2. Weight of nickel use in New Haven.

	Ni in use excluding non-resident (Mg)	Ni in use including non-resident (Mg)	Ni in use per capita (kg)
Building and infrastructure	40.9	40.9	0.33
Transportation	51.4	63.6	0.42
Industrial machinery	2.1	2.1	0.02
Household appliances and electronics	112.3	112.3	0.91
Metal goods and other end uses	113.9	113.9	0.92
Total	320.5	332.8	2.59

flows, accurate vehicle registrations, careful mining and processing statistics, etc. Given this information, one proceeds as follows. Assume one wishes to determine the stock σ of material μ in a repository such as the infrastructure of an urban area. During a specified time period, the repository receives an inflow I_μ and loses material in an outflow O_μ. This gain–loss sequence is repeated a number of times, often a number of years given the data commonly available. The in-use stock at the end of the time sequence is then given by

$$\sigma_m = \sum_{y=1}^{n} \left(I_{m(y)} - O_{m(y)} \right) \tag{7.1}$$

where σ_m is the stock of material at time 0, $I_{m(y)}$ is the inflow of material m during year y, and $O_{m(y)}$ is the outflow of material m during year y.

A significant limitation to the reliability of the top-down determination of stocks is that the results depend on a reasonably accurate knowledge of the average lifetime of the product (building, vehicle, appliance, etc.) that contains the material of interest. In some cases, and in some regions, this information can be quite precise (vehicle registration data, for example). It is generally unusual, however, for detailed public or corporate records of most lifetimes of interest to be available. As a consequence, most stock studies are forced to deal with limited data, perhaps drawn from different epochs and different regions and assumed to be "accurate enough".

No extensive compilation of available stock data for different repositories and different regions is currently available and is challenging to keep current. Building lifetimes appear to present a larger challenge than automobiles; they clearly range from a few years to more than a hundred. The lifetimes of "constructions" (bridges, ports, etc.) can vary from 10 years to a century depending on the product under study. Overall, it appears that uncertainties in final product lifetimes pose one of the two greatest challenges, the other being uncertainty in the materials content of the products themselves. All of this is not to say that determinations of in-use stocks are without value, but that their precision is likely to be moderate at best.

7.4 The Utility of Stocks Information

The results of in-use stock determinations provide insight into the ways in which material stocks develop over time and into the material compositions of stocks. In a useful global stock dynamics study, Wiedenhofer *et al.* (2019) estimated the mass of in-use stocks from the year 1900 to the present. The result is shown in Figure 7.3. It is immediately apparent that concrete constitutes half of all stocks, aggregate (crushed stone) perhaps another 30%, and bricks 5% to 10%. Therefore, all other materials — plastics, paper, wood, metals, glass, and so forth — together make up only 10% to 15% of all in-use stock. Because all stock exists to serve

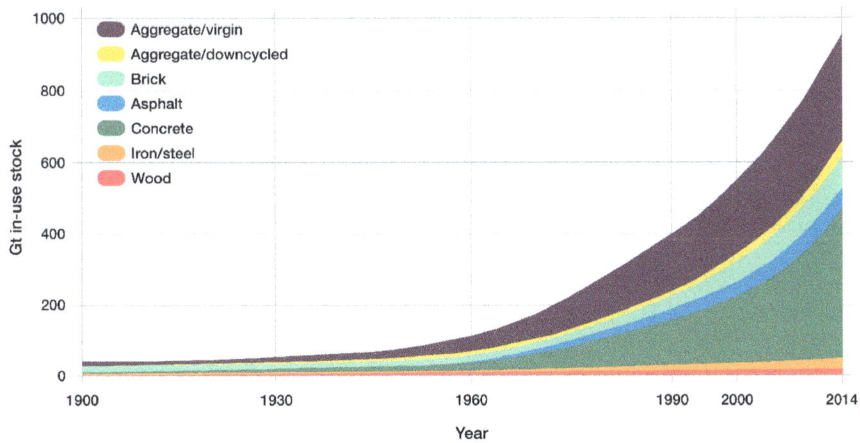

Figure 7.3. Material stocks in use (adapted from Wiedenhofer *et al.*, 2019).

human purposes, it is easy to conclude that the principal uses of concrete, aggregate, and bricks are vitally important to human well-being (in the forms, of course, of buildings, roadways, and simpler forms of infrastructure). Of course, plastics, paper, wood, metals, glass, and other materials have important functions to play in providing services that we desire, but they do so in amounts much smaller in mass than those of the more traditional materials, even though the amounts of those materials are quite dramatic in performing even relatively simple functions.

Less voluminous but perhaps equally useful can be more detailed stock analyses of specific materials or material groups. In an analysis of chromium in Japanese steel (Oda *et al.*, 2010), the authors derived year-by-year flows of alloy steels, subdivided into six subdivisions. The cycle for 2005 is shown in Figure 7.4(a). Austenitic stainless steel is the most widely used of the alloy steels, contributing more than 50% of the contained chromium to the chromium total. A comparison of input and output flows indicates that there are net additions to in-use stock in nearly every alloy steel category, with austenitic stainless steel again leading the way. When the chromium in other steel categories is also taken into consideration in the year-by-year analysis, the result is the determination over time of the in-use chromium in the various application groups (Figure 7.4(b)). Over the 25-year period of the analysis, the chromium in-use stock more than doubled. Such information is quite important to steel reprocessors, as successful separation and recovery of chromium-containing steels is much more profitable than if those steels are lost into the general large flows of the carbon steels.

In-use stocks may also be determined on a spatial basis. In a study that involved surveys both by images from space and from on-the-ground studies,

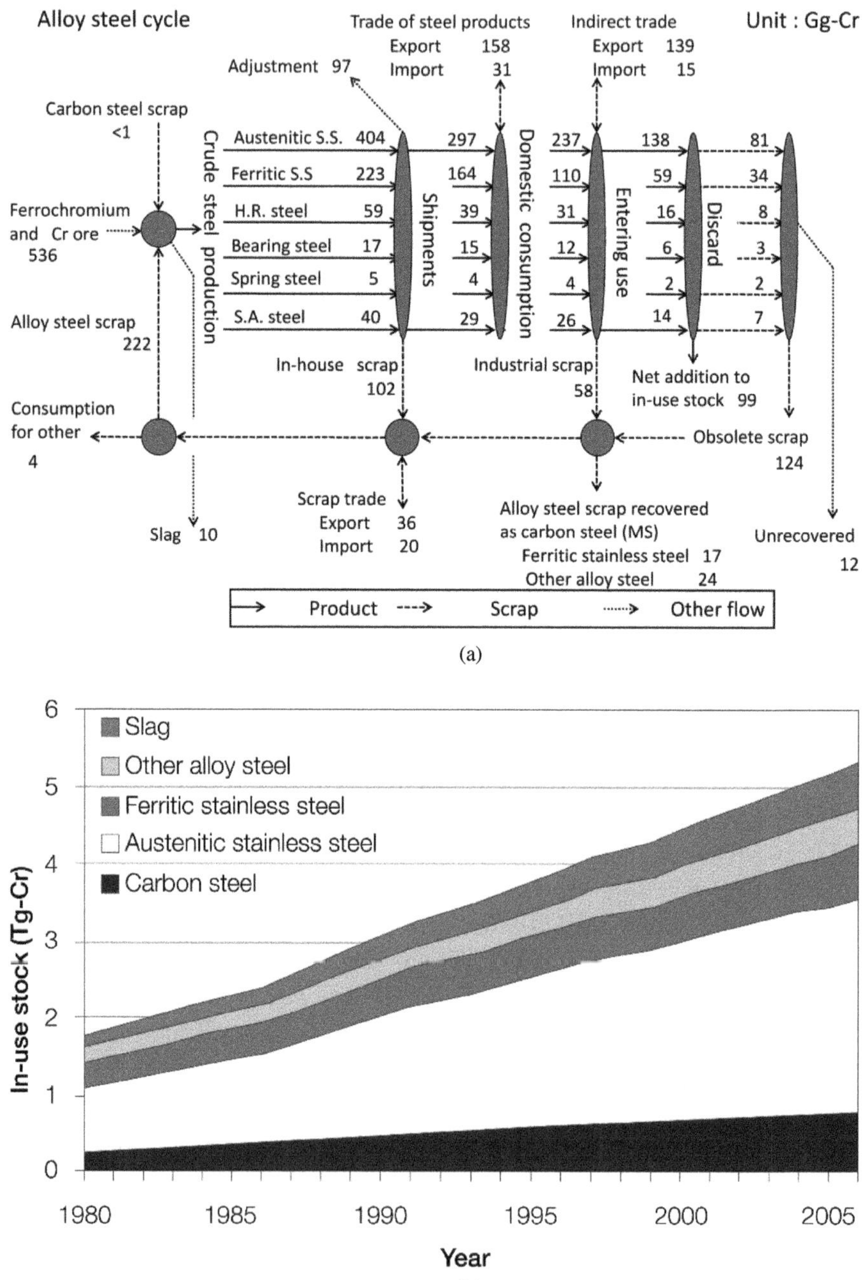

Figure 7.4. (a) The flows of chromium (Gg Cr/yr); (b) additions to in-use chromium stocks (Gg/Cr) in Japan in 2005 (Oda *et al.*, 2010).

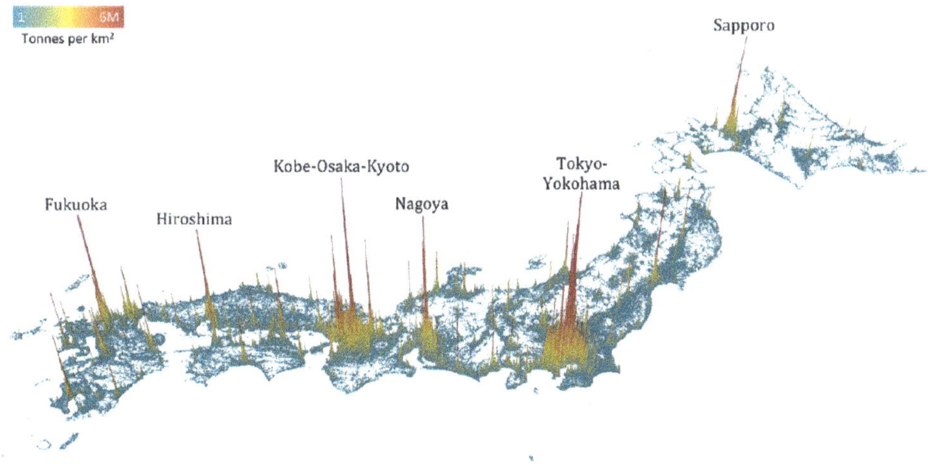

Figure 7.5. The spatial depiction of steel stock in Japan (Tanikawa *et al.*, 2015).

Tanikawa *ct al.* (2015) generated a Japan wide study of the total mass of material in major buildings. The results were presented in a dramatic diagram reproduced here as Figure 7.5. The metropolitan conglomerate of Tokyo–Yokohama and other major urban areas are very easy to distinguish, even without the labels provided by the diagram. In 2007, after the major tsunami off the Japanese coast caused major damage, especially in northern Japan, the stock information from the earlier study enabled a more accurate and quicker provisioning of the materials needed to rebuild those parts of the country that had suffered major damage.

7.5 Policy Aspects of Stock Analyses

Stocks reveal a number of different aspects about the long-term use of materials, the development of urban areas, and the prospects for technologies. Several examples of how this material can be both interesting and useful in urban planning and technology planning are given as follows.

Two parametric evaluations that have proven of significant interest deal with resource use as a function of wealth. Figure 7.6(a) shows that per capita copper stock demand is a roughly linear function of per capita income, although there is lots of scatter around the best fit line, with South Korea and other countries substantially higher and Norway and other countries substantially lower. Nonetheless, the correlation provides a rough estimate of copper demand as countries improve

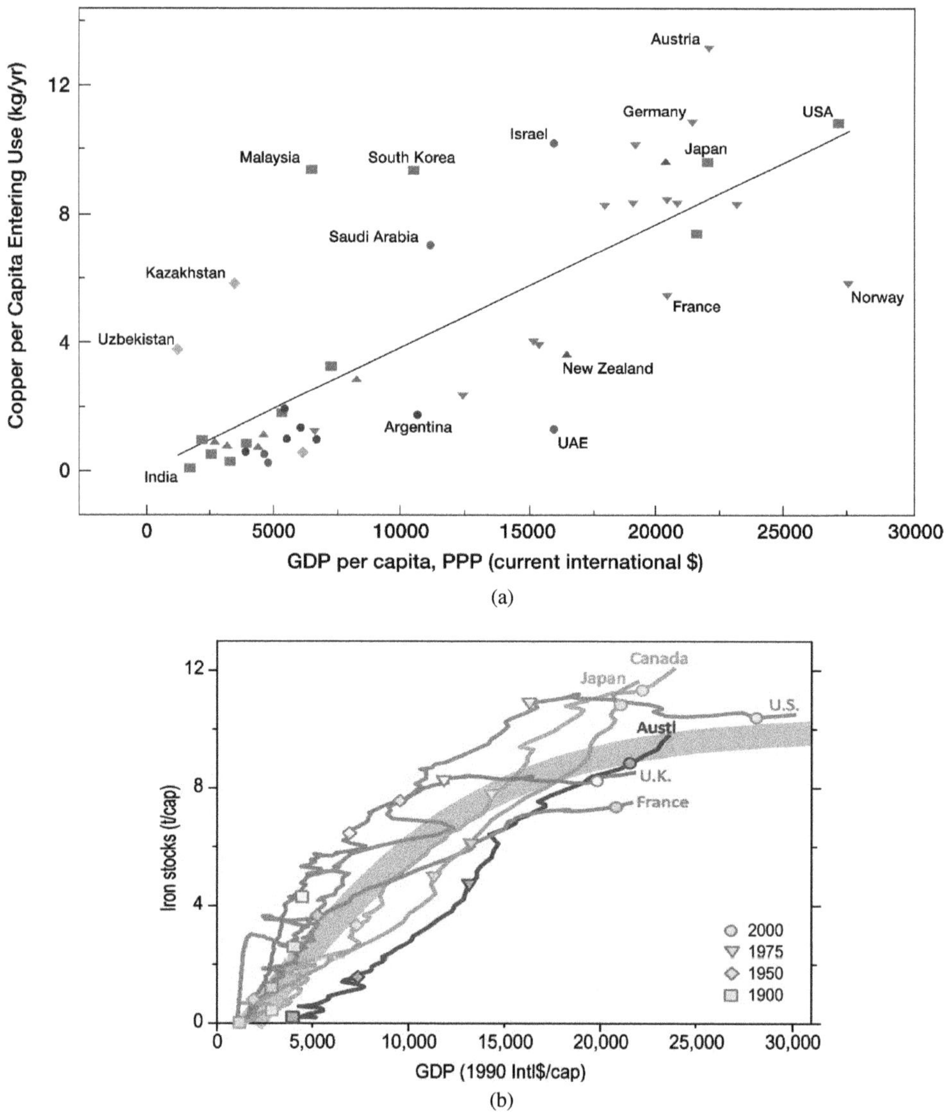

Figure 7.6. (a) Copper stock as a function of GDP/capita for different countries (adapted from Binder *et al.*, 2006); (b) the per-capita saturation of iron in-use stock (Müller *et al.*, 2011).

the average income of their citizens. The second study repeated the concept of Figure 7.6(a) but for iron in-use stock rather than annual copper flow into use. The result showed that some countries, including the U.S., the U.K., and France, had not only increased the per capita stock over time but also reached a saturation point of iron stock; see Figure 7.6(b). Presumably, these countries would only require iron in the future to accommodate the needs of new immigrants or to maintain existing facilities, a result that provides long-term insight into iron demand.

Although in-use stocks are generally beneficial, they can inhibit change even if change might be desirable. An example of this conundrum is the transition from fossil fuel energy to renewable power sources. From an environmental standpoint, such a change is desirable. However, it involves new investment, worker retraining, and technology development, and it may be simpler and cheaper in the short term to leave the current stocks in place. If fossil fuel sources decrease for one reason or another (as a matter of climate change governmental policy, for example), transitions to new and more desirable material stocks can be rapid, as shown by the growth of cellular communications in developing countries during the 2010s.

Further Reading

Bar-On, Y. M., Phillips, R., and Milo, R. (2018). The biomass distribution on Earth. *Proceedings of the National Academy of Sciences of the United States*, *115*, 6506–6511.

Binder, C. R., Graedel, T. E., and Reck, B. (2006). Explanatory variables for per capita stocks and flows of copper and zinc. *Journal of Industrial Ecology*, *10* (1–2), 111–132.

Cabrera Sorrenho, A., and Allwood, J. M. (2016). Material stock demographics: Cars in Great Britain. *Environmental Science & Technology*, *50*, 3002–3009.

Gerst, M. D., and Graedel, T. E. (2008). In-use stocks of metals: Status and implications. *Environmental Science & Technology*, *42*, 7038–7045.

Kong, X. *et al.* (2016). Changes in food web structure and ecosystem functioning of a large, shallow Chinese lake during the 1950s, 1980s, and 2000s. *Ecological Modeling*, *319*, 31–41.

Lin, C., Liu, G., and Müller, D. B. (2017). Characterizing the role of built environment stocks in human development and emission growth. *Resources, Conservation, and Recycling*, *123*, 67–72.

Müller, D. B., Wang, T., and Duval, B. (2011). Patterns of iron use in societal evolution. *Environmental Science & Technology*, *45*, 182–188.

Oda, T., Daigo, I., Matsuno, Y., and Adachi, Y. (2010). Substance flow and stock of chromium associated with cyclic use of steel in Japan. *ISIJ International*, *50*, 314–323.

Pauliuk, S., Milford, R. L., Müller, D. B., and Allwood, J. M. (2013). The steel scrap age. *Environmental Science & Technology*, *47*, 3448–3454.

Rostkowski, K. *et al.* (2007). "Bottom-up" study of in-use nickel stocks in New Haven, C.T. *Resources, Conservation, and Recycling*, *50*, 58–70.

Tanikawa, H., Fishman, T., Okuda, K., and Sugimoto, K. (2015). The weight of society over time and space: A comprehensive account of the construction material stock of Japan, 1945–2010. *Journal of Industrial Ecology*, *19*(5), 778–791.

Wiedenhofer, D., Fishman, T., Lauk, C., Haas, W., and Krausmann, F. (2019). Integrating material stock dynamics into economy-wide material flow accounting: Concepts, modelling, and global application for 1900–2050. *Ecological Economics*, *156*, 121–133.

Zhang, L., Yuan, Z., and Bi, J. (2012). Estimation of copper in-use stocks in Nanjing, China. *Journal of Industrial Ecology*, *16*, 191–202.

Chapter 8

Material Flow Analysis

Chapter Concepts

- Material flow analysis (MFA) is the methodology that is used to construct a metabolism for a chosen industrial material
- MFAs can be generated at different spatial levels and organizational types
- Dynamic MFAs construct an industrial metabolism over time, thereby revealing in-use material stock dynamics

8.1 Introduction

Material flow analysis (MFA) is an analytical response to the desire to study modern society in quantitative detail, with the realization that a great many aspects of that society — housing, food, transport, medicine, and so forth — are built on the backs of materials. Until the early 21st century, however, little quantitative information was available concerning rates of material use, material loss to the environment, efficiency of recycling, and other parameters of interest. Lord Rayleigh famously said in 1883 that "when you can measure what you are speaking about, and express it in numbers, you know something about it; when you cannot express it in numbers your knowledge is of a meager and unsatisfactory kind". MFA has evolved to provide such information related to material flows and stocks.

MFA is one of the central methodologies of industrial ecology. It is through MFA that an "industrial metabolism" (i.e., the flows of resources into and from a particular entity of human society) can be mapped and quantified, much as an accountant determines and quantifies monetary deposits and withdrawals into a bank account. Dynamic MFAs (i.e., those that treat a specific region or system over time) go further; they permit a determination of the in-use and "hibernating" stocks of materials in an industry or society (i.e., the material version of the accountant's "assets and liabilities").

Unlike the accountant, however, who deals only with stocks and flows generally well-reported in monetary terms, the MFA analyst faces a wide diversity of commodities — biomass, polymers, metals, minerals, and more, and transactions that often deal with inadequately described categories (e.g., "iron and aluminum alloys"), lumped categories (e.g., "plastics"), or resource flows that are seldom or never measured (many of the discard flows). MFA-related information quality may vary, from data to rough estimates to conjecture. The MFA analyst also needs to address flows that are of little importance to the accountant because they are not monetized, such as waste flows not captured or emissions to the environment. The MFA specialist must therefore be part detective, part archivist, part extractor of information from experts, and part bold estimator, in order to build the internally consistent database needed to achieve a useful material flow analysis.

In principle, MFA approaches can be applied to any material or combination of materials. In practice, metal stocks and flows have thus far proven to be particularly suitable for analysis, largely because they can often be relatively easily tracked and because data are commonly available for at least some parts of their life cycles. However, MFAs can also treat groups of materials, such as construction minerals (i.e., sand, crushed stone, or cement) or summed material flows into and from a country or region.

8.2 Flows, Stocks, and MFA Methodology

Industrial ecology is a concept with cycles at its very heart, and cycles are analyzed by means of materials and energy budgeting. Nearly everyone is familiar with the concept of a household or personal financial budget, whether or not she or he is conscientious about making and sticking to it. An approach very similar to that of financial budgeting is used to fashion budgets in industrial ecology. The situation can be appreciated with the aid of the diagram in Figure 8.1, which shows a tub receiving water from several faucets and having a number of drains of different sizes. When the water is supplied at constant (but probably different) rates by all the faucets and is removed at an equal total rate by drains with (probably) different capacities, the water level remains constant. When the tank is very large, however, and has some wave motion that makes it difficult to tell whether the absolute level is changing, an observer may not be able to know whether or not the system is in balance. In that case, she or he may try instead to measure the rate of

Figure 8.1. A simple conceptual system for material flow calculations. The water level in the tub over time is determined by the water flows in and out during each period.

supply from each of the faucets and the rate of removal in each of the drains over a period of time to see whether the sums are equivalent. A part of this technique involves the determination of the pool size (the total quantity of water in the tank) and either the rate of supply or the rate of removal. Determination of changes in the pool size then gives information about rates that are difficult to measure. The process of estimating or measuring the input and output flows and checking the overall balance by measuring the amount present in the reservoir constitutes the budget analysis.

Suppose that the input from one of the sources is increased; i.e., in our analogy, the flow from one of the faucets increases. Will the water level keep increasing? The answer depends on whether one of the drains can accommodate the additional supply. If no such drain is present, then the water level will indeed increase. Conversely, if the flow into a drain is enhanced for some reason, such as the removal of an obstruction, then the water level will decrease in the absence of a corresponding increase in the supply. Such a process will continue in this manner unless the flows through the drains adjust themselves to this new factor or unless the new factor results in other drains changing their functioning.

All budgets involve the same concepts mentioned earlier. One is that of the reservoir, in which material is stored. Examples include the shipping department where completed products are prepared for forwarding to customers, or the atmosphere where emissions of industrial vapors collect and react. A second concept is that of flux, which is the amount of a specific material entering or leaving a reservoir per unit of time. Examples include the rate of evaporation of water from a power plant cooling tower or the rate of transfer of ozone from the stratosphere to the troposphere. Third, we have sources and sinks, which are rates of input and loss of a specific material to or from a specific reservoir per unit of time. A system of connected reservoirs that transfer and conserve a specific material is termed a cycle. Industrial ecology budgets follow this pattern: determination of the present stock level (the concentration of a single material or a group of materials), a measurement or estimate of sources, and a measurement or estimate of sinks. A perfect determination of any two of these three components determines the other, as a consequence of the *conservation of mass* principle: material can be transformed, but not lost. Because any material of interest in an industrial facility or in the environment may have several sources and sinks, each source and sink must generally be studied individually.

The MFA process begins with defining the system to be studied and constructing a diagram of the material flows and stocks that occur within that system. This important first step requires detailed knowledge of the system: the points in the system at which material is designed to enter and leave, the points at which material can be stored for shorter or longer periods of time, the nature and location of the transformations that occur, and the locations at which waste products are generated and exported outside the system boundary. Unless the system is carefully and properly specified, the necessary data will not be identified nor acquired, the necessary data may not be pursued, and the resulting system specifications may be incomplete, inaccurate, or of little ongoing use.

A typical schematic diagram for industrial ecology material flows is pictured in Figure 8.2. The system boundary specifies what is to be studied (the element or other material), for what spatial entity (often a country or region), and for what time interval (often a year). The individual boxes are termed "processes"; they are identified by numbers or labels and the material flows among them are shown by

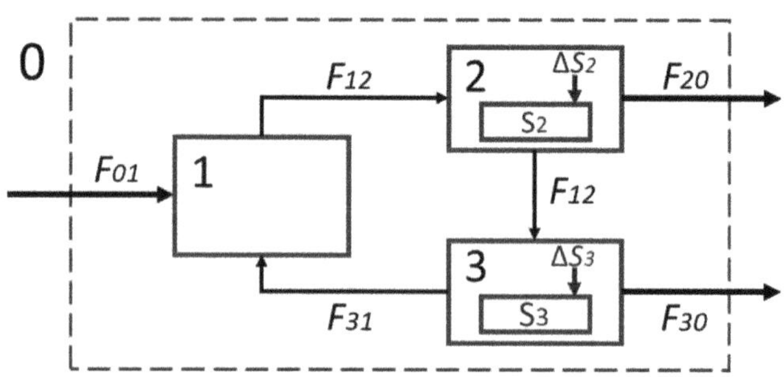

Figure 8.2. A process and system balance diagram for a simple system.

arrows. Each process may involve an internal stock (the small box within the larger one) and a flow into or from that interior box (the net addition to stock). This diagram therefore describes a system that incorporates three processes that exchange and perhaps store material, as well as flows across the system boundary to/from an external reservoir (labeled "0").

The system can be defined mathematically through a set of equations. First, the system variables can be specified:

Stocks: $S_2(t)$, $S_3(t)$
Flows: $F_{01}(t)$, $F_{12}(t)$, $F_{23}(t)$, $F_{20}(t)$, $F_{30}(t)$, $F_{30}(t)$
Net additions to stock: $\Delta S_2(t)$, $\Delta S_3(t)$

Then, writing a mass balance equation for each individual process, and the system as a whole, gives

Process 1: $F_{01}(t) + F_{31}(t) - F_{12}(t) = 0$
Process 2: $F_{12}(t) - F_{23}(t) - F_{20}(t) = \Delta S_2(t)$
Process 3: $F_{23}(t) - F_{31}(t) - F_{30}(t) = \Delta S_3(t)$
System: $F_{01}(t) - F_{20}(t) - F_{30}(t) = \Delta S_2(t) + \Delta S_3(t)$

Given the parameters thus far defined, various indicators of performance can now be derived:

Indicator #1: Efficiency = useful output (t) / total input (t)

Process 1: $\varepsilon_1(t) = F_{12}(t)/F_{01}(t)$
Process 2: $\varepsilon_2(t) = F_{20}(t)/F_{12}(t)$
System: $\varepsilon_s(t) = F_{20}(t)/F_{01}(t)$

Indicator #2: Emissions intensity = emissions/useful output

Process 2: $\varepsilon_2(t) = F_{20}(t)/F_{23}(t)$
Process 3: $\varepsilon_3(t) = F_{30}(t)/F_{12}(t)$
System: $\varepsilon_s(t) = F_{20}(t)/F_{31}(t)$

This MFA vision and methodology can be summarized by a list of attributes customarily possessed by any resulting analysis:

1. An MFA is the study of a clearly designed material flow *system*, not merely the study of a particular material flow.
2. An MFA includes a description of each flow in the system (e.g., the physical and chemical state of each material).

3. An MFA quantifies all flows of significance in the system. Conservation of mass constraints applies at each of the system nodes.
4. The presentation of MFA results is generally diagrammatic as well as numeric.
5. An MFA analysis includes a discussion (or, better yet, a detailed analysis) of the reliability of the results.

No matter what the cycle, the careful construction of a flow diagram is essential if an accurate picture is to be obtained.

8.3 Cycles and Data in MFA

An industrial ecology resource analysis may, in principle, deal with any spatial scale, though global and country-level MFAs are by far the most common. The effort may also deal with many reservoirs, although something like a half-dozen is probably most common. A generic metal cycle, shown in Figure 8.3, is an example. There are six (or seven, depending on how the cycle is constructed) sequential processes (or reservoirs or life stages), linked by the arrows. First, the ore (rock containing metal-rich minerals) is mined (i.e., Mining, "Mi") and the minerals separated from the rock, followed by Smelting, or "S" (to separate the metal minerals from the rock debris) and Refining, or "Rf" (to purify the nickel) — the "S" and "Rf" steps are sometimes combined. At the Fabrication ("F") stage, the metal is fashioned into intermediate products such as rolls and sheets, which are then employed in the Manufacture ("Mf") of products such as appliances or vehicles.

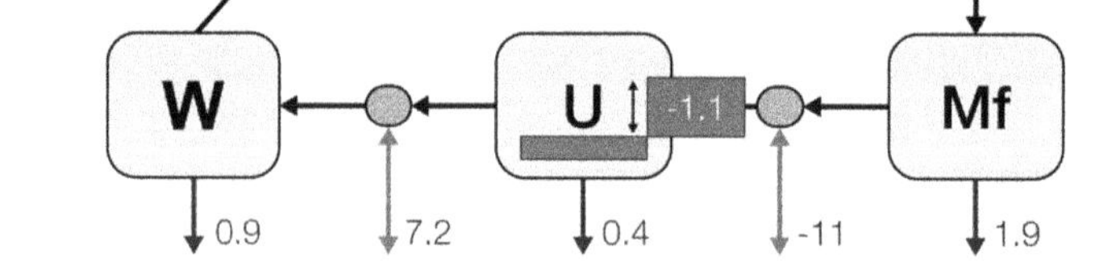

Figure 8.3. An example material flow analysis for a metal. Dark arrows indicate process flows, circles indicate "markets", and light arrows indicate loss flows. Min = mining, S = smelting, Rf = refining, F = fabrication (of intermediate products), Mfg = manufacture (of final products); U = use, W = waste management. All flows are measured in convenient metal mass units, such as Mg/yr.

Products are in Use, or "U" (generally for a number of years), and then are discarded to waste management ("W"), where they may be recovered for reuse, sold as scrap (presumably for reuse), or lost. No process is perfect, however, and losses at each stage are shown as gray arrows.

The cycle in Figure 8.3 is characterized by processes that are linked through "markets" (the intermediate circles), each indicating trade with other regions (light gray arrows) at the respective life stages, and losses to the environment (dark gray arrows). The scrap market plays a central role in that it connects waste management with Refining and Fabrication. The Use process incorporates a stock repository that may increase or decrease (black arrow). The cycle is surrounded by entities lying outside the system boundary: trade partners (other regions), the lithosphere from which ore extraction takes place, and repositories for nickel in tailings and slag (i.e., production wastes) and in landfilling.

Example 8.1

Imagine that you are assigned to quantify the flows and stock of a particular material whose system is pictured in Figure 8.3. All import and export flows are measured (in unspecified units), as is the flow leaving the in-use stock. Compute the missing flows.

Answer:

Process Mi: 55 – 11 = 44; Process S, Rf: 44 + 7 – 2 – 5 = 44;
Process F: 28 – 3 = 25;
Process Mf: 25 – 10.1 – 1.9 = 13; Process U: 13 – 11 – 0.2 – 1.1 = 0.7;
Process Rc: 0.7 + 7.2 – 0.9 = 7.0

Material exports (negative flows from the system under analysis; i.e., sales) indicate a manufacturing process that is producing and selling products. Imports of material occur only in the recycling stage (except the original input) and are used to minimize raw material extraction.

A frequent problem in MFA is getting an accurate assessment of the flow from in-use stock. Typically, the stock of a particular element will be distributed among several types of uses, each of which has its own in-use lifetime. An element contained in home appliances will typically be in use for two or three decades, while the same element in a smartphone may leave use in a year or two. This fact introduces a variable time delay into the system, meaning that material often becomes available for urban mining a decade or more after its initial entry into the system. This delay must be considered in any efforts to quantify the amount of material likely to be available for reuse, and the forms in which it will be discarded.

Product lifetimes are not routinely measured, and thus need to be determined on the basis of expert judgment, field surveys, or the like. Such estimates often are the only way to proceed. In some cases, however, reasonably good data are available, as in the following example.

Assume that element E has four principal uses, and that the flows into use are tracked. Over a six-year period, those flows are listed in the following table:

Flows into use (Mg E/yr)

Use	2015	2016	2017	2018	2019	2020
1	45	60	70	60	75	80
2	30	35	40	30	40	50
3	15	17	20	15	18	25
4	10	10	12	15	16	20

Problem: Find the total outflow from these uses in 2021 if the lifetimes of use are 4, 6, 3, and 1 years.

Solution: For Use 1 with a lifetime of four years, the amount of E released from in-use stock is that which entered in 2017 (four years earlier): 70 Mg E. Similarly, Use 2's flow from use is that which entered in 2015 (30 Mg E); for Use 3: inflow in 2018 = 15 Mg E; and for Use 4, 20 Mg E.

The total outflow in 2021 for all of element E's major uses is thus 70 + 30 + 15 + 20 = 135 Mg E.

All of this material flow analysis seems like a lot of effort. Why do we wish to do this? There are many reasons for wanting to quantify these flows, including the following:

- Manufacturers wish to know how their operations connect with other product life stages.
- Market analysts search for this information to become better informed.
- Taxes, tariffs, and other fees are paid on the amounts and types of imports and exports.
- Environmental scientists are interested in rates of material loss (and related impacts).
- Sustainability researchers want to know the effectiveness of recycling.

Where does the information for generating a material flow analysis come from? Some examples include the following:

- *Government reports on material use, use sectors, and loss*: (The United States Geological Survey's Mineral Commodity Summaries is widely used, for example.)

- *Import/export data*: The UN Comtrade database is widely employed for such information although its level of detail may in some cases be problematic.
- *Corporate reports*: Corporations make public selected material flow information in annual reports, conference presentations, speeches, and the like.
- *Market analysts*: A number of private companies collect and synthesize data and comments from corporations and other experts.
- *Market surveys*: (For example, what percentage of a specific population discarded personal electronics in the last calendar year?)
- *Informed estimates*: It is often possible to informally ask industry experts specific questions about the flows and markets of interest for an MFA analysis.

How can one be sure that stocks or flows with the same name but reported by different sources are really reporting on the same flow? Various national and international standards can be useful in this regard if they are carefully explained in related documents, but poorly described data remain a problem.

None of the earlier suggestions promise surefire success in procuring the desired information. Data may be missing or not useful due to corporate secrecy, government secrecy, not disaggregated enough to be useful, or simply that nobody measured it.

How can missing data be dealt with? Among the many possibilities are the following:

- Employ the principle of conservation of mass. (If information for all but one flow into a node is available and you believe no significant in-use stock change is occurring, the missing flow can be approximated by addition and subtraction.)
- Use data from other countries, regions, etc. as proxies.
- Restrict the scope of the analysis. (A common approach is to determine flows up to and including in-use reservoirs but to not address in-use lifetimes and recycling flows.)
- Restrict the scale of the analysis. (For example, produce a European-level analysis instead of a country-level analysis. Global data are often more available and more accurate than regional data, which in turn are often more accurate than country-level data.)
- Infer flows of your target material from related information. (For example, alloy flow information (e.g., tool steel) provides information about the individual elements in the alloy (e.g., tungsten and cobalt in tool steels).)

An example of a carefully prepared MFA analysis is shown in Figure 8.4, which shows a flow diagram for the zinc cycle of Asia in 2010. Several features

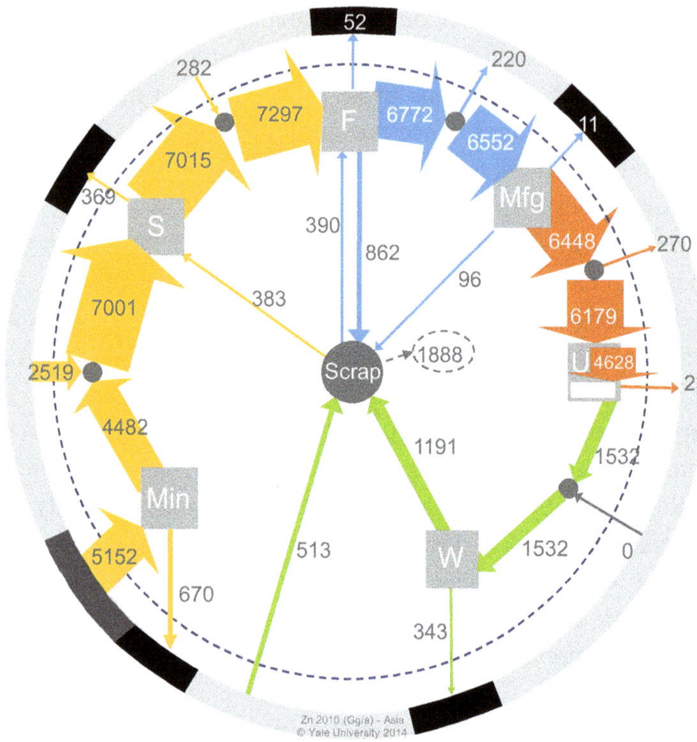

Figure 8.4. The 2010 zinc cycle for Asia. The line widths are proportional to the magnitudes of the zinc flows from one node of the diagram to the next. The colors indicate flows of zinc during ore processing (yellow), fabrication (blue), manufacturing (tan), and discard, recycling, and loss (green). Min = mining, S = smelting, F = fabrication, Mfg = manufacturing, U = use, W = waste management. The units are Gg/yr (Meylan and Reck, 2017).

are easy to deduce, including the largest input flow (yellow arrow, from domestic Chinese mines), the large reuse of scrap (blue arrow), and the large rate of loss at the end of use (green arrow). Trade with other regions occurs between life stages at "markets" (i.e., the small black circles between the flows). A small dashed-line insert shows input to the Scrap node to reflect an input flow that is required to balance input and output, but for which available data do not determine definitive assignment to one of the larger flows.

An extension of a static MFA analysis is the "dynamic MFA", which consists of a sequence of cycles, usually year by year for a decade or more. A benefit of such an analysis is that it provides a quantitative estimate of in-use stock — the cumulative difference of flows into and out of use. Examples of in-use stock determinations utilizing MFA approaches are illustrated in Chapter 7.

It is common that the data on material flows, having been collected from many different sources, are not completely consistent. This challenge can be dealt with in two ways. The first is to make one's best assumption as to which of the variables is more likely to be correct and to adjust the other(s) to achieve numeric consistency. The second is to admit that the problem exists but to call attention to it rather than make an assumption as to likelihood (the latter of which, in effect, disguises the discrepancy from the reader). Calling attention to an MFA data problem can be done in various ways, but perhaps the most common is to add a distinctive flow and reservoir to the diagram, as shown by the dashed line oval from the Scrap sector in Figure 8.4 to complete the resource balance but to indicate where the discrepancy occurs.

In some cases, as in Figure 8.4, the existence of flows crossing a boundary may be well known, and the principal challenge is to quantify them (and the related stocks as well). In others, as in the atmospheric chlorofluorocarbons (CFCs), the importance of some of the sources and sinks may not be known and must be deduced by quantifying those flows for which information is available. In all cases, however, the process begins by specifying the metabolic cycle, its metabolites, and its enzymes, in as much detail as possible.

A high level of detail is desirable in an MFA. The additional work needed to generate such a product is substantial, but the result can have considerable value. Such an achievement is shown in Figure 8.5 — the global cycle of the important

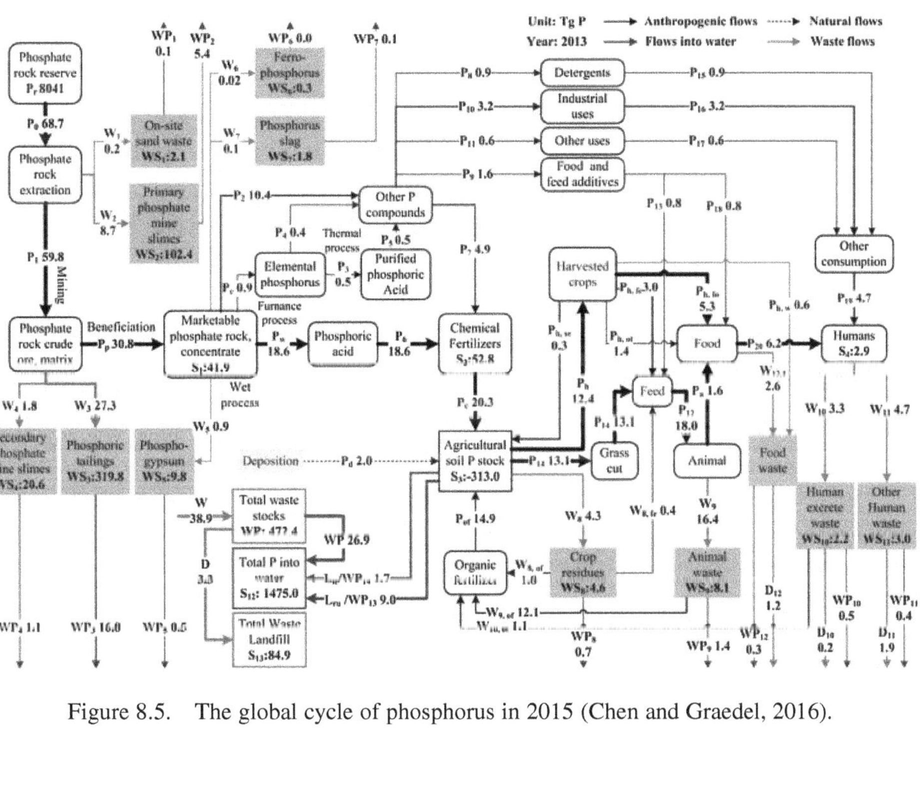

Figure 8.5. The global cycle of phosphorus in 2015 (Chen and Graedel, 2016).

nutrient phosphorus. Inputs, flows, and losses are clearly shown, and the opportunities for flow intercept are clearly visible.

In industrial ecology, the concepts of budgets and cycles are applied not only to the anthropogenic use of resources but also sometimes to the combination of anthropogenic and natural processes. The results help evaluate present metabolic needs and to estimate those that may be required in the future. Similarly, it may be possible to study specific resources as they pass through various technological organisms, and thus evaluate resource supply, use, and loss. Substance-centered cycles complement the product cycles addressed by life cycle assessment (Chapters 11 and 12).

8.4 Scope of Materials Being Addressed in MFA

As the protocols and approaches of MFA became reasonably well structured in the early part of the 21st century, the applications were largely to three metals: iron, aluminum, and copper. In decades since, a large number of other elements have been studied by MFA approaches. This has been most successful for elements that are used primarily or entirely by themselves.

Petrochemicals and plastics differ from metals in that they cannot readily be recovered, reprocessed, and reused. The traditional end-of-life approach has been either to burn them for energy production or to down-cycle them into products such as park benches. Perhaps partly because of this "one time through" design, material flow data for these materials as a group, or for individual plastics, have not been widely available. Nonetheless, MFAs for plastics as a group were published in peer-reviewed journals as early as 1998. These MFAs have usually been for specific plastics, generally the consumer plastics polyethylene, polypropylene, polystyrene, and polyurethane. Polyvinyl chloride, which sees use both commercially and in consumer goods, has also been addressed. Comprehensive (grouped plastics) analyses that emphasize flow to final use sectors of the economy rather than specific polymers have also been published.

Fibers constitute a diverse group of materials that includes forest products and clothing. The former has a wide variety of uses, from the very long term (buildings and construction) to the very short term (paper). In recent years, MFAs for forest products have been generated for lumber, paper, or both. Wood flow dominates in some countries and regions, while paper flow dominates in others. Data for clothing fibers such as cotton or wool are much less available and clothing fiber MFAs appear not to exist.

Most modern construction employs minerals: cement, crushed stone, sand, and ornamental stone such as marble. The amounts used are very large; larger even than food or fossil fuels. Most of these materials have low value and are mined and used locally, which tend to limit the availability of data. Unlike other minerals, however, data for cement seem well enough established to enable cement analyses to be generated. In fact, estimates of the annual production of concrete are generally produced by utilizing cement data in combination with average ratios of cement to crushed stone and sand. Because construction minerals are locally sourced and have very long in-use lifetimes, MFA cycles have not yet been constructed.

Composite materials are those in which materials are intertwined to achieve physical properties that are unavailable from less complex material combinations. The common example for many years has been fiberglass, in which glass fiber rods are laid within sheets of plastic so as to produce a lightweight product with good mechanical properties. More recently, carbon fiber-reinforced polymers (CFRPs) have seen rapidly increasing use, especially in the automotive and aircraft industries. Consequently, CFRP manufacture rates are now roughly equal to those of some metals such as molybdenum (about 40 Gg/yr). With CFRP recycling apparently nearing potential commercial reality, detailed MFA analyses of composites seem likely to become increasingly common in the future.

Food and related materials have been quantified in some detail for decades, particularly by the UN Food and Agriculture Organization (FAO), although the information is not typically presented in MFA form. An exception has been for phosphorus, in view of its vital importance for agriculture. Detailed phosphorus MFAs have been derived for the world, countries, and cities.

It is well known that in most cases energy is required to pump and purify water, and that water is often required to produce energy (e.g., hydroelectric dams, cooling water for nuclear power plants). As a consequence, water-energy MFAs are reasonably common. Water for non-energy uses is typically not conserved but rather discarded at the end of use. Thus, non-energy water has not customarily fit the pattern of a typical MFA. With the increasing reprocessing and reuse of water, however, MFAs will soon become possible at small scales such as buildings, towns, and regions.

In later chapters of this book, these groups of materials are addressed in more detail, and examples of MFA studies of those groups are illustrated.

8.5 Displaying Material Flow Analysis Results

The flows of material in an MFA system can be simply depicted as lines with labels and numbers, as seen in Figure 8.3. An alternative approach that visually enhances such a diagram is the Sankey diagram originated from Irish engineer

Matthew Sankey in 1898. In Sankey diagrams, the magnitudes of resource flows are indicated by line width. Sankey's initial diagrams were for steam flow in equipment. Many variations and expansions have followed.

Sankey diagrams have become so common in MFA that they have been termed "the visual language of industrial ecology". These diagrams have a unique ability to illustrate many features of an MFA analysis on a single diagram. Color is often utilized therein to aid in referring to the magnitudes of particular flows, to indicate additional information such as life cycle stages, or to illustrate a particular property of the analysis (see Figure 8.4 for an example).

Software to generate Sankey diagrams became routinely available in 2004, when Helmut Rechberger *et al.* at the Technical University of Vienna developed a software program to minimize the effort involved in generating MFA analyses for a variety of systems. The current version of this "STAN" open source software is freely available for download and use. It includes features such as data reconciliation and error propagation. It also indicates the degree of uncertainty in non-measured flows. An example of its graphical output (the numeric output is also available) is shown in Figure 8.6, the result of a detailed MFA collaboration involving aluminum industrialists, researchers, and analysts worldwide. The progress of the aluminum from the mine through to final use is easily followed through the widths of the flow lines. The amount of scrap recovered and reused also stands out. Import and export flows are clearly depicted as well.

8.6 Determining of Uncertainty in MFAs

As mentioned earlier, MFA results are uncertain to various degrees. Some flows are relatively well-determined, such as the import flow of mineral concentrates. Others, such as the flows of materials contained in products, are generated by averaging information on major product types. In some cases, such as losses to landfills are determined by difference or by informed estimation.

Given reliability variations from one flow to another on the same diagram, an exemplary MFA will attempt to define the levels of uncertainty for each flow. In a recent example characterizing MFA uncertainty, Lupton and Allwood displayed estimated uncertainty for each flow by varying the color on a Sankey MFA diagram notable for its clarity and beauty (Figure 8.7). The line widths provide visual estimates of flow rates. In this otherwise excellent display, the authors follow a common but unfortunate practice of using a metric (megatons, Mt) that merges an SI unit (Mega) with a non-SI unit (tons, or is it tonnes, or short tons, or long tons?). Such an approach is an invitation to confusion. This book advocates using

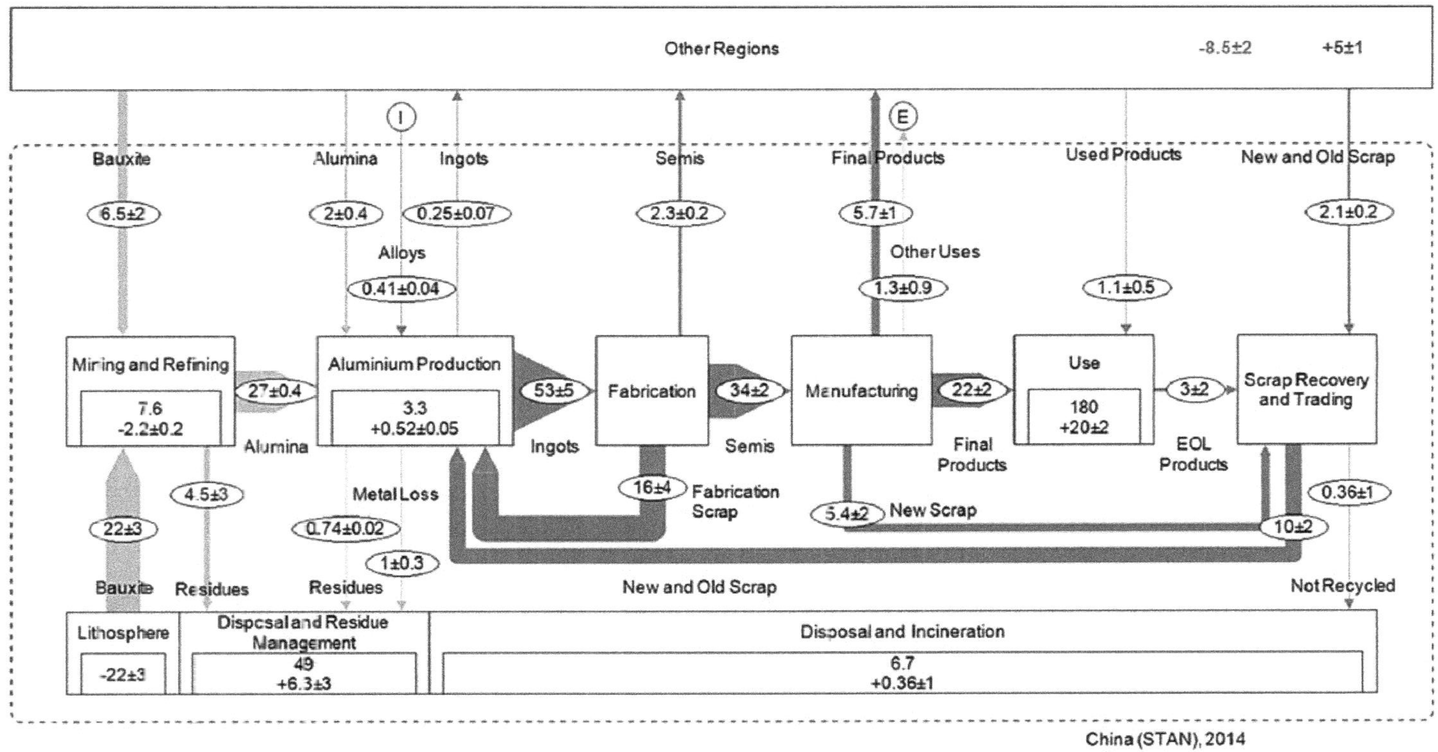

Figure 8.6. The MFA analysis of aluminum stocks and flows in China in 2014 as rendered by the STAN software (Bertram *et al.*, 2017).

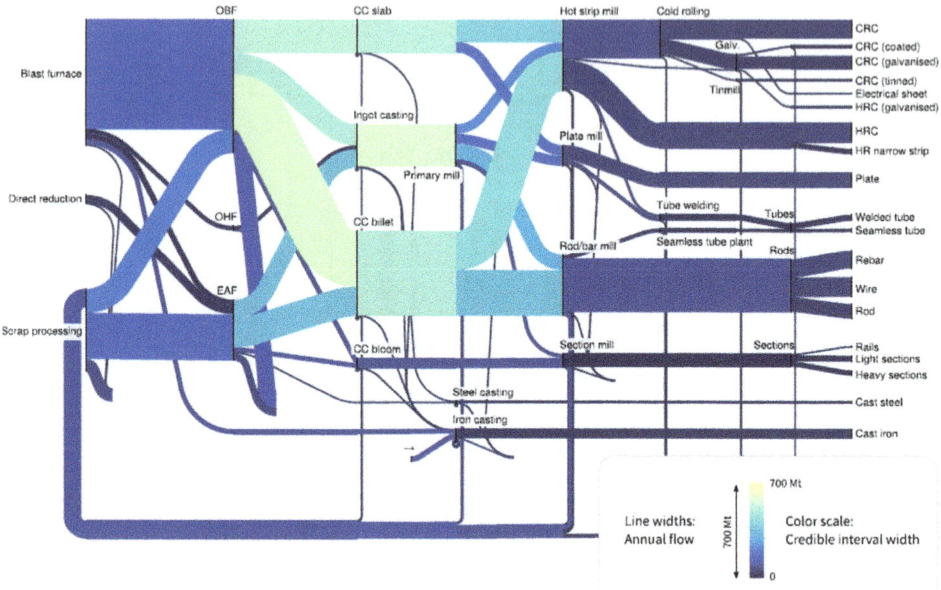

Figure 8.7. An example of a Sankey diagram resulting from a material flow analysis (this is for iron on a global basis). In this diagram, the line width indicates the flow magnitude, while the color indicates the level of uncertainty in the flow (Lupton and Allwood, 2017).

nothing but SI units in industrial ecology, as listed and defined in one of the appendices.

Acknowledgement

Aspects of the mathematical description of the dynamic MFA system of Figure 8.2 are abstracted from an open-source presentation by Professor Stefan Pauliuk of the University of Freiburg, Germany.

Further Reading

Bertram, M. *et al.* (2017). A regionally-linked, dynamic material flow modelling tool for rolled, extruded, and cast aluminium products. *Resources, Conservation and Recycling*, *125*, 48–60.

Chen, M., and Graedel, T. E. (2016). A half-century of global phosphorus flows, stocks, production, consumption, recycling, and environmental impacts. *Global Environmental Change, 36*, 139–152.

Eckelman, M. J., and Chertow, M. R. (2009). Using material flow analysis to illuminate long-term waste management solutions in Oahu, Hawaii. *Journal of Industrial Ecology, 13*, 758–774.

Graedel, T. E. (2019). Material flow analysis from origin to evolution. *Environmental Science & Technology, 53*, 12188–12196.

Kawecki, D., Schneeder, P. R. W., and Nowack, B. (2018). Probabilistic material flows analyses of seven commodity plastics in Europe. *Environmental Science & Technology, 52*, 9872–9888.

Lupton, R. C., and Allwood, J. M. (2017). Hybrid Sankey diagrams: Visual analysis of multidimensional data for understanding resource use. *Resources, Conservation and Recycling, 124*, 141–151.

Meylan, G., and Reck, B. K. (2017). The anthropogenic cycle of zinc: Status quo and perspectives. *Resources, Conservation and Recycling, 123*, 1–10.

Stan2Web. (2012). Retrieved from: https://www.stan2web.net.

Chapter 9

Material Recycling and Reuse

Chapter Concepts

- Recycling that preserves the inherent properties of a material ("functional recycling") is currently compromised by applications that are inherently dissipative or by applications for which recycling technology has not yet been developed
- Functional recycling rates for many materials are very low
- Few products are designed from the start with recycling in mind; this renders efficient recycling of materials and parts extremely difficult and expensive

9.1 The Availability of Recyclable Material

Why can't materials that are incorporated in products of various kinds be reused when the use of those products is finished? This seemingly obvious inquiry can be addressed, at least to some extent, with the diagram shown in Figure 9.1(a), which is directed at the principle applications in which an industrial material is employed. In the figure, the applications are divided into four categories: (1) "in-use dissipated", (2) "currently unrecyclable", (3) "potentially recyclable", and (4) "unspecified" (generally small-scale uses whose low volumes do not justify tracking them). The *in-use dissipated* category includes uses that may seem beneficial (fertilizer ingredients, fireworks, etc.) but offer little or no prospects for material recovery and reuse. Some other applications, such as the use of rare earth elements in polishing powders, could be recyclable if a technological approach had been developed, but often no suitable technologies exist. In the *potentially recyclable*

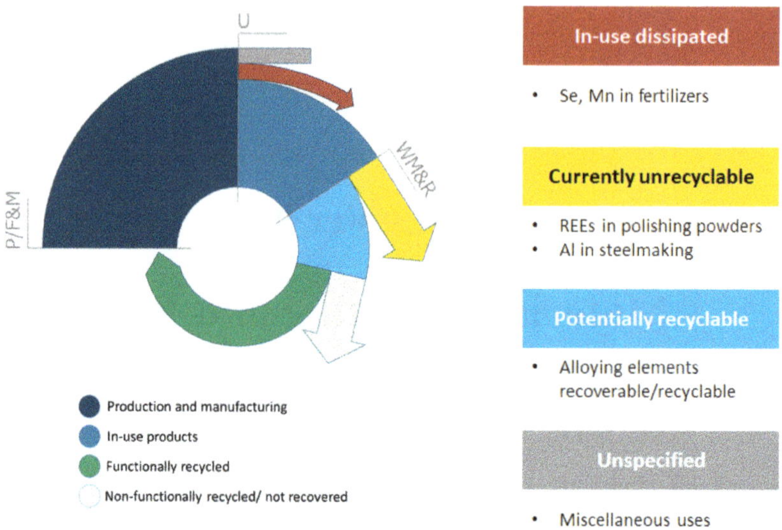

(a)

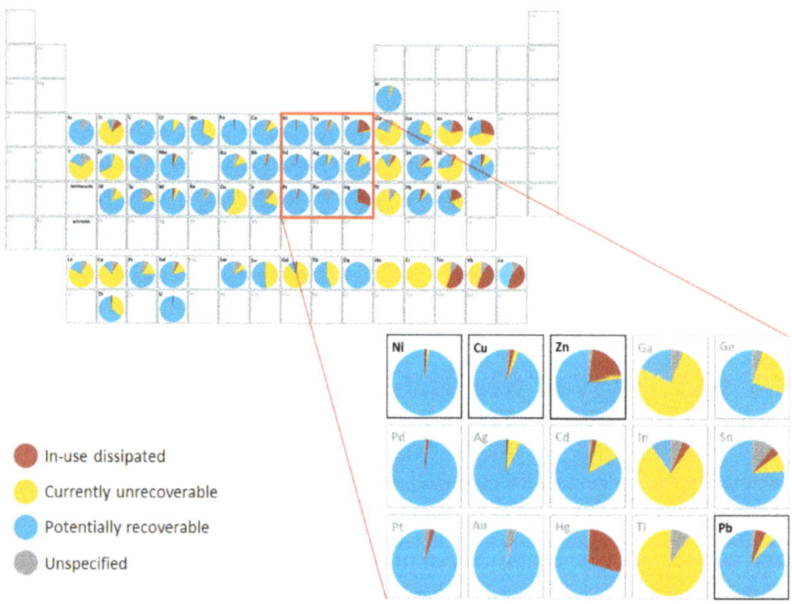

(b)

Figure 9.1. (a) Material streams for materials in products: dark blue — production and manufacturing; gray — unspecified minor uses; red — in-use dissipative; medium blue — in-use products; light blue — potentially recyclable; yellow — currently unrecyclable; green — functionally recycled; white — non-functionally recycled or not recovered. (b) Percent distributions among the four material streams for a selection of elements (Ciacci *et al.*, 2015).

category, recycling methods exist, although are sometimes not employed for reasons of cost, inconvenience, or lack of sufficient incentives.

Two examples help to illustrate the challenges of availability for recycling. Figure 9.1(b) shows the percentages of uses of several elements. Consider the diagram for zinc (Zn) (top center). Nearly two-thirds of all zinc is used as a corrosion protection coating for steel ("galvanized steel"). Over time, around 20% of those coatings are dissipated during use; the remainder is typically recovered in steel recycling. Almost all other uses, such as die casting and alloying, are also recoverable. Gallium (Ga), to zinc's immediate right on the diagram, is mostly used in integrated circuits and other electronic applications. In principle, those products are recoverable and one might assume that a recycling technology has been developed, but none has emerged to date. Smaller fractional uses of gallium in alloys and magnets can, however, be recycled. Similar studies for other elements indicate a wide range of recycling potential that can only be improved by more insightful product design and more extensive development and deployment of new recycling technologies.

In the ideal world, material available through recycling would satisfy the demand for the same material and no new resource extraction would be needed. However, materials enter service and remain there for extended periods, often decades. This situation is termed the "delaying effect of stocks", a consequence of which is that in a world of increasing demand even perfect recycling is not enough

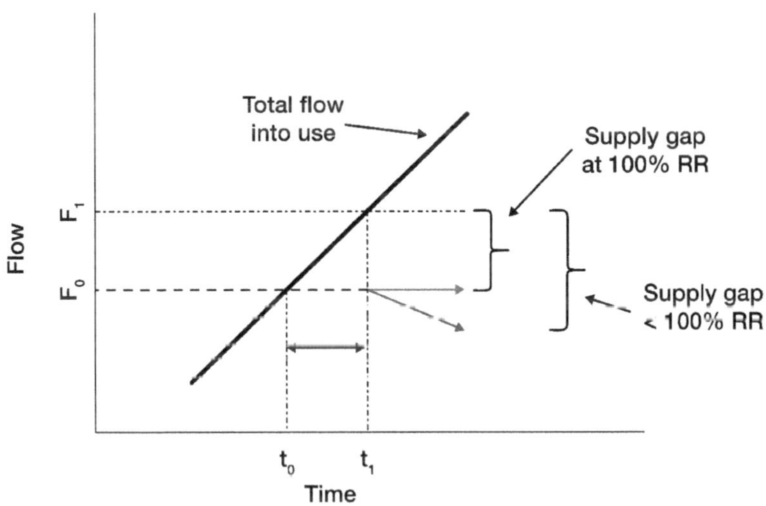

Figure 9.2. The "delaying effect of material in product stocks in use". In a world of increasing demand, even perfect recycling is not enough to meet supply.

Figure 9.3. Map of disconnected power cables (bold black lines) in the city center of Linköping, Sweden (Krook *et al.*, 2011).

to meet supply (Figure 9.2). At time $t = 1$, the material that entered in-use stocks at time $t = 0$ cannot meet the new higher demand even if completely recycled (green line). If recycling is less than 100% efficient (red line), as is always the case, the demand/supply gap is even worse. Moreover, some materials may not immediately undergo reprocessing and reuse. Personal electronics are famous for being retained in a bedroom drawer for as long as a decade — these are often termed "hibernating stocks". A related category is "comatose stocks" — material that is stored in such a way that it may never be recovered, such as power distribution cables disconnected from service but left in place because the benefits of recovery do not offset the effort and expense involved provide an example (Figure 9.3). Finally, there are stocks that are designed never to be recovered and reused, such as the foundation pilings under tall buildings and harbor structures; these might be termed "abandoned stocks".

Imagine, however, that a decision has been made to discard a product containing potentially recyclable material. Many steps may be involved in actually carrying out technologically appropriate recycling, as discussed in the following in some detail.

9.2 The Recycling Process

Assume that a material (contained in a product) is not subject to any of the constraints to recycling and reuse discussed in Section 9.1 and that the material has been discarded. A stepwise sequence is involved in successful recycling, as shown in Figure 9.4, but failure to capture components or improper processing occurs to varying degrees all along the sequence. Given the estimated current probabilities for successful processing at each stage, the efficiency of the overall process turns out to be quite low. Improving this situation requires efforts at all stages of the recycling process, but also in the original product design process. Environmental product design will be discussed in detail in Chapter 31, but some of the main points are summarized as follows:

- If possible, capture a product before discard and seek to use it elsewhere (this is termed "relocation").
- If relocation is not feasible, seek to remanufacture the product so as to return it to its original condition and capabilities or, better yet, upgrade it to the most recent capabilities of similar products.
- If remanufacturing is not practical, disassemble the product and reuse the components. This step will be enabled by identifying opportunities for efficient

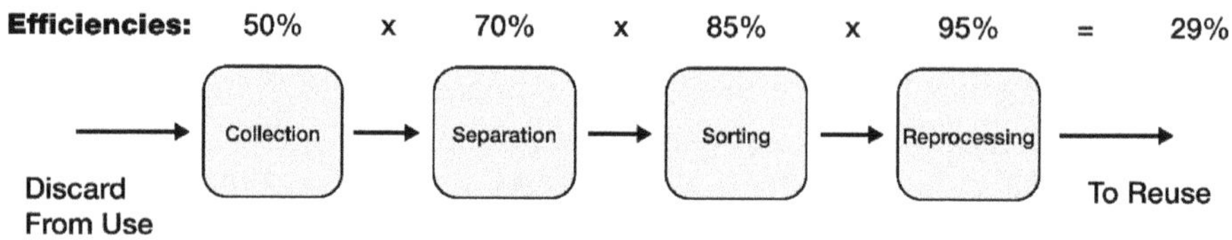

Figure 9.4. The stepwise recycling sequence from discard to newly available materials. Typical percentage efficiencies at each step are indicated. (Adapted from diagram developed by C. Hagelüken, Umicore Corporation.)

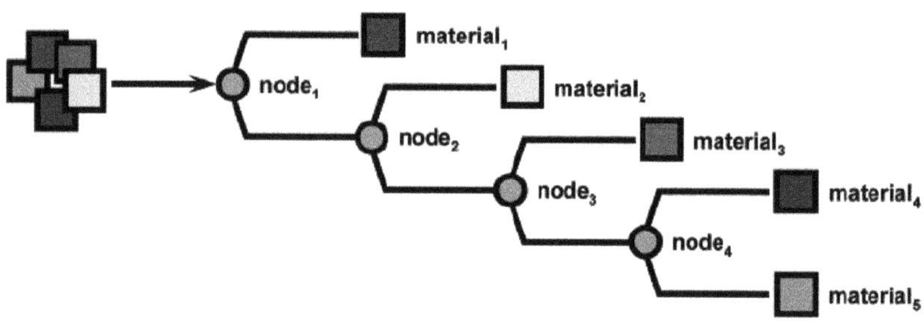

Figure 9.5. A four-node separation sequence for disassembly of a generic product (Dahmus and Gutowski, 2007).

identification of the components and researching opportunities for their redeployment. Disassembly is best addressed at the product design stage by minimizing the steps needed for disassembly (Figure 9.5).

- Components and assemblages that cannot readily be disassembled, or where doing so is not economically or practically feasible, may or may not be shredded, but in any case, are sent on to sorting facilities, followed by treatment in chemical or metallurgical reactors.

A further consideration relates to basic thermodynamics and is related to the fact that a great many modern products are mixtures of materials. If such a mixture is supplied to an industrial process, some portion of the materials may be difficult or impossible to recover. The situation is illustrated in Figure 9.6 for metal mixture processing in smelters and electric arc furnaces. Elements in green are fully recoverable if recycled by an appropriate process, but those in yellow may require special processing and those in red are very likely to be lost. Elements present in small quantities are particularly likely to be unrecoverable, at least without unreasonable cost and effort. For example, in copper or nickel smelters, rare earth elements (REEs) are very likely to be lost. Thus, minimizing elemental loss during recycling

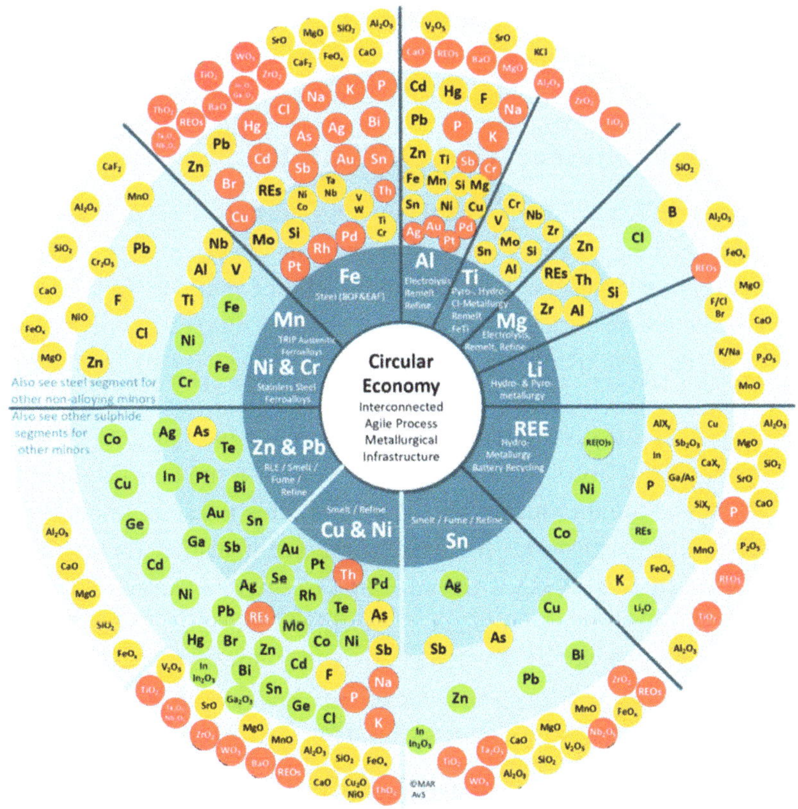

Figure 9.6. The "metal wheel" that shows the likely destinations of elements in various metallurgical reactors (the elemental reactors shown in the innermost circle divisions). Elements in green are likely to be recovered in elemental form, those in yellow in compound form, and those in red lost or incompatible with the base metal.

imposes a challenge for process designers who must minimize or avoid mixing metals into products which recycling technologies cannot yet easily separate them.

An opposite challenge arises if a product made in the past contains elements or compounds whose properties while in use were thought sufficiently valuable to offset their toxicity, but must now be managed as a waste stream. Cadmium plating of aircraft landing gear as a corrosion preventer is an example. In these cases, the current approach is to encase these materials permanently in hazardous waste disposal sites, but Paul Brunner of the Vienna Institute of Technology and his colleagues call for designing and implementing "safe final sinks": long-term storage locations where such materials can be stored without hazard, and perhaps be re-mined in the future if the need arises.

9.3 Quantifying the Recycling of Materials

Recycling rates have been defined in a number of different ways over the years, a practice that resulted more in confusion than in enlightenment. The situation stimulated the United Nations International Resource Panel to appoint a committee in 2009 to establish a standard protocol for recycling rate determination and to then provide the best available recycling rate values by element under that protocol.

After meetings and discussion, the diagram shown in Figure 9.7(a) was generated. It clearly labels flows through the life cycle of a metal. (The approach has since been adapted to other materials.) The committee made an important distinction between recycled material returned to the scrap market for the element in question [flow (g)] and recycled material lost to the recycled flow of other elements [flow (f)]; the former was labeled "functional recycling" because the physical and chemical properties of the element in question were retained for reuse, and the latter labeled as "non-functional recycling" because the element, while not lost, was no longer capable of contributing its unique properties to a second use. The "end-of-life functional recycling rate" was then defined as

$$\text{EOL-RR (functional)} = g/d \tag{9.1}$$

With that definition, the committee then assigned the "best-estimate" functional recycling rate of the elements of the periodic table to one of five percentage ranges, as shown in Figure 9.7(b) (for more details, see Reck and Graedel, 2012).

It is easy to see in the figure that 15–20 elements have rates above 50%. (The estimates are for 2008 and the committee judged most of them to be "not much above 50%".) Perhaps more dramatic are the more than 30 elements with essentially no functional recycling at all. Only a few elements were assigned values in between the two extremes. Thus, a majority of the elements employed in technology were used once and then lost to technology forever, a sad fate given the energy and effort expended to acquire them in the first place.

Recycling statistics have never been very good, as no regulations require them to be collected. Consequently, the best estimates of EOL-RR values remain those of the International Resource Panel in 2009. This would seem to call for a more

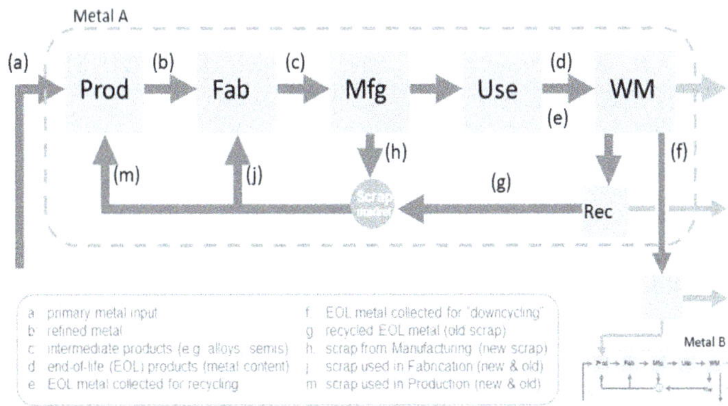

(a)

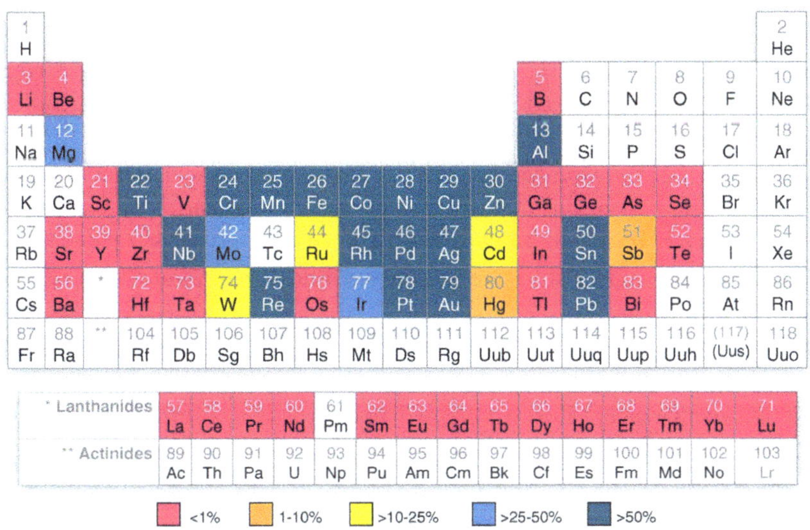

(b)

Figure 9.7. (a) Flows related to the life cycle of metals and the recycling of production scrap and end-of-life products. Boxes indicate the main processes (life stages) (International Resource Panel, 2011). (b) End-of-life functional recycling rates of 60 elements, with the individual elements categorized into one of five ranges (Reck and Graedel, 2012).

structured data-driven approach to routinely quantifying recycling rates. It is to be hoped that such an approach could be put in place in the future; at present, all end-of-life recycling rates must be considered "informed estimates based on minimal data".

Further Reading

Ciacci, L., Reck, B. K., Nassar, N. T., and Graedel, T. E. (2015). Lost by design. *Environmental Science & Technology*, *49*, 9443–9451.

Dahmus, J. B. and Gutowski, T. G. (2007). What gets recycled: An information theory based model for product recycling. *Environmental Science & Technology*, *41*, 7543–7550.

Gaustad, G., Olivetti, E., and Kirchain, R. (2012). Improving aluminum recycling: A survey of sorting and impurity removal technologies. *Resources, Conservation, and Recycling*, *58*, 79–87.

Graedel, T. E., Harper, E. M., Nassar, N. T., and Reck, B. K. (2015). On the materials basis of modern society. *Proceedings of the National Academy of Sciences of the U.S.A.*, *112*, 6295–6300.

Helbig, C., Thorenz, A., and Tuma, A. (2020). Quantitative assessment of dissipation losses of 18 metals. *Resources, Conservation, and Recycling*, *153*, 104537.

Inghels, D. and Bahlmann, N. D. (2021). Hibernation of mobile phones in the Netherlands: The role of brands, perceived value, and incentive structures. *Resources, Conservation, and Recycling*, *164*, 105178.

Kral, U., Kellner, K., and Brunner, P. H. (2013). Sustainable resource use requires "clean cycles" and safe "final sinks". *Science of the Total Environment*, *461–462*, 819–822.

Krook, J., Carlsson, A., Eklund, M., Frändegård, P., and Svensson, N. (2011). Urban mining: Hibernating copper stocks in local power grids. *Journal of Cleaner Production*, *19*, 1152–1156.

Powell, J. T. and Chertow, M. R. (2018). Quantity, components, and value of waste materials landfilled in the United States. *Journal of Industrial Ecology*, *23*, 466–479.

Reck, B. K. and Graedel, T. E. (2012). Challenges in metal recycling. *Science*, *337*, 690–695.

Reuter, M. A., van Schaik, A., Gutzmer, J., Bartie, N., and Abadías-Llanus, A. (2019). Challenges of the circular economy: A material, metallurgical, and product design perspective. *Annual Reviews of Materials Science*, *49*, 253–274.

United Nations Environment Programme, International Resource Panel (2011). *Recycling Rates of Metals: A Status Report*. Retrieved from: https://www.resourcepanel.org/reports/recycling-rates-metals.

Chapter 10

Environmentally Extended Input–Output Analysis

Chapter Concepts

- Input–output analysis uses a sector-based model of an economy to model the links between production and consumption and to analyze how demand for goods and services causes activity throughout the economy
- Environmentally Extended Input–Output (EEIO) analysis incorporates sector-specific information on environmental burdens to link consumption with measures of emissions, resource use, and environmental impacts
- EEIO modeling has become one of the most versatile and popular tools within industrial ecology, spawning numerous innovations, including the use of physical input–output tables (PIOTs) to understand material flows through the economy

10.1 Understanding Flows at an Economy Scale

Technological societies as a whole are comprised of many factories, not one, and also shops of various kinds, medical facilities, service industries, residences, etc. Complete materials and energy flow data for such systems are not routinely available, but they are essential to a detailed understanding of their resource demands and environmental impacts. On the other hand, transactions between the various entities of society that involve monetary exchanges generate data that is captured by governments.

In the 1930s, economist Wassily Leontief developed economic input–output (I–O) analysis based on industrial sector transaction tables to address these interactions. His approach has proven enormously successful in the field of economics, and it is currently the practice in many countries to routinely generate I–O tables of their national economies. Beginning in the 1970s, economists (including Leontief) and industrial ecologists began to develop methods for including environmental information into I–O tables, so that they could be used to understand how economic activity influences emissions, use of natural resources, and other environmental concerns. This gave rise to the field of environmentally extended input–output (EEIO) analysis, one of the primary tools of industrial ecology.

10.2 The Mathematics of Input–Output Analysis

Here, we will demonstrate the structure of input–output models with a simple two-sector economy, as shown in Figure 10.1(a). The variables are as follows:

z_{ij} = monetary flows from sector *i* to sector *j*
y_i = monetary flows from sector *i* entering use (final demand)
x_i = total production (output of all components of sector *i*)

In this arrangement, the columns (e.g., $z_{11} + z_{21}$) represent the inputs to Sector 1, while the rows (e.g., $z_{11} + z_{21} + y_1 = x_1$) represents the outputs of Sector 1. In I–O, all values are expressed in monetary units, and the number of sectors (depending on the country) may be as many as several hundred.

Import and export flows (also expressed in monetary units) can be included by adding an additional row and an additional column, as shown in Figure 10.1(b).

Each of the z_{ij} factors contribute a portion of the total production, so we can write

$$z_{11} = a_{11}x_1 \text{ and } a_{11} = z_{11}/x_1 \tag{10.1}$$

where the a_{ij} terms are called technical coefficients representing these proportions.

For the two-sector model, the production equations then form a system of linear equations:

$$\begin{aligned} x_1 &= a_{11}x_1 + a_{12}x_1 + y_1 \\ x_2 &= a_{21}x_2 + a_{22}x_1 + y_2 \end{aligned} \tag{10.2}$$

or, expressed in matrix notation:

$$\vec{x} = \mathbf{A}\vec{x} + \vec{y} \tag{10.3}$$

(a)

To / From	**Sector 1**	**Sector 2**	**Final demand**	**Total output**
Sector 1	z_{11}	z_{12}	y_1	x_1
Sector 2	z_{21}	z_{22}	y_2	x_2

(b)

To / From	**Sector 1**	**Sector 2**	**Exports**	**Final Demand**	**Total Output**
Sector 1	z_{11}	z_{12}	e_1	y_1	x_1
Sector 2	z_{21}	z_{22}	e_2	y_2	x_2
Imports	i_1	i_2			

(c)

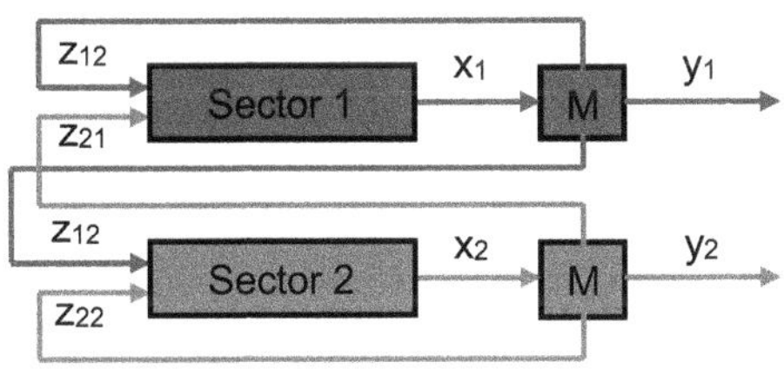

(d)

To / From	**Sector 1**	**Sector 2**	**Final demand**	**Total output**
Sector 1	z_{11}	z_{12}	y_1	x_1
Sector 2	z_{21}	z_{22}	y_2	x_2
Env. Flow	F_1	F_2		

Figure 10.1. (a) A two-sector input–output transaction table. (b) A two-sector input–output transaction table including imports and exports. (c) A representation of an input–output table as a material flow diagram. (d) A two-sector input–output transaction table with appended sectoral environmental flows.

Source: D. Müller.

where x is the production vector, **A** is the technical coefficients matrix, and y is the final demand vector. The relationships among sectors, final demand, and total output are shown graphically in Figure 10.1(c).

The inverse of a square matrix **A** is a matrix $\mathbf{A}^{-1}$, so that

$$\mathbf{A} \cdot \mathbf{A}^{-1} = \mathbf{A}^{-1} \cdot \mathbf{A} = \mathbf{I} \tag{10.4}$$

where **I** is the identity matrix, and $\mathbf{I}_2$ is

$$\mathbf{I}_2 = \begin{bmatrix} 1 & 0 \\ 0 & 1 \end{bmatrix} \tag{10.5}$$

To solve Eq. (10.5), subtract $\mathbf{A}\vec{x}$ from both sides:

$$\vec{x} - \mathbf{A}\vec{x} = \vec{y} \tag{10.6}$$

Grouping terms gives

$$(\mathbf{I} - \mathbf{A})\vec{x} = \vec{y} \tag{10.7}$$

Multiply both sides by $(\mathbf{I} - \mathbf{A})^{-1}$ gives

$$(\mathbf{I} - \mathbf{A})^{-1}(\mathbf{I} - \mathbf{A})\vec{x} = (\mathbf{I} - \mathbf{A})^{-1}\vec{y} \tag{10.8}$$

Applying Eq. (10.6), this becomes

$$\vec{x} = (\mathbf{I} - \mathbf{A})^{-1}\vec{y} \tag{10.9}$$

where the term $(\mathbf{I} - \mathbf{A})^{-1}$ is called the Leontief inverse and x is the activity vector that describes the total output of each sector of the economy in order to fulfill the final demand $\vec{y}$, including all of the inter-industry requirements. We can now use Eq. (10.9) to calculate what happens when final demand shifts to a new $\vec{y}^*$. This formulation of the monetary inputs and outputs of the sectors of an economy is one of the central pillars of modern economics.

Example 10.1

Consider a simple two-sector economy, where the sectors are Agriculture and Food Processing, and the currency is units. In this economy, the Agriculture sector outputs 10 units to itself (e.g., seeds for planting), 100 units to Food Processing, and 40 units to final demand (e.g., raw food that can be eaten directly). The Food Processing sector outputs 20 units to Agriculture (e.g., to feed farmers), 10 units to itself (e.g., to feed factory workers), and 170 units to final demand. **(a)** Construct an I–O table for this economy.

Now suppose that the population increases and the final demand for processed food increases by 50%. **(b)** Use your I–O table from Part **(a)** to calculate how the total output of each sector will change in response to the increase in demand.

Answer:

Part (a). Following the structure of Figure 10.1(a), we just need to assign which flows are inputs and which flows are outputs. Then, the total outputs from each sector simply sum across each row:

From \ To	**Agriculture**	**Food processing**	**Final demand**	**Total output**
Agriculture	10	100	40	150
Food processing	20	10	170	200

Part (b). The first step is to calculate the inter-sectoral coefficients a_{ij} using Eq. (10.2). a_{11} is the proportion that the flow from Agriculture to itself (10 units) makes up of the total output from Agriculture (150 units), or 10/150 = 0.07.

Filling out the remainder of **A** gives

$$\mathbf{A} = \begin{bmatrix} a_{11} & a_{12} \\ a_{21} & a_{22} \end{bmatrix} = \begin{bmatrix} 10/150 & 100/150 \\ 20/200 & 10/200 \end{bmatrix} = \begin{bmatrix} 0.07 & 0.67 \\ 0.1 & 0.05 \end{bmatrix}$$

Then, we can construct the Leontief inverse in two steps:

$$\mathbf{I} - \mathbf{A} = \begin{bmatrix} 1-0.07 & 0-0.67 \\ 0-0.1 & 1-0.05 \end{bmatrix} = \begin{bmatrix} 0.93 & -0.67 \\ -0.1 & 0.95 \end{bmatrix}$$

$$(\mathbf{I} - \mathbf{A})^{-1} = \begin{bmatrix} 1.16 & 0.81 \\ 0.12 & 1.14 \end{bmatrix}$$

(Calculating the inverse of a matrix can be done easily using Microsoft Excel or mathematical software packages.)

Now, let's consider the new final demand:

$$\bar{y}^* = \begin{bmatrix} 40 \\ 340 \end{bmatrix}$$

Using Eq. (10.9), the new total output becomes

$$\vec{x}^* = (\mathbf{I} - \mathbf{A})^{-1}\vec{y}^* = \begin{bmatrix} 1.16 & 0.81 \\ 0.12 & 1.14 \end{bmatrix} \begin{bmatrix} 40 \\ 340 \end{bmatrix} = \begin{bmatrix} 323 \\ 392 \end{bmatrix}$$

Compared to the total output of Part (**a**), the value for Food Processing nearly doubles, but the value for Agriculture also increases dramatically by more than 160%, because even though its own final demand remains constant, it has to supply the Food Processing sector.

Thinking through Example 10.1 raises several practical questions that get to the major assumptions of I–O modeling. First, if production in both sectors increases so dramatically, will there not be economies of scale? In fact, no; I–O models are linear, so the same quantity of inputs is needed to produce a unit of output, regardless of the scale of production. This is called constant returns to scale. Further, the I–O models assume that the 'recipes' for what each sector needs as inputs (represented by **A**) remain constant, called a fixed structure of inputs. Second, won't some food items be affected more than others, depending on the processing required? Because I–O models are aggregate representations of the economy, they are only resolved to the sector level and can't tell the difference between individual items within a sector. This assumption is called industry homogeneity — each sector has only one recipe, and its output has just one price. In reality, of course processing fruit juice is physically very different than processing chickens, with different amounts of energy and other inputs required, and different prices of the final products. Input–output modeling isn't useful for differentiating between items within the same sector, but it is a fast and convenient method for analyzing economy-wide consequences of consumption.

10.3 Adding Environmental Extensions

The I–O tables of Section 10.3 can tell us much about how the economy responds in monetary terms, but in industrial ecology, we want to model environmental flows. This is done by adding environmental extensions or 'satellite tables' to our I–O table that contain flows of interest. For example, we might want to include information on emissions of pollutants, generation of solid waste, inputs of raw materials, or any number of social or biophysical flows. How much are these flows affected per unit of consumption?

Figure 10.1(d) shows an extension of our two-sector economy, now with information on environmental flows F_1 and F_2, representing the totals for that sector. Any number of rows can be added, corresponding to each environmental flow of interest. In the same way that we normalized monetary flows z_{ij}, we normalize the environmental flows F_i by the total monetary output in each respective sector *i*:

$$f_i = F_i / x_i \qquad (10.10)$$

Then, the emissions (scalar) associated with a new final demand vector y^* can be calculated as follows:

$$E = \vec{f}\,\vec{x}^* = \vec{f}(\mathbf{I} - \mathbf{A})^{-1}\vec{y}^* \qquad (10.11)$$

Example 10.2

Let's return to our simple two-sector economy in Example 10.1. Suppose the Agriculture sector uses 20 TJ of energy per year, while the Food Processing sector uses 1,000 TJ.

(a) What are the energy requirements per unit of output?
(b) If we again double the final demand for processed food, how much additional energy will be required?

Answer:
Part (a). Following Eq. (10.10), the energy required per unit of output is as follows:

$$\vec{f} = \begin{bmatrix} f_1 & f_2 \end{bmatrix} = \begin{bmatrix} F_1/x_1 & F_2/x_2 \end{bmatrix} = \begin{bmatrix} 20/150 & 1000/200 \end{bmatrix} = \begin{bmatrix} 0.13 & 5 \end{bmatrix}$$

Part (b). Using Eq. (10.11), the new energy required is as follows:

$$E = \vec{f}\,\vec{x}^* = \begin{bmatrix} 0.13 & 5 \end{bmatrix} \begin{bmatrix} 323 \\ 392 \end{bmatrix} = (0.13 \times 323) + (5 \times 392) = 2002 \text{ TJ}$$

The total energy use in the base case was 1020 TJ, so this is a near doubling of the energy required to run the economy.

Oftentimes, we want to find the environmental impact of a single purchase. For this, it is useful to have information on the total emissions per unit of expenditure in each sector, denoted $\hat{e}_i$, calculated by setting the final demand $\hat{y}_i$ to just one unit of expenditure in sector *i*.

$$\hat{e}_i = \vec{f}(\mathbf{I} - \mathbf{A})^{-1}\hat{y}_i \tag{10.12}$$

Or, calculated as a vector of emissions per unit expenditures $\vec{e}$:

$$\hat{e} = \vec{f}(\mathbf{I} - \mathbf{A})^{-1}\mathbf{I} = \vec{f}(\mathbf{I} - \mathbf{A})^{-1} \tag{10.13}$$

Example 10.3

Again building on Example 10.1, calculate the vector $\hat{e}$ of total emissions per unit of expenditure.

Answer:
We have already done this within our earlier matrix calculations, but here it is explicitly

$$\hat{e} = \vec{f}(\mathbf{I} - \mathbf{A})^{-1} = \begin{bmatrix} 0.13 & 5 \end{bmatrix}\begin{bmatrix} 1.16 & 0.81 \\ 0.12 & 1.14 \end{bmatrix} = \begin{bmatrix} 0.76 & 5.80 \end{bmatrix}$$

So, if we want to produce an additional 5 units of processed food, the additional energy required will be 5 units × 5.80 TJ/unit = 24 TJ.

Unless you become an industrial ecology researcher, you may make these kinds of EEIO calculations directly, but it is important to understand how they are calculated. It is very common to see EEIO-derived emissions factors built into various carbon, water, and other environmental and ecological footprinting calculators; any factor that is described per unit of expenditure, such as [kg CO_2e per $] for carbon calculators, comes from EEIO modeling.

10.4 Physical Input–Output Tables

Researchers have developed many innovative approaches to use EEIO modeling to answer various large-scale questions about the impacts of consumption and global flows of various resources. There are too many to explore in this textbook, but one important development that we will describe is using monetary input–output tables (MIOTs) to create physical input–output tables (PIOTs). PIOTs are of great interest to industrial ecologists because they can be used in conjunction with stock analysis (Chapter 7) and material flow analysis (MFA) (Chapter 8) to understand the physical basis of the economy. PIOTs can be generated for a single material, such as aluminum, or for the entire mass of materials flowing through each sector.

As an example, we illustrate the steps involved in transforming the U.S. MIOT into a PIOT for aluminum in the U.S. economy in 2007, as reported by Chen *et al.* (2016). The process begins by compiling the MIOT, often done by national statistical agencies, as shown in Figure 10.2(a). In Step 2 (Figure 10.2(b)), the subject material (aluminum) is introduced into the I–O table by utilizing material flow information to identify those sectors that have direct flows of ore, scrap, process losses, and imports/exports, and to link them in physical units. In Step 3 (Figure 10.2(c)), the incoming aluminum is spread across the inter-industry network by identifying which sector outputs contain aluminum. A client who pays an hourly fee to a lawyer, for example, transfers only money, not aluminum (or any other material). Accordingly, industrial ecologists must transform the economic table by eliminating non-material transactions. Step 4 (Figure 10.2(d) determines the amount per unit of monetary output based on a detailed investigation of material composition and prices for each sector output. At each step through the inter-industry network, mass balance must be observed to constrain the calculations. Finally, Step 5 closes the life cycle loop by adding scrap collection flows originating from in use stock discards. The product of these steps is the PIOT (Figure 10.2(e)).

10.5 The Status and Utility of EEIO Tables

EEIO analysis has advanced tremendously over the past decade. Industrial ecology adaptations of I–O tables have been accomplished for many countries, regions, and the planet, and these underpin all types of macro-level calculations of emissions, resource use, and environmental impacts. PIOTs can be used with MFA to understand how resources are moving through the economy, and thus how changes in demand or the material content of certain items will percolate through the entire supply chain. One of the simpler examples (though in reality far from simple) is the cobalt intensity diagram of Figure 10.3, from Nuss *et al.* (2019) where several of the major flows from a particular US input sector to corresponding US output sectors are shown as cobalt intensities (in metric tons). Cobalt is considered by many to be a critical material (see Chapter 23), so the information in the figure helps companies in those sectors understand how their supply chain is dependent on stable, affordable cobalt supplies. PIOTs can also be used to understand when materials will come out of service and be available for recovery as a secondary resource, and how much. By linking with statistics on waste generation using so-called waste input–output accounting, PIOTs can be used to measure the overall circularity of our economy (see Chapter 37) and the longevity of each material life cycle, over all cycles of use (see Poncelet *et al.*, 2022). Those

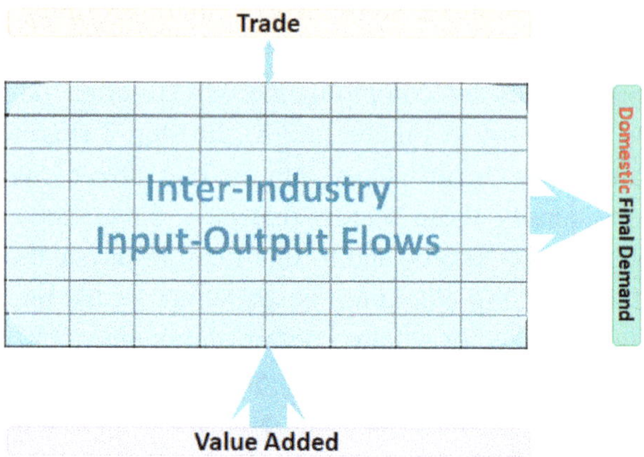

(a)

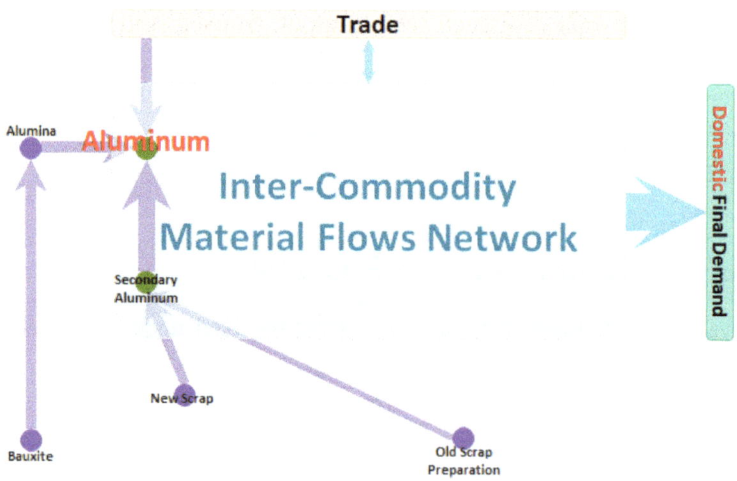

(b)

Figure 10.2. Converting the US 2007 input–output table of monetary flows between industrial sectors to physical flows of aluminum: (a) Step 1: Compile the MIOT (393 inter-industry input–output flows) in which the aluminum life cycle is embedded. Green nodes and links are determined using the U.S. input–output table (nodes for trade and domestic final demands are not shown); (b) Step 2: Introduce aluminum into the network. Purple nodes and links are determined based on a traditional MFA study so as to reflect the comprehensive aluminum life cycle (Chen *et al.*, 2016); (c) Step 3: Spread aluminum input paths across the inter-industry network; (d) Step 4: Convert monetary flows into aluminum flows assuming that the monetary flows are proportional to aluminum flows; (e) Step 5: Close the life cycle loop by adding scrap collection flows. The result is the environmentally extended input–output material flow network of aluminum for the US in 2007.

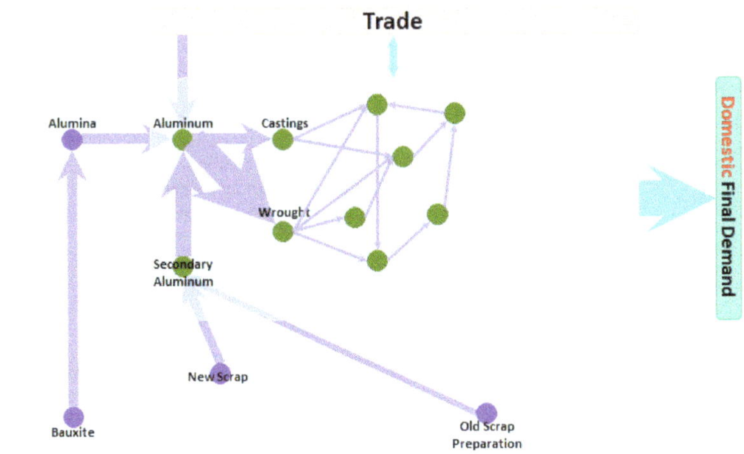

(c)

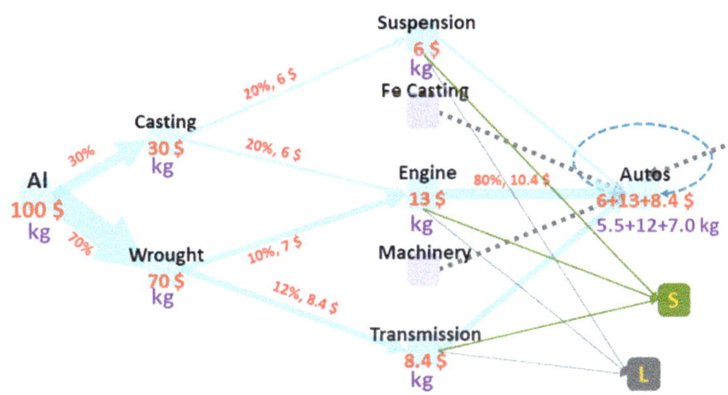

(d)

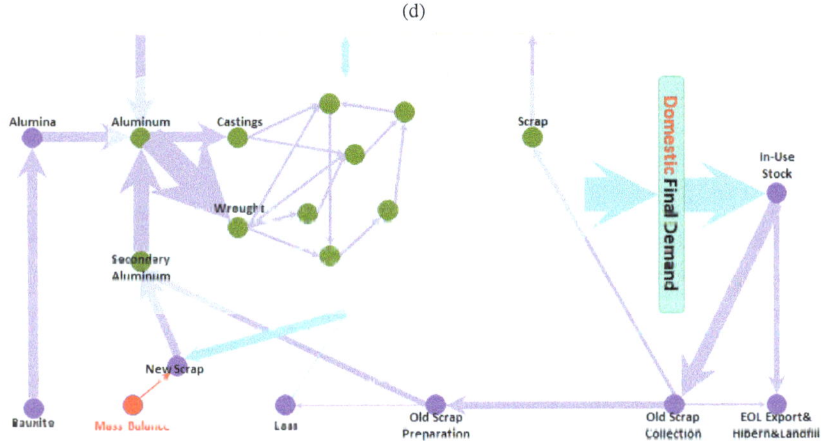

(e)

Figure 10.2. (*Continued*)

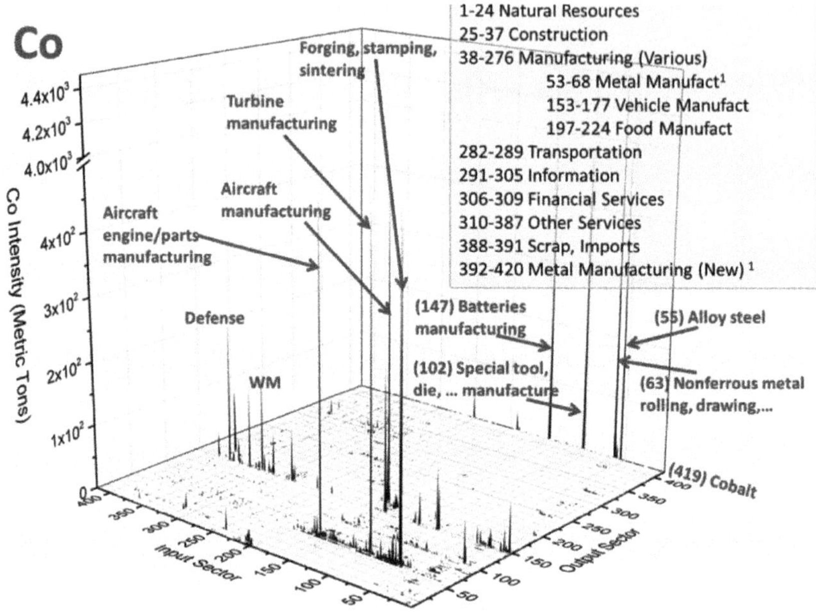

Figure 10.3. A three-dimensional visualization of the physical input–output table of cobalt in the US, 2007. The horizontal axes are the input sectors (the "sending" sectors) and the output sectors (the "receiving" sectors) of the US economy. The vertical axis is the sector to sector cobalt flow intensity in mass units (Nuss *et al.*, 2019).

authors found that most metals last less than a century in the economy, and some only a handful of years, indicating that there is much work to be done in improving our product designs and material management systems to achieve sustainable use of the Earth's resources.

Further Reading

Chen, W.-Q., Graedel, T. E., Nuss, P., and Ohno, H. (2016). Building the material flow network of aluminum in the 2007 U.S. economy. *Environmental Science & Technology*, *50*, 3905–3912.

Joint Research Centre, Institute for Prospective Technological Studies, Huppes, G., Tukker, A., and Heijungs, R. (2011). *Environmentally Extended Input-Output Tables and Models for Europe*. Publications Office. https://op.europa.eu/en/publication-detail/-/publication/1edb6271-5b07-40fa-ae6b-55bce1c1c220.

Kitses, J. (2013). An introduction to environmentally-extended input-output analysis. *Resources*, *2*, 489–503.

Lenzen, M., Kanemoto, K., Moran, D., and Geschke, A. (2012). Mapping the structure of the world economy. *Environmental Science & Technology*, *46*, 8374–8381.

Nuss, P., Ohno, H., Chen, W.-Q., and Graedel, T. E. (2019). Comparative network analysis of metals use in the United States economy. *Resources, Conservation and Recycling*, *145*, 448–456.

Poncelet, A. C. *et al.* (2022). Losses and lifetimes of metals in the economy. *Nature Sustainability*, *5*, 717–726.

Shafie, F. A., Omar, D., Karupannan, S., and Gabarrell, X. (2013). Urban metabolism using economic input output analysis for the city of Barcelona. In Zubir, S. S. and Brebbia, C. A. (Eds.), *The Sustainable City VIII*, Vol. 1. Southampton, UK: WIT Press.

Chapter 11

Introduction to Life Cycle Assessment

Chapter Concepts

- Products move through a life cycle, including the extraction of raw materials, manufacturing and assembly, transportation, use, and waste management
- Product life cycles are interconnected through various materials and processes
- Emissions and environmental impacts can occur at any point in the life cycle of a product; it is often difficult to predict the largest contributors
- Environmental life cycle assessment (LCA) provides a formalized, quantitative approach to analyzing the environmental impacts of a product or process

11.1 The Concept of the Life Cycle

The environmentally related activities of the 1970s and 1980s focused on the manufacturing facility: reducing energy use, waste generation, and various emissions. This was a necessary concentration of effort to address the poor environmental performance of manufacturing at the time, but it considered industrial facilities as isolated entities, completely separate from their suppliers and customers. As we have learned in prior chapters, industrial ecology applies a systems perspective, wherein any industrial activity lives within an industrial ecosystem and is highly interconnected to other nodes (facilities, sectors, or users). Any action taken in one node of the industrial ecosystem will influence other nodes, and addressing environmental challenges is best done by considering the overall system.

Focusing greening efforts just at the level of a manufacturing facility ignores critical links in the industrial ecosystem. Any facility needs material and energy inputs for its operations, and often times these resources are extracted and processed elsewhere. The manufactured products may themselves have environmental or human health impacts when used, as well as when they are eventually discarded. In the 1990s, the scope of interest for greening manufacturing was enlarged and the concept of a product 'life cycle' began to take hold. Again making use of the biological analogy, physical products began to be considered from 'cradle to grave', with associated use of energy, resources, and generation of emissions occurring at every point along the life cycle. Just as we can follow an organism on its physical journey through the world, watching as it interacts with its environment and other organisms, the concept of the product life cycle invites us to follow a product through its own physical journey, watching as its constituent material parts come together from all corners of the globe, as the product is transported, sold, used, discarded, and lives out the rest of its days in a landfill or perhaps transformed by fire in a waste incinerator.

The components of a product life cycle can be defined in various ways depending on the goals and level of detail desired, but the four numbered stages in Figure 11.1 are typical: (1) acquisition/extraction and processing of the necessary resources, (2) manufacture and assembly, (3) use, and (4) end-of-life (EOL), with transport occurring between each of the stages. From any particular point in the product life cycle, activities that occur in previous stages are considered *upstream*, while activities that occur in later stages are considered *downstream*. From the use stage perspective (where most of us interact with products as consumers), upstream corresponds to the production and supply chain of a product, while downstream corresponds to end-of-life.

There are also several opportunities for the cycling of flows among life cycle stages. The generation of reusable discards in manufacturing stimulates a flow of 'prompt scrap' or 'pre-consumer' recycling back to material processing (upper right on the diagram), such as paper mill scrap being reprocessed into pulp. After use ('post-consumer'), products may still have functionality, stimulating a reuse flow, either for the same purpose or another, such as used cardboard packing boxes being shared or sold. Finally, if the product can no longer be reused directly but the materials still have value for manufacturers, there is a recycling flow of post-consumer scrap for reprocessing into new products.

Consider, for example, the life cycle of paper. Trees are cultivated, harvested, debarked and chipped, and processed into the raw material, pulp. Next, pulp is used in a paper mill to produce all manner of paper products. That paper is then purchased by consumers for a variety of uses. Some paper may only stay in the use stage for one day before it is promptly discarded, while other paper, such as in

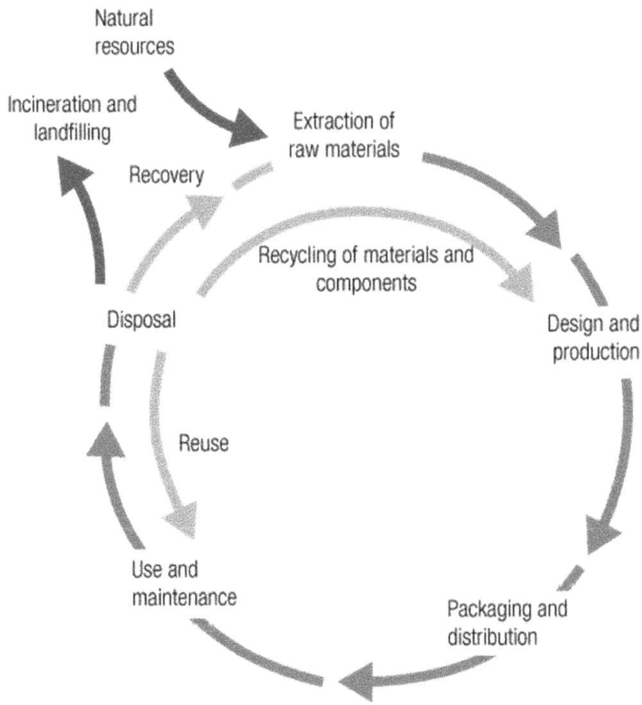

Figure 11.1. Life cycle stages of a generic product (UNEP Life Cycle Initiative).

a book, will (one hopes) stay in circulation for many years. Finally, paper that reaches the end-of-life phase enters the waste management system, where it could be recycled, incinerated, landfilled, composted, or lost as litter directly to the environment.

We certainly use a lot of paper: an estimated 80–90 million metric tons every year in the US. How does all of this paper use affect the environment, and more importantly, where should we focus our efforts to reduce these environmental impacts? Traditionally, we have focused on manufacturing, for example by improving the energy efficiency of papermaking equipment. By taking a life cycle perspective, we can see many other potential drivers of environmental degradation: deforestation and erosion from the overharvesting of trees, water pollution from the pulp plant, particulate matter emissions from diesel trucks that transport the finished products, or greenhouse gas emissions when waste paper is incinerated.

Which one of these drivers or processes causes the most environmental harm? Should we focus on managing forests better? Should we develop different chemical processes for bleaching paper? Should we invest in new ways to reuse or recycle paper? Or perhaps, all the transportation emissions from paper production, delivery, and waste collection are the biggest problem? We want to improve the environmental performance of paper, but how do we know where to focus our efforts?

Actually, this is a very difficult question to answer. First, we need a way to quantify each type of emission so that we can compare them. But there are different types of emissions that cause different types of environmental problems! We can't just compare the mass of a water emission like a bleaching agent to that of an air emission like carbon dioxide, and these emissions are happening in different locations anyway. So, second, we need a way to determine which emissions affect which environmental systems, where, and how much. Third, it would be useful to be able to compare the severity of the impacts on these different systems.

11.2 The Genesis of Life Cycle Assessment (LCA)

Life cycle assessment (LCA) is a modeling tool that has been developed over the past several decades in order to address the challenges described above. The goal of the LCA is to quantify or otherwise characterize all flows of resources and emissions over the entire life cycle of a product, to specify their potential environmental impacts, and to consider alternative designs or approaches that can change those impacts for the better.

The genesis of LCA was in early 'Resource and Environmental Profile Analysis' reports undertaken by companies in the late-1960s, most notably by Coca-Cola to examine different packaging options. This was before much of our current scientific understanding of emissions, and so these early studies tended to focus on energy use. The field of energy analysis developed rapidly, particularly after the global energy crisis in the 1970s, providing macro- and process-level quantification of energy use in important industrial sectors like steel production and giving rise to the first estimates of embodied energy for common materials. As emissions data and emissions factors began to be developed, for example for sulfur dioxide (an important precursor to acid rain), early LCAs began to add these aspects as well as waste flows such as the total mass of solid waste generated. By the 1980s, many companies and researchers were using LCA, but the lack of a standardized framework meant that the studies were not comparable and would lead to confusion. That changed in the 1990s with various workshops and efforts at coordination, with the Society for Ecological Toxicology and Chemistry (SETAC) playing a leading role with its publication of *Guidelines for Life-Cycle Assessment: A 'Code of Practice'* in 1993. This was followed by the development and publication of technical standards by the global International Organization for Standardization (ISO, in its French acronym), which provided the basic theoretical framework that we use today. Additional standards have been developed as LCA methods have become more sophisticated and reached specialized application, such as carbon footprinting.

11.3 LCA Framework and the ISO Standards

The formal structure and terminology of LCA was promulgated by ISO primarily through Standards 14040 and 14044 (both most recently revised in 2006). According to these standards, LCA contains four phases:

(1) *Goal and Scope*, where the overall study design is explained and documented;
(2) *Life Cycle Inventory Analysis,* or life cycle inventory (LCI), where flows of resources and emissions are quantified throughout the life cycle;
(3) *Life Cycle Impact Assessment* (LCIA), where the use of resources and emissions of substances into the environment are linked to physical changes and/or damages; and
(4) *Interpretation*, where results are discussed relative to the original study goal and scope, particularly with respect to issues of data quality and uncertainty.

The ISO LCA framework is pictured in Figure 11.2. First, the goal and scope of the LCA are defined and documented. An inventory analysis and an impact assessment are then performed based on this goal and scope. Often the goal and scope must be adjusted because of data gaps or other issues. Finally, the interpretation that follows guides an analysis of potential improvements and uncertainty, which typically feeds back to influence the other three stages, so that the entire process is iterative. It is perfectly normal to redo an LCA several times over the course of a project, as adjustments are made to scope, data sources, or other assumptions and methodological choices.

There is perhaps no more critical step in beginning an LCA evaluation than to define as precisely as possible the goal of the study — the precise questions to be answered and for what purpose — followed by choosing the evaluation's scope — what materials, processes, or products are to be considered and how broadly will alternatives be defined. We will focus on the goal and scope phase in the remainder of this chapter and will return to the other stages in detail in the following chapters.

Consider, for example, the question of releases of chlorinated solvents during a typical dry-cleaning process. The purpose of an LCA on this topic might be to identify effective ways to reduce environmental impacts from dry cleaning. But, if

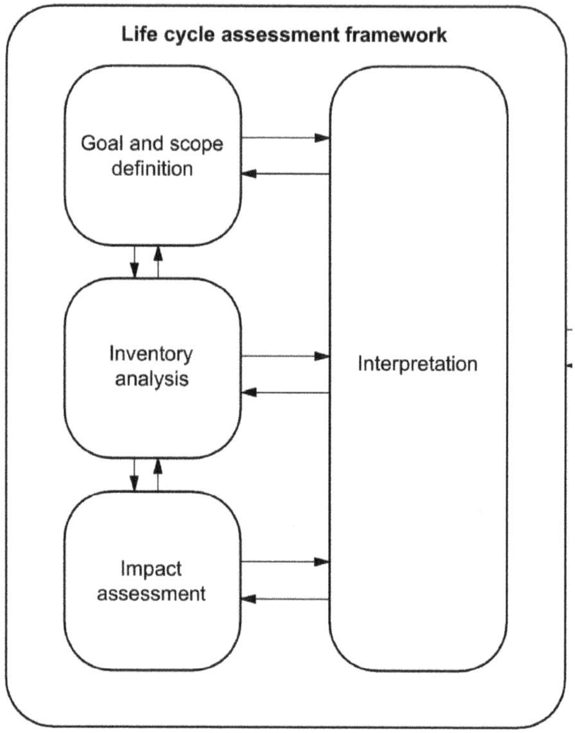

Figure 11.2 Diagram of LCA phases according to ISO 14040: 2006.

the goal and scope of the study are overly narrow, the study might assume that the basic process design is fixed and only consider good housekeeping techniques, end-of-pipe controls, administrative procedures, and perhaps alternative petrochemical solvents. If, however, the goal and scope are defined broadly, the study could include alternative service options. For example, some data indicate that a substantial number of items are sent to dry cleaning establishments not for dry cleaning *per se* but simply for pressing. Accordingly, offering an independent pressing service might reduce emissions considerably. Taking an even broader, systems-level view of the problem, the study might focus on design and consider innovative fiber alternatives or new process technologies that do not require chlorinated solvents at all.

Several methodological and practical considerations influence the goal and scope, including the following: (1) who is sponsoring and who is performing the analysis, and how much control can they exercise over the implementation of options; (2) what resources are available to conduct the study; and (3) what is the most limited scope of analysis that still provides for adequate consideration of the systems aspects of the problem. The resources that can be applied to the analysis should also be assessed. Most traditional LCA methodologies provide the potential for essentially open-ended data collection — and, therefore, virtually unlimited expenditure of effort. As a general rule, the depth of analysis should be keyed to the degrees of freedom available to make meaningful choices among options, and to the importance of the environmental or technological issues leading to the evaluation.

For example, an analysis of using different plastics in the laptop computer would probably not require a complex analysis, because the constraints imposed by the existing design and its market niche make the options that are available to a designer quite limited. On the other hand, a government regulatory agency contemplating rules on single-use plastics used in large amounts in numerous and diverse manufacturing applications would want to conduct a comprehensive analysis, because the degrees of freedom involved in finding substitutes could be quite numerous and the environmental impacts of substitutes could be significant.

11.4 Setting LCA Study Goals

A common LCA objective is to derive information on how to improve environmental performance of a product or to compare multiple products. If the exercise is conducted early in the design phase, the goal may be to compare two or three alternative designs. If the design is finalized, or the product is in manufacture, or the process is in operation, the goal can probably be no more than to achieve modest changes in environmental attributes at minimal cost and minimal disruption to existing practice. LCA can also be carried out at large scales, comparing not just a single product or process but entire organizations or systems, for example, an entire manufacturing facility, corporation, city, or even country. In such cases, the goal might be to evaluate alternative operational designs or to determine what aspects of consumption are causing the largest environmental impacts, in order to inform policy changes. Either way, the objective of a study should match the scale of the problem and the options that the various stakeholders have for implementing change.

According to the ISO standards for LCA, the Goal statement should state and explain the following:

- the aim or objective of the study;
- the intended use of the results or how they will likely be applied;

- the initiator (or commissioner) of the study;
- the practitioner of the study;
- the stakeholders of the study (interested parties);
- the intended users of the study (target audience); and
- if the results will be used in comparative assertions and released publicly (if yes, this will trigger an obligatory third-party peer review).

It is useful to recognize that life cycle assessment is an iterative process and that the goal may need to be revisited as the LCA proceeds.

11.5 Defining the Functional Unit of Analysis

In many cases, we conduct an LCA in order to compare multiple options, so that we might choose the environmentally preferable one. As a regular consumer, life is full of choices that have environmental implications, and it is often very challenging to know what to do. Is seafood better than red meat? Should I use a sponge or a paper towel or a cloth? Is recycled content paper really better than virgin paper? These types of comparative questions occur for large-scale production systems as well, where the choices affect many people. Should a company build a wood-framed building or a reinforced concrete building? Should the state invest in renewable energy or natural gas-fired power plants? Should society strive to make plastic out of petrochemicals or bio-based chemicals?

In all cases, we need a fair way to compare the options. This is harder than it sounds. Take the example of the sponge versus paper towel versus cloth. Physically, these are very different products. It doesn't make sense to compare one of each type, especially because sponges and clothes can be used multiple times and washed, whereas the paper towel is designed for single use. If there is a liquid spill, they each have different levels of absorbance, so you might need one sponge (washed out 10 times) versus five pieces of paper towels versus two pieces of cloths that then must be washed in the laundry.

LCA recognizes that products have different levels of performance. So, rather than the physical items themselves, it is the core function of the products that is used as the basis of comparison. The quantity of this function you want to compare is called the *functional unit*. As the name implies, the functional unit should specify the common function of the product alternatives, a quantity, and a unit of measure. In the earlier example, the functional unit could be "the capacity to absorb one liter of liquid". The physical quantity of each product alternative that is needed to fulfill the functional unit is called the *reference flow*. This would be the one sponge, five paper towels, or two pieces of cloth. In a comparative study, the common functional unit should always be determined first, and the reference flows derived from that functional unit.

The choice of the functional unit in large part determines what options can be considered for comparison in an LCA, and so requires consultation and expert judgment. In the example earlier, if the functional unit is set as one liter of absorptive capacity, this excludes options that don't rely on absorption, such as a wet vacuum or even a thirsty pet! Ideally, the functional unit should be set so that the comparison can include all options that could plausibly be taken. The broader the functional unit, the greater the diversity of options that can be considered, including system-level options that may not be immediately apparent.

A functional unit can also be not specific enough. Again using the earlier example, suppose the liquid that is spilled is a strong acid. Paper towels might disintegrate, and personal safety is also a concern. A more appropriate functional unit would then be "the ability to clean up one liter of pH 2 liquid while maintaining personal safety". Time is another aspect of performance that can be specified in a functional unit. One option for dealing with a liquid spill is simply to wait for it to evaporate (assuming the liquid is volatile enough). But this might take hours and may not actually be a plausible option. So, in order to exclude this option, a new functional unit could be "the capacity to absorb one liter of liquid in under five minutes". It is also possible that a product system has more than one function, such as a window that provides a physical barrier but also transmits light. When considering options, both of these functions could be considered in specifying alternative reference flows.

Let's consider another example. An early topic of LCA was the question of shopping bags, a classic question of paper versus plastic versus reusable cloth bags. Comparing these bags one-to-one-to-one is incorrect because each bag provides a different level of performance, a different quantity of stuff that it can hold. Therefore, possible functional units could be "the ability to carry 10 kilograms of stuff" or "the ability to carry 15 liters of stuff". This might require one typical kraft paper grocery bag or two typical low-density polyethylene (LDPE) plastic grocery bags or a large cloth bag, which would be the reference flows. Since the cloth bag can be reused (this assumes the paper and plastic bags are single-use), its reference flow needs to be adjusted by the number of expected uses. If the average large cloth shopping bag is used 100 times, then its reference flow in this case would be 1/100 bags, with some consideration for cleanliness.

What if we want more breadth? Specifying that the stuff must be carried implies that you have gone to a store to purchase items. If the functional unit instead states "the delivery of 10 kilograms per 15 liters of stuff to my home", now we can begin to consider larger systems-level questions, such as how to get to the store or e-commerce and drone delivery options. Or, what if we want more specificity? Paper bags perform poorly for wet items, and plastic bags perform poorly for sharp items. If the stuff in question has physical characteristics that preclude certain options, then performance around these characteristics should be specified in the functional unit as well.

Correctly specifying functional unit and reference flows is fundamental to any comparative LCA, but can be challenging because it requires us to consider products in a different way than we are used to doing, to see them for what they do rather than what they are. Ideally, the choice of functional unit should be made with input from multiple stakeholders and revised during the study if necessary. For product systems that have been studied extensively, such as water supply or fuels, consensus functional units have been agreed upon over time.

11.6 Determining an Appropriate Scope of Analysis

As with the Goal statement, the description of the Scope of an LCA must address specific topics according to the ISO standards. These include the following:

- the functional unit,
- the system boundary,
- major methodological choices, such as mode of analysis and allocation procedure
- data sources and data quality requirements, including temporal, spatial, and technology coverage,
- impact assessment methods and categories of impacts considered,
- uncertainty analysis,
- other major assumptions.

We will address questions of data and methods in the following chapters on the Inventory, Impact Assessment, and Interpretation phases, and focus the remainder of this chapter on the system boundary.

The potential complexity of comprehensive LCAs is nowhere better illustrated than by the challenge of defining appropriate boundaries of the study. There have been many potential considerations and heuristics developed over the years in order to guide LCA practitioners, but in the end, the choice of system boundary should depend on the questions that the study is trying to answer. If the goal of an LCA includes a comparison with other studies, then it is important that study boundaries be consistent. (Indeed, in many sectors there is a general agreement about a minimum scope of inclusion for this reason.) But if the initiator of an LCA has specific questions, especially if the results are being used for internal decision-making or research purposes, then the system boundary is more flexible and is often more targeted. And remember, LCA is an iterative process; your choice is not fixed. Your initial study boundary may need to change as the project evolves, and during the Interpretation phase, it is common to alter boundary assumptions in order to see the effect on overall results.

11.6.1 *Life cycle stages*

The first basic scoping question is which life cycle stages to include. There are three common approaches and a host of variations; the best choice depends on the goal of the study. Consider again the life cycle diagram in Figure 11.1. Often, we want to know what is the impact of making a product, including raw material extraction, processing, manufacturing, and assembly. An LCA that focuses just on the production stages is called *cradle-to-gate* ('cradle' being the natural source of raw materials and 'gate' referring to the boundary of the manufacturing facility). This scope can be further truncated to exclude the supply chain and focus only on what happens during manufacturing and assembly stage, which is called *gate-to-gate*. The supply chain is really a sequence of gate-to-gate processes, and these can be strung together to form a cradle-to-gate scope if desired. Finally, if the use stage and end-of-life stage are included, covering the entire product life cycle, this is called *cradle-to-grave*, borrowing a metaphor from the human life cycle. If end-of-life consists mostly of reusing or recycling, then this is sometimes called *cradle-to-cradle*.

A cradle-to-grave study is clearly the most comprehensive but is often not necessary or even possible. For example, a manufacturer will have access to information and a degree of control over manufacturing and assembly stage and also on the extraction and processes stage (the supply chain) of the various inputs required, but often not, however, on exactly how the product is used or treated at end-of-life. This is especially true if the product is an intermediate such as a polymer resin and is used in a wide variety of products. For these reasons, most manufacturer LCAs are cradle-to-gate. This scope is probably the most common and has different forms depending on the product type. For smaller consumer items, this would include the production of the finished good, usually with packaging. For buildings and infrastructure, this would include the production of all the materials as well as construction and commissioning.

If the location of use is known, then a cradle-to-gate scope can be extended to include transportation, distribution, and/or sale. For example, for fossil fuels (whose industrial life cycle begins at an oil or gas well), a cradle-to-gate study is called 'well-to-gate'. If distribution to a fuel station is included, the scope becomes 'well-to-pump'. If filling up the fuel tank of the vehicle is included, the scope becomes 'well-to-tank'. For consumer goods, this scope including everything upstream from the point of sale might be called 'cradle-to-shelf'.

At the other end of the life cycle, LCA comparisons of waste management options usually have the starting point of waste generation and don't need to consider how various products were used before they entered the waste management system. Or sometimes the use stage is the only stage of interest, for example, when different operational modes are being compared, and the production and waste management stages are assumed to be identical across options.

The inclusion or exclusion of life cycle stages largely determines what kinds of questions can be answered with an LCA. This can be demonstrated clearly with the example of food. As one of humanity's common basic needs, food and food systems have been analyzed extensively using LCA (and other IE tools). LCA studies including the cultivation and harvesting stage can inform choices around crop types, farming practices, or land use, and answer questions such as, is organic farming better than conventional farming? Alternatively, many people are interested in the benefits of local food systems and questions of food miles and access to healthy food, and to answer these questions, transportation and distribution stages must be included in the study scope. Other researchers focus on new food technologies such as alternative proteins or lab-grown 'meat', which are typically analyzed and compared to conventional options using a cradle-to-gate scope. Still, other food researchers focus on the topic of food waste, which can occur at multiple points and requires a much broader system boundary, including waste during harvesting (extraction stage), from food processing facilities (manufacturing stage), from spoilage during transportation and storage or from unsold and discarded food, or from waste during preparation or from uneaten food (use stage). Then understanding the environmental consequences of food waste necessitates including the end-of-life stage, depending on if the food waste is composted, landfilled, fed to animals, and so on.

Whatever the topic of an LCA, the choice of which life cycle stages to include should be clearly stated and justified based on the underlying goal of the study and the questions it is trying to answer. Especially for comparative studies, it is critical to include any life cycle stages that differentially influence the various options. Excluding relevant life cycle stages is one of the most common mistakes in LCA. For example, an LCA on sustainable packaging that is only cradle-to-gate and leaves out the end-of-life stage will ignore the fact that packaging materials have very different physical fates, that paper in a landfill will degrade, that paper and plastic in an incinerator will release carbon dioxide (and potentially other air pollutants), that different materials have different recovery and recycling rates; in short, the study will miss important physical flows and have misleading results.

11.6.2 *Physical versus economic modeling of product systems*

So far, we have considered product life cycles in isolation, but in reality, they are highly interconnected. A piece of paper is not only made of wood pulp but also has chemicals added for brightening or coating, whose production also must be considered in the product system. But look beyond just what is physically in the paper — each life cycle stage requires a whole variety of inputs. Just to make the pulp requires processing chemicals, fuel for heating, machinery, electricity to run the machinery, and the physical building that houses all the equipment, which itself is made up of steel, concrete, and other building materials. A pulp plant also requires workers, who require transportation for commuting, food, clothing, protective equipment, on-site bathrooms, telecommunication systems, and computers. The business of the plant also needs insurance, banking services, government permits, advertising, and a host of other services. And in turn, any one of these inputs into the pump plant also requires an entire supply chain of interlinking goods and services. Trace the life cycle for just about any product or service, and the picture quickly becomes quite complex, often requiring us to model processes all throughout the economy. If we want to understand the environmental impacts of a piece of paper (our reference flow in this case), then we need to consider the network of processes that are required to produce it. This is called the *product system*.

There are two basic approaches to modeling a product system in LCA. The first approach is to model the system in physical terms, as a network of physical processes that produce goods and services. This is called *process LCA*. This is typically how engineers are trained to understand the world, as each process can be measured and modeled in terms of physical flows of energy and mass. The second approach is to model the system in monetary terms, as a network of economic sectors that produce goods and services. This is typically how economists are trained to understand the world, as each process can be measured and modeled in terms of monetary flows. We learned about this approach in Chapter 10 on environmentally extended input–output models. When applied to studying product systems, this approach is called *economic input–output LCA* (IO-LCA).

Process LCA and IO-LCA each have their own advantages and disadvantages, which we will discuss over the next several chapters. One of the most important differences, however, is how the choice of which to use affects the system boundary of a study.

11.6.3 *Level of detail boundaries*

How much detail should be included in an LCA? An analyst frequently needs to decide whether effort should be expended to characterize the environmental impacts of trace constituents such as minor additives in a plastic formulation or small brass components in a large steel assembly. With some modern technological products containing hundreds of materials and thousands of parts, this is far from a trivial decision. This quandary also applies to considering processes in the life cycle, such as infrastructure. For example, almost all product life cycles include transportation of some form or another, and it is common to include emissions from trucks, trains, planes, and ships, but is it also necessary to consider the environmental impacts of road materials? What about from the production of the machinery needed to build the road? What about the factory needed to build the machinery? What about the transportation of the factory worker? And so on.

In process LCA, where each physical process is modeled separately, it is essentially impossible to include interconnections of every physical input at every level, either because data aren't available or there is simply not enough time. So, a system boundary must be drawn that cuts off some inputs and interconnections; the challenge is to do so in a way that ensures that the results are still accurate, by minimizing what is called *cut-off error*. One rubric for drawing a system boundary is called a *cut-off rule*. If a material or component or process comprises less than a certain percentage of inputs, say 5% by weight or energy or some other physical basis, it can be considered outside the scope of an LCA. The assumption behind the cut-off rule is that weight or energy content is an accurate proxy for environmental impacts. This assumption does not always hold, particularly when there are inputs or components with particularly severe environmental impacts. For example, the lead-acid battery in an automobile weighs less than 5% of the vehicle, but the toxicity of lead implies that it should be included if the goal of the study includes toxicity concerns. There are many examples of substances that have small physical flows but high potency, such as ozone-depleting fire suppressants or radioactive materials, so the LCA practitioner must be very careful not to exclude these important flows from the system boundary, as they might actually drive differences among the results. Another type of cut-off rule is to exclude any flow that contributes less than a certain percentage of environmental impacts. To make this determination requires some initial analysis, but then inconsequential but time-consuming flows can be safely left out of the LCA.

Another rubric for setting the system boundary of a process LCA is to leave out certain categories of inputs, with the implicit assumption that they do not contribute much to environmental impacts and their exclusion will not meaningfully affect the results. This exclusion rubric is commonly applied to inputs of labor, machinery, or infrastructure. Note that this is distinct from the exclusion of life cycle stages. Rather, the exclusion of, say, machinery, from an LCA would apply to all life cycle stages, including machinery needed for extraction, production, transport, use, and end-of-life. The risk here is that the excluded category actually is important. For example, for an LCA that compares different freight transportation options, it would be prudent to include infrastructure (but still perhaps exclude labor). Whatever the rubric taken, any flows that fall outside of the system boundary of a process LCA need to be explicitly documented in the study scope.

While each individual flow that is cut off in a process LCA might be small, their cumulative effect can be significant. In contrast, IO-LCA completely avoids the cut-off problem. Because IO-LCA models the monetary flows and emissions over the entire economy, all of the interconnections among processes are included and in theory, nothing is left out. IO-LCA has other disadvantages that we will discuss in later chapters, but one major advantage is its comprehensiveness.

11.6.4 *Spatial and temporal specification and boundaries*

For every LCA, it is essential to know where and when the product system is being considered. This information is best provided alongside the reference flow, such as 'one kilogram of tomatoes grown in the Netherlands and shipped to the UK in 2020'. The specification of location and time period in the study scope serves as a benchmark to judge the quality of the data used in the life cycle inventory (see Chapter 12). Tomatoes grown in the Netherlands may require different types and quantities of inputs than those grown in Spain, and of course shipping to the U.K. is closer than to, say, South Africa, requiring less fuel. And one hopes that both cultivation and shipping have improved in resource efficiency over time. The locations and timing of production and consumption indicate exactly which product system is under study.

In addition, the study scope should specify not just where and when emissions are taking place, but also the boundaries of the environmental impacts that are caused. A characteristic of environmental impacts is that their effects can occur over a very wide range of spatial and temporal scales. The emission of large soot particles mostly affects a local area, those of nitrogen oxides generate acid rain over hundreds of kilometers, and those of carbon dioxide influence the entire planetary energy budget. Similarly, emissions causing photochemical smog have a temporal influence of only a day or two, the disruption of an ecosystem several decades, and the stimulation of global climate change several centuries. Which spatial and temporal boundaries to choose again depends on the goal of the study, but in any case, these boundaries should be clearly stated in the scope of an LCA. If, for example, an LCA is being conducted by a national agency with the goal of assessing domestic health impacts, then a national spatial boundary would be suitable. But if the goal is to understand the environmental impacts of a new technology, then considering effects on global ecosystems would be appropriate. Unsurprisingly, the larger and longer the system boundaries are, the more time-consuming and scientifically uncertain an LCA becomes. Exactly how the spatial and temporal boundaries of environmental impact are modeled is part of the life cycle impact assessment phase of LCA.

11.6.5 *Choosing boundaries*

It should be apparent that the choice of LCA boundaries can have enormous influence on the time scale, cost, results, meaningfulness, and tractability of an LCA. The best guidance that can be given is that the boundaries should be consistent with the stated goals of the study. The most common mistake made in setting the system boundary is simply omitting important information. A vague system boundary makes it difficult or impossible to interpret results properly or compare them to other product systems or LCA studies. Clear documentation is essential for the system boundary and throughout the goal and scope phase of an LCA, and is a primary reason for the existence of international standards.

Further Reading

Guinée, J., *et al.* (2002). *Handbook on Life Cycle Assessment — Operational Guide to the ISO Standards*. Dordrecht, The Netherlands: Klewer Academic Publishers.

Guinée, J., *et al.* (2011). Life cycle assessment: Past, present, and future. *Environmental Science & Technology, 45*(1), 90–96.

Matthews, S., Hendrickson, C., and Matthews, D. *LCA Textbook*. Retrieved from: https://www.lcatextbook.com/.

United Nations Environment Programme (UNEP) (n.d.). Life Cycle Initiative. Retrieved from: https://www.lifecycleinitiative.org/.

Chapter 12

Life Cycle Assessment in Practice

Chapter Concepts

- The fundamental building block of a life cycle assessment is the unit process, each with discrete inputs and outputs, and linked together to form a product system
- The life cycle inventory is the complete list of material and energy flows for a product system and is made up of foreground data (collected directly) and background data (linked through existing datasets)
- Life cycle impact assessment characterization models quantify how flows in the inventory cause environmental impacts, either in terms of environmental changes (midpoint) or damages (endpoint)
- Proper interpretation of the results includes consideration of uncertainty and reflection on whether the LCA has adhered to its goal and scope; iteration is common

12.1 Building a System Diagram

Once the goal and scope of an LCA have been established, including the definition of functional unit and the reference flows for each of the options being considered, an LCA practitioner can begin the modeling process. The first step is to construct, ideally in cooperation with the design and manufacturing team, a *system diagram* that includes all the life cycle stages included in the study, and to break each stage down into *unit processes* that represent each major step or transformation.

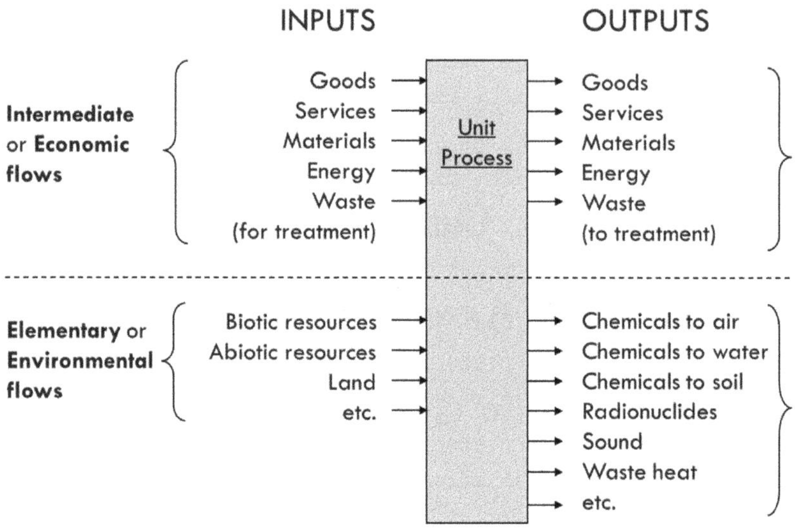

Figure 12.1. A generic diagram of a unit process (adapted from Geyer).

Each unit process within the product system is commonly represented as a block, with arrows representing flows of materials, energy, or emissions (Figure 12.1(a)). There are two types of flows that we want to track. *Economic flows* are those that are linked to other unit processes (that is, they are inputs from or outputs to other parts of the economy). Inputs of electricity and processed materials or outputs of solid waste sent for treatment are all examples of economic flows. In contrast, *environmental flows* are those that are linked directly with nature (that is, they do not connect to other unit processes in the product system). Water taken directly from a river for process cooling or pollutants released directly to the atmosphere are examples of elementary flows. It's not necessary to label flows in this way in a system diagram, but this distinction will become important when we look at the following computational structure of life cycle assessment (LCA).

Unit processes are then linked together by various flows; the resulting system diagrams are thus commonly referred to as block diagrams or process flow diagrams (PFDs) and are extensively used in the various fields involving process engineering. System diagrams will also show processes that are excluded from the study, either because they are not relevant to the goal, or for some other methodological or practical reason. Figure 12.2 shows a system diagram for the life cycle (cradle-to-grave) of an electric scooter (see Hollingsworth *et al.*, 2019). We will return to this example throughout the chapter to reflect on each phase of an LCA.

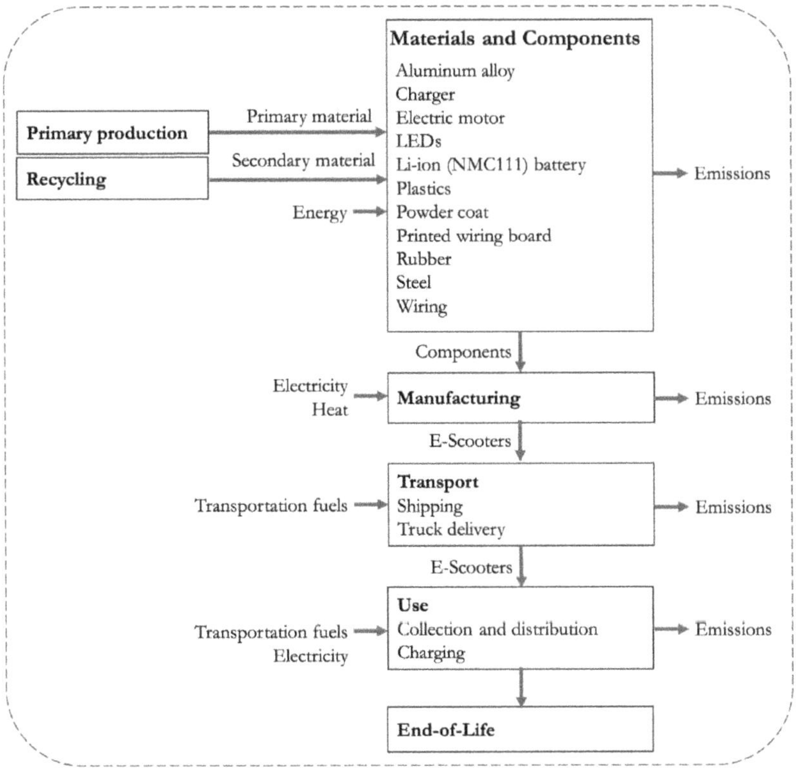

Figure 12.2. A system diagram of the life cycle of an electric scooter (Hollingsworth *et al.*, 2019).

12.2 Compiling the Life Cycle Inventory

12.2.1 *Acquiring foreground data*

Once the system diagram of the LCA has been established, the analyst proceeds to collecting the necessary data on input and output flows, which is called the life cycle inventory phase of LCA (see Figure 11.1). This phase typically begins with examining the main product or process reference flow, either by taking measurements directly or by working with manufacturers or public information. What is the product made of, and what are the quantities of different materials? This information could be found by deconstructing the product and weighing the pieces, or from a manufacturer's specifications sheet. If the product is assembled from components supplied by others, it may be necessary to deal with suppliers to get a complete picture. How much electricity is used to make the product? If the factory is accessible, this value could be measured directly with individual meters, or derived from the factory's electricity bills. How far does the product travel to the user, and how? This value can be estimated from online maps, calculated based on internal details from a manufacturer's logistical operations, or averaged from freight transport data reported to the government. Whether measured directly, calculated based on engineering models, estimated, or averaged, these product- or process-specific inputs and outputs are called *foreground data*, as they are in the foreground of the product system, "within sight" of the LCA practitioner. Foreground data can include both economic and environmental flows.

In the case of the e-scooter study, researchers disassembled and weighed a scooter, estimated shipping distances from the manufacturer, estimated electricity needed for charging based on manufacturer specifications, used data from the e-scooter sharing platform to understand how much the scooters were actually being used, and conducted a survey of employees to understand the extent to which scooters needed to be relocated after hours. This mix of direct measurement, technical specifications, estimation, company data, and personal interviews is typical of the foreground inventory analysis phase of LCA.

12.2.2 *Linking foreground data to background data and LCI databases*

Foreground data represent only *direct* inputs and outputs at individual stages of a product life cycle, but they don't yet tell us anything about the product system as a whole, in particular all of the indirect flows that occur upstream in the supply chain or downstream in waste management. For example, the amount of electricity used annually in a building doesn't reveal how that electricity was generated, how much fuel was used (if non-renewable), what emissions were generated and where, the quantity of water used at the power plant, etc.

The second step of compiling the life cycle inventory is to connect each economic flow in the foreground either upstream or downstream to other unit processes in the product system. It is not practical for an LCA practitioner to investigate every process and supplier along every branch of a supply chain; there are so many interconnections in our global economy that the study would never be completed!

Luckily, there are already inventory data for thousands of unit processes that have been investigated in the past, based on decades of LCA research. So, rather than an LCA practitioner collecting data directly from each individual material supplier, the product-specific economic flows in the foreground can simply be linked to existing LCI data sets, called *background data.* Background LCI data sets live in LCI databases that are compiled by researchers, government agencies, or commercial entities. There are dozens of LCI databases globally; taken together, they represent a giant intellectual contribution to the field. A list of common LCI datasets can be found on the openLCA Nexus page. In our example, the e-scooter researchers linked the quantity of aluminum needed for the scooter with a background dataset name "Aluminum alloy, $AlMg_3$" from the ecoinvent 3.3 LCI database. All LCA studies should clearly document how foreground data were matched to background data so that these matching choices can be checked and validated.

Once all of the foreground economic flows have been linked to background LCI data sets, an algorithm (described in the next section) traces each supply chain all the way back to the initial extraction of resources from nature and counts all of the emissions, scaling each unit process according to the quantities needed to produce the final reference flow. This system-wide set of resource inputs and emissions is called the life cycle inventory and may cover thousands of individual substances or flows.

The life cycle inventory forms the basis of life cycle impact (LCI) assessment (described in Section 12.3), but can also provide valuable environmental insights, especially about resource inputs. For instance, the life cycle inventory can tell us how much energy is used throughout the life cycle of a product, both directly and indirectly — this is called *primary energy demand* or *cumulative energy demand.* We can also see how much water is withdrawn from natural sources throughout the life cycle — this is a form of *water footprint.* Total land use, total emissions of a specific pollutant, total solid waste generated, all of these can be extracted from the life cycle inventory and used as a basis to compare product options or to target environmental improvements.

One fascinating inventory analysis study considered the mass of inputs required to manufacture a 2-gram microchip (see Williams *et al.*, 2002). The authors found that the chemical input needed to make the chip (including the upstream use of fossil fuels) weighed 1.7 kilogram, or 630 times the mass of the chip itself! Water used for fabrication contributed an additional 32 kilogram of material use. Chip fabrication relies on high purity materials and highly controlled process environments, and based on these results, we say that the chip has a large material footprint. This type of analysis has been carried out for numerous products, and it is always striking to see that the upstream use of resources is typically many times larger than the resources that physically make it into the final product.

Figure 12.3. Photographic visualization of the embodied materials in a car, represented next to a normal sized car but is about three times the size and about 25 times the mass (from the LCAart project, lcaart.com).

Figure 12.3 is an artistic (but dimensionally accurate) visualization of this result for a building.

12.2.3 *Computational structure of life cycle inventory analysis*

Each of the unit processes in an LCI database has input and output flows, and remember that these can be grouped into economic flows (linked to other unit processes) and environmental flows (directly to and from nature). Consider a group of n unit processes, each of which can be expressed as a vector $\vec{p}_i$ that has economic flows a and environmental flows b:

$$\vec{p}_i = \begin{pmatrix} a_1 \\ \vdots \\ a_r \\ b_1 \\ \vdots \\ b_s \end{pmatrix} \tag{12.1}$$

There are r possible economic flows and s possible environmental flows. If we line up all n unit processes side-by-side, we can list all of the flows in the database, separated into two adjoint matrices **A** and **B**, with economic flows on top and environmental flows on the bottom:

$$\begin{pmatrix} a_{11} & a_{1n} \\ & \\ a_{r1} & a_{rn} \\ b_{1n} & b_{1n} \\ & \\ b_{s1} & b_{sn} \end{pmatrix} = \left(\frac{\mathbf{A}}{\mathbf{B}}\right) \tag{12.2}$$

A is called the *technology matrix* and describes how the unit processes are linked together through economic flows. Each unit process has a reference flow that describes its primary economic output and each economic flow a_i has a corresponding unit process that produces it, so that **A** is square and $r = n$. **B** is called the *environmental matrix* and describes how the unit processes are linked to nature. There may be any number of environmental flows, so there is no requirement that **B** be square.

The basic matrix mathematical problem when calculating the life cycle inventory is to trace how each of the foreground inputs that make up the reference flow (the product or process under study) causes activity in upstream or downstream unit processes. These foreground inputs can be expressed as a final demand vector $\vec{Y}$:

$$\vec{Y} = \begin{pmatrix} y_1 \\ \vdots \\ y_r \end{pmatrix} \tag{12.3}$$

We want to "activate" each unit process in the technology matrix **A** in order to deliver these economic flows. In other words, we want to find the vector of activity coefficients X such that

$$\mathbf{A}\vec{X} = \vec{Y} \tag{12.4}$$

Solving for $\vec{X}$ gives

$$\vec{X} = \mathbf{A}^{-1}\vec{Y} \tag{12.5}$$

So, the economic flows of each unit process are scaled by the corresponding activity coefficient in order to deliver the economic flows that constitute the final demand. At the same time, the environmental flows of each unit process are also scaled by the same activity coefficients. We can thus express the system-wide environmental flows as

$$\vec{E} = \mathbf{B}\vec{X} = \mathbf{B}\mathbf{A}^{-1}\vec{Y} = \begin{pmatrix} b_{11} & \cdots & b_{1n} \\ \vdots & \ddots & \vdots \\ b_{s1} & \cdots & b_{sn} \end{pmatrix} \begin{pmatrix} x_1 \\ \vdots \\ x_n \end{pmatrix} = \begin{pmatrix} \sum_{i=1}^{n} b_{1i} \cdot x_i \\ \vdots \\ \sum_{i=1}^{n} b_{1i} \cdot x_i \end{pmatrix} \tag{12.6}$$

$\vec{E}$ is the life cycle inventory vector, it represents the quantity of each type of environmental flow summed over all of the unit processes in the product system. For example, $\vec{E}$ could contain the total amount of freshwater withdrawn as a resource input or the total amount of carbon dioxide emitted from the product system. For more details on the computational structure of LCA, see Heijungs and Suh (2002).

12.2.4 *Allocation and system expansion*

One complication in compiling a life cycle inventory is when a unit process has more than one reference flow, that is, more than one salable product output. Actually, this is quite common in industry; for example, refineries can produce a dozen or more different fractions of chemicals using crude oil as an input. The refinery has resource inputs and emissions, so how best to assign, or *allocate* those flows to the multiple product outputs?

The ISO standards for LCA provide specific guidance on this topic, giving a list of options, detailed as follows and shown in Figure 12.4.

1. *Avoid allocation by sub-dividing the unit process*: In some cases, a unit process might have discrete inputs and output for each of its outputs. For example, a food processing plant could grind oats into oatmeal and corn into cornmeal using different machinery. As long as each of the inputs and emissions can be assigned to one and only one of the output products, then we can break the multi-output unit process into multiple single-output unit processes.
2. *Avoid allocation by system expansion*: This is the most complex option, one that forms an entire sub-discipline within LCA. The idea here is to choose a primary output to which all of the flows are allocated, but then also assign it credits (negative flows) because the other output(s) can act as substitutes and

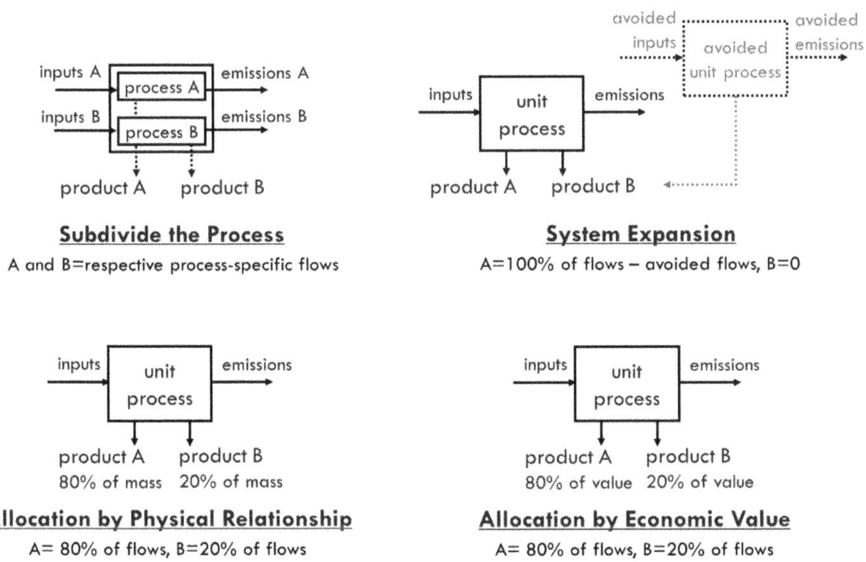

Figure 12.4. Approaches to allocation.

avoid production. In this way, the system is expanded to include the avoided production as a credit. For example, a co-generation plant produces both electricity and steam heat. An alternate way of generating that heat is with steam boilers. Using system expansion, we would assign all of the fuel inputs and emissions to the electricity, but then subtract the avoided emissions from the steam boilers because they no longer need to operate.

3. *Allocation on the basis of physical relationships*: In this case, we assign all flows of the multi-unit process to each product output according to its proportion of some total physical quantity, typically mass or energy content. So, under mass allocation, if in a two-output process, Product A has 95% of the mass, then it is allocated 95% of each flow, and Product B is assigned 5%.
4. *Allocation on the basis of economic value or some other relationship*: The final option is to use a non-physical proxy for allocating flows, the most common choice being economic value. This option resonates with many businesses because industrial facilities are often built with a primary product or source of revenue in mind, while other co-products only marginally influence profitability. The approach is very similar to the earlier allocation example; under economic allocation, if in a two-output process, Product A has 95% of the economic value, then it is allocated 95% of each flow, and Product B is assigned 5%.

Different industries have developed their own guidelines about which approach to allocation should be followed for LCA studies in their sector. The choice of allocation can greatly affect results, and for this reason, it is critical that an LCA practitioner clearly document how allocation was done.

12.3 Life Cycle Impact Assessment

The life cycle inventory flows tell us only the *magnitude* of resource inputs and emissions for each option. The third phase of LCA, called life cycle impact assessment (LCIA), allows us to quantify *effects* of these flows on the environment and public health. Whereas the inventory phase is largely about accounting, the assessment phase relies on models from environmental science, chemistry, physics, ecology, toxicology, and other fields in order to create causal chains that link an emission in a particular location to the physical changes (called *midpoints*) or damages (called *endpoints*) that it causes all over the world.

Of course, there are multiple ways that our activities affect the planet: global warming, depletion of the ozone layer, and toxicity are all critical concerns. These are called *impact categories*. One of the key features of LCA is that it considers multiple impact categories simultaneously, in order to provide more complete information for decision-making. This also can avoid *burden shifting*, where we lessen impacts in one category while unintentionally increasing impacts in another. Different models are used to assess different types of impacts; an LCIA method defines which impact categories are included and how each one is assessed.

LCIA methods are typically developed by scientific research organizations within government labs and universities, with an intended focus on geography. Table 12.1 shows an example for the United States, called Tools for Reduction and Assessment of Chemicals and other environmental Impacts (TRACI). While other LCIA methods might consider different sets of impact categories and use different models to do so, most methods cover the same themes of resource use, atmospheric change, air quality, water quality, and toxicity.

The LCIA procedure is a four-step process: *classification* and *characterization* are required by the ISO standards, while *normalization* and *weighting* are optional.

1. *Classification*: Each flow is classified according to whether it causes a certain type of impact. For example, nitrous oxide (N_2O) causes both global warming and ozone depletion, and so is assigned to both impact categories. Other flows might not cause effects in any of the impact categories and are left unassigned.

Table 12.1. Impact Categories Covered by the TRACI 2.1 LCIA Method.

Impact category	**Description**	**Indicator**
Global warming potential (GWP)	Indicates the radiative forcing (heat-trapping potential) associated with greenhouse gases (GHGs) and other climate forcers, using a 100-year time horizon	CO_2 eq
Acidification potential (AP)	Indicates the potential acidification of water and soils due to acid rain, formed by reaction of emissions such as sulfur and nitrogen oxides with water in the atmosphere	SO2 eq
Eutrophication potential (EP)	Indicates the potential for algae growth and reduced amounts of oxygen in water due to nutrient pollution	N eq
Ozone depletion potential (ODP)	Indicates the potential for destruction of ozone in the stratospheric ozone layer	CFC-11 eq
Photochemical smog formation	Indicates the potential for formation of tropospheric (surface-level) ozone due to reactions of nitrogen oxides and volatile organic compounds (VOCs) in the presence of sunlight	O_3 eq
Human health particulate matter	Indicates the potential for emissions and secondary formation of harmful particulate matter (PM)	$PM_{2.5}$ eq
Human health cancer	Indicates the potential for chemical emissions to cause cancer in humans	CTUh
Human health non-cancer	Indicates the potential for chemical emissions to cause non-cancer disease in humans	CTUh
Ecotoxicity	Indicates the potential for chemical emissions to cause damages to non-human organisms	CTUe
Resource depletion	Indicates the potential for depletion of non-renewable resources, in particular fossil fuels	MJ eq

2. *Characterization*: Each classified flow is assigned a value, or characterized, according to how potent a substance it is for each type of impact. This is called a characterization factor (CF) and is expressed relative to an indicator substance. In the case of global warming potential, we measure CFs relative to carbon dioxide, and so all greenhouse gases are characterized in units of carbon dioxide equivalents (CO_2 eq). Then the mass of each classified flow E_j from the life cycle inventory is multiplied by its respective characterization factor $CF_{i,j}$ and summed up over all j flows to determine the total impact I_i for each impact category:

$$I_i = \sum_j CF_{i,j} \cdot E_j \qquad (12.7)$$

3. *Normalization*: Because LCA compares options across multiple sustainability indicators, results typically have different units, many of which are unfamiliar to non-scientists. It can be useful to normalize results across impact categories by some set of reference values R_i, for example as a proportion of a person's average annual impacts, in order to tell which type of impact is larger on a relative basis. The normalized results N_i are then:

$$N_i = I_i / R_i \tag{12.8}$$

4. *Weighting*: Valuation is the process of assigning weighting factors to the different impact categories based on their perceived relative importance. For example, an assessor, an international standards organization, or a stakeholder panel might choose to regard climate change impacts as twice as important as acidification, and apply weighting factors w_i to the normalized impacts accordingly. Then the impact assessment results can then be expressed as a single weighted score W:

$$W = \sum_i w_i \cdot N_i \tag{12.9}$$

Single scores are simple and attractive for comparing products and tracking progress. But, the application of weighting factors is controversial because doing so involves making social, political, and ethical choices. As a result, LCIA evaluations that are suitable for a particular culture or location or time are unlikely to be useful in other circumstances. Because of this limitation, weighting is often omitted from LCIAs. (Doing so, however, is equivalent to making an implicit weighting with all categories valued the same.)

12.4 Interpretation

This last and perhaps most important phase of LCA is interpretation, where results get translated into knowledge and recommendations. If Option A has lower impacts than Option B across all impact categories, then this supports a clear preference. But it frequently happens that one product option is better for some categories but worse in others. The LCA practitioner must discuss these different trade-offs in the context of the study Goal, leaving the final decisions to the client. Overall, the purpose of the Interpretation phase is to provide context to the study so that readers can fairly interpret the results.

12.4.1 Draw conclusions and recommendations

In many cases, an LCA is motivated by a desire to improve designs and practices. These studies will include a *contribution analysis* to tell which processes or hot spots are driving results for different impact categories, in order to suggest where to focus efforts. If a particular material is driving impacts, then designers might investigate low-carbon alternatives or ways to reduce how much material is used. If a particular upstream process is the main culprit (such as electricity generation), then managers might work with suppliers to source a greener option (e.g., through renewable energy contracts). This list of recommendations constitutes the core output of an LCA study.

In developing a list of recommendations based on the inventory and impact assessment results, it is important for the assessor to be inclusive, and to range widely. Some recommendations will be very specific (i.e., avoid the use of toxic metals), while others are much more diffused (i.e., minimize the diversity of packaging materials). Both types are important to include. The highly specific recommendations are easier to generate, and their accomplishment is more easily measured. The diffuse recommendation may be more difficult to deal with, but may in some cases be very important; their inclusion may crucial to actually effecting change.

12.4.2 ***Evaluate the data used in the LCA***

LCA is not an exact science — there are various assumptions and choices that are a part of every study, as well as inherent uncertainty in model parameters such as material quantities and emissions factors. An essential step in the LCA interpretation phase is to evaluate the quality, completeness, and consistency of the data. The goal is to ensure that each identification of significance is backed by adequate, reliable information. This is particularly important if alternative product designs are being evaluated because the designs need to have consistent bases for comparison.

A data quality assessment is an investigation of how well the input data reflect the reality of what is being modeled. For instance, if a study uses emissions factors that are five years old but the actual technology has improved dramatically, then this must be clearly documented and its implications discussed. Several dimensions of data quality can be investigated; a typical list of data quality indicators includes the following:

- *Reliability* — how were the data gathered, from direct measurement, secondary sources, estimation, etc.
- *Completeness* — how well do the data represent the range of production sites for the process or product
- *Temporal correlation* — how long between when the data were collected and the time of the study
- *Geographical correlation* — how well does the location of the data source match the location of the study
- *Technological correlation* — how well do the data represent the actual process or product being modeled

For each data set used, each of these data quality indicators might be scored "1" to "5", or described qualitatively to make up the data quality assessment.

At the data completeness step, one wishes to confirm that all product life stages have been addressed, as well as all relevant environmental impacts. This information should be verified as meeting the system boundaries established at the beginning of the study, and that the significant raw materials and releases have been incorporated. Finally, at the data consistency step, the LCA practitioner checks that consistent choices have been made in terms of data quality, source databases and background data sets, allocation, LCIA methods, and other methodological choices across the options being considered.

12.4.3 *Sensitivity and uncertainty analysis*

LCA studies often include a *sensitivity analysis*, where important assumptions are changed in order to see how much the overall results respond. If an LCA is found to be very sensitive to a single parameter, then a caveat should be added to the recommendations and the parameter might be investigated more in order to make sure that the best possible information is being used. In the study on e-scooters, sensitivity analysis was conducted on several parameters, including the daily use and product lifetime. Testing the full range of values for these parameters produced global warming results that spanned a factor of "3". This magnitude of model sensitivity isn't necessarily bad or particularly unusual; it just means that care must be taken in generalizing the results. In addition to testing different parameter values, a sensitivity analysis might also test other choices, such as functional unit, allocation scheme, LCIA method, and so on, in order to make sure that the recommendations coming out of the study are robust and not unduly influenced by modeling choices.

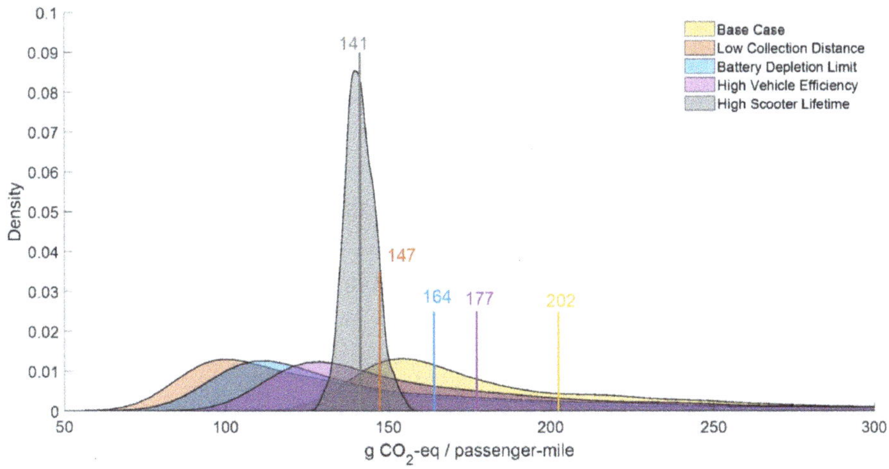

Figure 12.5. Monte Carlo simulation results for electric scooter GHG emissions per passenger mile under different modeling assumptions (Hollingsworth *et al.*, 2019).

LCA studies might also include a quantitative *uncertainty analysis* using statistical methods. The most common approach is called the Monte Carlo simulation, which aims to combine the uncertainty of each data point into an uncertainty estimate for the overall results. In this approach, each data point is assigned a probability distribution (e.g., a normal distribution) that defines its possible range of values. Then the entire model is run hundreds or thousands of times, with each run sampling a different set of values for each parameter. The distribution of model results can then be used to create confidence intervals, compare options, and run statistical tests. Figure 12.5 shows Monte Carlo simulation results for the e-scooter study under several different modeling scenarios. The mean values of each study are shown by the vertical bars. These values alone *could* be used to communicate results, but also showing the spread of the results adds a new dimension for interpretation and conveys where there is significant overlap among options.

12.5 LCA Software

In this chapter, we have presented the fundamentals of LCA methods and mathematics, but because of the complexity of the LCA process, most practitioners rely on software that holds the background inventory data, allows for easy inputting of foreground data and metadata, and runs the algorithms (including uncertainty analysis). LCA software comes in two general flavors. The first is comprehensive models meant for in-depth analysis or research, where the user has control over each parameter and can build custom, complex models. The second is screening models meant for non-specialists that consist of pre-calculated impact results for common materials and processes. This is the approach taken by online calculators, product design, and architectural software, and many sector-specific LCA tools.

Table 12.2 shows an excerpt of the Inventory of Carbon and Energy, with embodied energy and embodied carbon values per kilogram of selected building materials. All the user needs to do is figure out how much of each material is needed for the product in question, multiply by each respective factor, and sum to find the cradle-to-grave results. This approach is fast and easy if all one needs is results for reporting or the choice of materials is fixed, but understanding exactly where and why certain materials are causing impacts instead requires comprehensive LCA software.

Table 12.2. Excerpt of Pre-calculated Inventory and Impact Factors of Selected Building Materials from the Inventory of Carbon and Energy (University of Bath, UK and Circular Ecology).

Material	Energy MJ per kg	Carbon kg CO_2 per kg
Aggregate	0.083	0.005
Aluminum (including 33% recycled)	155	8.24
Bitumen (general)	51	0.38–0.43
Bricks (common)	3	0.24
Concrete block (medium density)	0.67	0.07
Concrete (1:1.5:3)	1.11	0.16
Copper (including 37% recycled)	42	2.6
Glass	15	0.85
Gold	100,000	10,000
Iron (general)	25	1.91
Lead (including 61% recycled)	25.2	1.57
Plywood	15	1.07
Polyurethane insulation (rigid foam)	102	3.48
PVC (general)	77.2	2.41
Stainless steel	56.7	6.15
Steel (average recycled content)	20.1	1.37
Timber (excluding sequestration)	8.5	0.46
Vinyl flooring	65.6	2.92
Wool carpet	106	5.53

LCA software packages are continuously being refined, as are the LCI data and LCIA methods that are loaded into them. When choosing an LCA software platform, be sure to check which versions of datasets and methods and being used, and that they are providing good coverage for the types of flow and environmental impacts that are part of your goal and scope.

12.6 LCA Limitations and Outlook

LCA has enjoyed widespread popularity as an ideal way to quantify environmental impacts at a systems level, and it has been applied in thousands of studies. But despite its widespread use, it is important to keep in mind its limitations. Most generally, LCA is meant to provide information to support decisions. Its core competency is providing a broad life cycle perspective, but there are other perspectives and tools that can complement it, such as risk analysis, exposure assessment, or cost-benefit analysis.

As with any model, the quality of LCA results depends on the quality of the data inputs. LCI datasets continue to grow but simply cannot keep pace with all of the millions of items that are produced worldwide, using a variety of processes and in different locations, all of which are continuously changing in technology and efficiency. As we learned about in this chapter, in addition to which data sets to use, the LCA practitioner has to make various decisions about how to carry out the modeling, especially around the scope and setting of the study. As a result, LCAs by different assessment teams can produce different, though perhaps equally defendable, results. This is one reason why documenting assumptions in LCA is so important, as it allows readers to figure out what is driving the differences among studies.

There are many efforts to harmonize LCA, including developing reference datasets of government data, more detailed technical guidance (such as the International Reference Life Cycle Data System), standards around individual impact categories that are more specific than the ISO standards, guidelines for specific types of products (called Product Category Rules, or PCRs), and others. These efforts are critically important for LCA to be used for absolute measurements (as in carbon accounting) and policy settings where consensus is vital, as opposed to being used for internal design decisions.

LCA methods also continue to evolve. One of the greatest opportunities is to improve the spatial and temporal resolution of both inventory data and impact assessment methods. Current datasets are often resolved to the national or sometimes regional scale, but in reality, the location and timing of emissions matter in terms of downwind or downstream exposures. Allowing LCIA methods to incorporate this type of information will bring LCA results more in line with the sophisticated environmental models that underpin policy and will make results more useful to communities who have specific questions about their particular situation and for whom generic results are not that useful (see Hellweg and Milà i Canals, 2014). A second opportunity has to do with the mathematical structure. LCA is a linear model: double the flows and you get double the impacts. But for some types of environmental impacts, notably toxicity, responses are non-linear. Lower concentrations of a pollutant might have no effect (or even be essential to life). Capturing these threshold effects consistently across all types of flows has been difficult, but is an active area of research.

Despite these limitations, LCA remains an extremely useful and flexible tool, and a core one within industrial ecology. Its large community of practitioners and researchers continue to generate data and add new methods and capabilities, including the integration of important environmental and social impacts. This chapter has covered the basics but there are many resources available (some included in References) for those who want to learn more.

Further Reading

Bare, J. C. (2012). Tool for the Reduction and Assessment of Chemical and Other Environmental Impacts (TRACI), Version 2.1 — User's Manual (EPA/600/R-12/554).

Guinée, J. *et al.* (2002). *Handbook on Life Cycle Assessment — Operational Guide to the ISO Standards*. Dordrecht, The Netherlands: Kluwer Academic Publishers.

Heijungs, R., and Suh, S. (2002). *The Computational Structure of Life Cycle Assessment*. The Netherlands: Kluwer Academic Publishers.

Hellweg, S., and Milà i Canals, L. (2014). Emerging approaches, challenges and opportunities in life cycle assessment. *Science*, *344* (6188), 1109–1113.

Hollingsworth, J., Copeland, B., and Johnson, J. X. (2019). Are e-scooters polluters? The environmental impacts of shared dockless electric scooters. *Environmental Research Letters*, *14* (8), 084031.

Matthews, S., Hendrickson, C., and Matthews, D. (2014). *LCA Textbook*. Retrieved from: https://www.lcatextbook.com/.

openLCA Nexus (n.d.). Retrieved from: https://nexus.openlca.org/databases.

United Nations Environment Programme (n.d.). Life Cycle Initiative. Retrieved from: https://www.lifecycleinitiative.org/.

Williams, E. D., Ayres, R. U., and Heller, M. (2002). The 1.7 kilogram microchip: Energy and material use in the production of semiconductor devices. *Environmental Science & Technology*, *36* (24), 5504–5510.

Chapter 13

Economics Tools: Techno-Economic Analysis, Life Cycle Costing, and Valuation

Chapter Concepts

- Industrial ecology often employs economic analyses to complement environmental and social sustainability tools
- Techno-economic assessment and life cycle costing provide a life cycle approach to the financial analyses of products, processes, and projects
- Valuation can be used to convert environmental and social benefits impacts into economic terms in order to facilitate communication and comparison

13.1 A Life Cycle Perspective on Costs

One of the main goals of industrial ecology is to apply a life cycle perspective to understanding physical products and systems so that we can make design and consumer decisions that are holistically better for the environment. Tools like life cycle assessment (LCA) can aid us in quantifying emissions and resource concerns and identify hotspots in a product life cycle that are targets for improvement. But of course, environmental performance is just one of many considerations that can inform decisions. Another vital consideration is economic costs and benefits.

In this chapter, we will briefly introduce several economics tools that are widely applied in industrial ecology, either in parallel to environmental analyses or as integrated approaches. As with LCA, these tools apply a life cycle perspective to ensure that we are seeing the whole picture, and not making economic decisions now that shift costs to future generations. Techno-economic analysis

(TEA) is a process-based modeling tool for quantifying the total costs of production, typically carried out from a manufacturing perspective when considering investment in a new product or process. The closely related tool of Life Cycle Costing (LCC) has a wider scope, covering the entire life cycle costs of a physical asset like a bridge. Both TEA and LCC are akin to the inventory phase of LCA but account for flows of money rather than flows of mass or energy. Economic valuation, on the other hand, is akin to the impact assessment phase of LCA, but in units of economic damages or benefits rather than environmental and health consequences.

Economics tools are an important component of an industrial ecologist's toolbox, especially in interfacing with industry and policymakers. Modeling both environmental and economic impacts ensures that recommendations are more likely to produce win–win–win scenarios that are good for people, prosperity, and the planet.

13.2 Techno-Economic Analysis

As long as humans have been making products for barter or sale, there has been a calculus of whether it is worthwhile. The modern manifestation of that calculus is called techno-economic analysis (TEA), widely used by process and manufacturing engineers to quantify costs associated with production.

Costs fall into two major categories. Capital costs are those associated with the construction of the production facility, including construction materials, equipment, land, labor, as well as services such as financing, design, legal services, permitting, and insurance. Once the plant has been constructed and commissioned, operating costs are those incurred while the plant is running, including material inputs, energy and water utilities, and labor. Adding capital and operating costs together gives an estimate of the cradle-to-gate costs of production, and therefore whether the process will be economically competitive.

But how to add capital and operating costs? A central tenet of economic analysis is the 'time value of money', which holds that cash flows that occur at different times have different values. This is because money today can be placed into a risk-free investment that produces interest or has a rate of return. Thus, the future value (FV) of money spent today (present value, or PV) depends on the prevailing rate of return i and the number of compounding periods n into the future:

$$FV = PV \cdot (1 + i)^n \tag{13.1}$$

Conversely, the present value of money spent in the future can be found by rearranging the terms as

$$PV = \frac{FV}{(1+i)^n} \tag{13.2}$$

Initial capital costs are assumed to be spent today (Year 0) while operating costs (and later capital costs for multi-year construction projects) are spent in the future. So, in order to add them together, we discount costs in each future year over the plant lifetime N to their present values and add to capital costs in Year 0 to obtain the net present value NPV:

$$NPV = \sum_{n=0}^{N} \frac{FV_n}{(1+i)^n} \tag{13.3}$$

This is called a discounted cash flow analysis and is the basis of all manner of investment decisions. The net present value divided by the total production volume gives the unit cost of production, which in turn sets a minimum price for the product in order for the plant to be profitable. For fuels, this is called the minimum fuel selling price (MFSP, per liter or gallon); for electricity, it is called the levelized cost of electricity (LCOE, per MWh or kWh).

The starting point of a TEA, as in a process-based LCA, is a detailed process flow diagram (PFD), including the desired throughput. Figure 13.1 shows a PFD

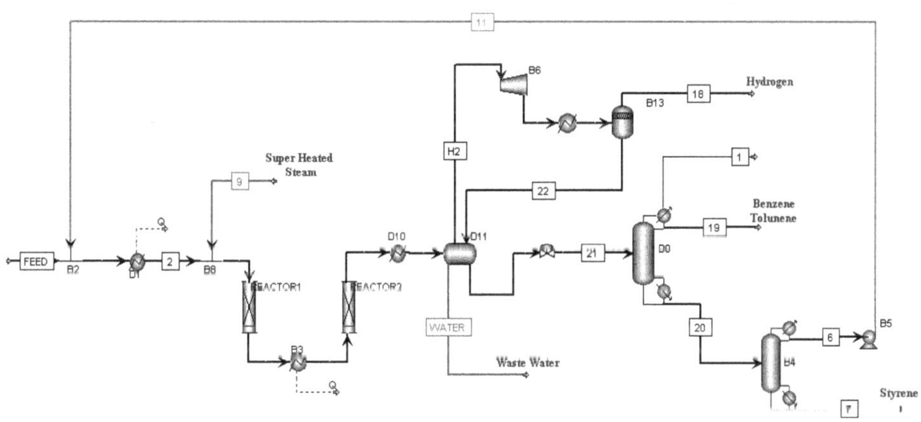

Figure 13.1. PFD of styrene production from ethylbenzene in AspenPlus (Parvatker and Eckelman, 2019)

for a plant that produces styrene from the dehydration of ethylbenzene. Specialized process simulation software has been developed that can then calculate the required size of each piece of equipment (for capital costs) and the material, energy, and water requirements (for operating costs). Labor and other operating costs are determined based on similar plants.

Because its process models are highly detailed, TEA can identify exactly which aspects of production most contribute to costs, and test how costs can be reduced through process modifications. TEA can also model how production costs will be influenced by external factors, for example, if the price of a certain input falls or if a policy incentive like a production tax credit is introduced. Manufacturing operations can be designed to be more or less efficient; TEA is used as a process design tool to optimize designs based on profitability. For example, many chemical plants operate with a conversion efficiency for the feedstock of less than 100%. Higher efficiencies are physically possible (with additional separation and recycling stages), but may not be economically worthwhile.

13.3 Life Cycle Costing

While TEA focuses on cost estimation for process manufacturing, the related tool of life cycle costing (LCC), life cycle cost analysis (LCCA), or total cost of ownership (TCO) considers ownership costs over the entire lifetime of a generic physical asset. TEA and LCC both quantify capital and operating costs, and apply cash discounting methods in order to compare options and inform investment decisions; the main differences are the ownership perspective and the cradle-to-grave scope.

As an example of how LCC is used, consider the choice between two technology options for a home water heating system: a high-efficiency heat-pump system with substantial capital costs of $2,000 but low operating costs of $200 per year versus a less efficient natural gas-fired system with lower capital costs of $800 but higher operating fuel costs of $600 per year. A typical water heater has a lifetime of 10 years. Table 13.1 shows a LCC discounted cash flow model for the comparison using a discount rate of 5% (and ignoring any salvage value or energy price inflation):

Even though the up-front (or 'first') cost of the heat-pump system is 2.5 times that of the gas one, over the total lifetime of the water heater, the heat-pump system is 35% cheaper. Psychologically, humans tend to focus on first costs when making investment choices, so life cycle costing is essential for prudent decision-making.

Table 13.1. Example LCC Comparison Table for Home Water Heating Systems.

Year	Heat-pump nominal	Heat-pump discounted	Gas nominal	Gas discounted
0	$2,000	$2,000	$800	$800
1	$200	$190	$600	$571
2	$200	$181	$600	$544
3	$200	$173	$600	$518
4	$200	$165	$600	$494
5	$200	$157	$600	$470
6	$200	$149	$600	$448
7	$200	$142	$600	$426
8	$200	$135	$600	$406
9	$200	$129	$600	$387
10	$200	$123	$600	$368
LCC		**$3,544**		**$5,433**

The earlier example is but one case of one of the most common questions in life cycle studies: is it worth it to invest more in the materials and manufacturing stage in order to achieve benefits in the use stage? Similar LCC studies are routinely conducted to evaluate investments in green buildings or renovations, design and maintenance of infrastructure, or switching to a new technology such as electric vehicles. The very same core question motivates many LCA studies as well; literally thousands of LCA articles have been published on embodied versus operational carbon in buildings, for example, in evaluating insulation products. By conducting LCA and LCC on the same system in parallel, industrial ecologists can provide additional value to consumers and policymakers than by just considering environmental impacts.

13.4 Economic Valuation

So far, we have learned about cost analysis in terms of actual costs that are borne by a manufacturing company or an asset owner. But there are others in the larger system who bear costs as well. If a manufacturing facility gives off air pollution that impacts the health of downwind residents, their hospital bills are certainly real, but they are not paid by the facility owner. Or, if that pollution causes acid rain that kills a protected forest, no cash costs are incurred necessarily, but there is certainly a loss to society and to the natural world. These types of costs are called 'externalities' — they are the costs that occur external to a process or asset that are not reflected in the financial decisions of firms or in the prices of goods and services. For example, a recent meta-analysis found that the mean externality associated with coal-fired electricity generation is 14.5 cents per kWh (Sovacool *et al.*, 2021), which is higher than the market price of electricity in many countries. Considering energy and transport, the authors estimate global externalities of nearly $25 trillion, or 29% of global GDP! Clearly, these external costs are large, and we should make every attempt to quantify and include them in decision-making. This process is called 'economic valuation' and is one of the main objectives of the field of environmental economics.

Air pollution is an important source of externalities, but there are many others. If a road construction project causes traffic delays, externalities will include lost time, additional fuel costs, as well as air pollution impacts from idling vehicles. Residents who live nearby might also experience noise, light, and vibration disturbances during construction. New construction in a coastal wetland area might negatively impact recreational activities like fishing. It might also degrade 'ecosystem services' like erosion prevention that natural systems provide to us for free so that we have to spend cash to build engineered systems like coastal barriers to do the same job.

Quantifying external costs is complicated and depends very much on which externalities are included. Some external costs like crop losses are straightforward to quantify because there are established market prices. Others such as health damages can be inferred from large data sets on hospital admissions. But what is the cost of losing something that has no monetized value, like a beautiful vista or a species that goes extinct?

Over the years, environmental economists have developed a wide range of tools to aid in economic valuation for different types of externalities. One important technique is hedonic price analysis, which uses the price of a proxy to infer intrinsic value, such as how home prices change based on proximity to a park or a metro station can be used to value green space or public transit. Another technique is willingness-to-pay (WTP), which asks people how much they would be willing to pay to avoid a loss. For example, economists have found that people are willing to pay between $5 and $100 to protect an individual species from extinction. Taken across millions of households, this means that the value of that species to humanity could be in the billions of dollars, such that it is worth funding conservation programs accordingly. These non-market valuation techniques don't capture the intrinsic value of species or natural systems independent of their perceived value to humans, but they are still important tools in capturing external costs.

Over the years, public agencies and large institutions like development banks have begun to require an analysis of externalities for any new project and have published standard tables of external costs for economic valuation. Table 13.2 shows published values for transportation projects from the most recent U.S. guidelines. We can see, for example, that the cost of particulate matter pollution per metric tonne is quite high, as this is the leading cause of human health damage from air pollution, causing millions of premature deaths annually worldwide. Especially with the publication of standard, consensus factors by public agencies and international institutions, social costs are increasingly being included in LCC studies.

The last value in Table 13.2 is of particular note. This is the external cost associated with carbon greenhouse gas emissions, what is called the 'social cost of carbon' (SCC). Climate change will affect human health in various ways, from increased risk of natural disasters and heat waves to higher prevalence of vector-borne diseases and crop failures leading to malnutrition. Economists use climate models to see how these various effects may manifest and then assign social costs based on the number of people affected. As with the transportation example, SCC is now being included in investment decisions or is being levied as an actual cost, either internally or through carbon markets and government policies. By internalizing the cost externalities of greenhouse gas emissions, the world will have a stronger financial incentive to decarbonize and meet our collective climate goals.

The place where industrial ecologists most use economic valuation is in life cycle impact assessment (LCIA), specifically in the calculation of endpoint characterization factors. Recall that these factors quantify environmental impacts in terms of damages, in contrast to midpoint characterization factors that quantify physical changes in the environment. The most common endpoint categories are

Table 13.2. External Costs for US Transportation Projects (US DOT, 2022).

Externality	**Published value (in 2019$)**
Vehicle travel (average) per hr	17.90
NO_x emissions per Mg	15,700
SO_2 emissions per Mg	40,400
$PM_{2.5}$ emissions per Mg	729,300
CO_2 emissions per Mg	50

health damages and species loss. By assigning monetary values from economic valuation studies to each of these categories, LCA results can be expressed in financial terms, which may add to their policy relevance and influence.

Further Reading

National Renewable Energy Laboratory (NREL), U.S. Department of Energy (DOE). (n.d.). *Techno-Economic Analysis*. Retrieved from: https://www.nrel.gov/analysis/techno-economic.html.

Parvatker, A. G., and Eckelman, M. J. (2019). Comparative evaluation of chemical life cycle inventory generation methods and implications for life cycle assessment results. *ACS Sustainable Chemistry & Engineering*, *7*(1), 350–367.

Sovacool, B. K., Kim, J., and Yang, M. (2021). The hidden costs of energy and mobility: A global meta-analysis and research synthesis of electricity and transport externalities. *Energy Research & Social Science*, *72*, 101885.

U.S. Department of Transportation (DOT) (2022). *Benefit-Cost Analysis Guidance for Discretionary Grant Programs*. Retrieved from: https://www.transportation.gov/sites/dot.gov/files/2022-03/Benefit%20Cost%20Analysis%20Guidance%202022%20%28Revised%29.pdf.

U.S. Environmental Protection Agency (EPA) (2019). *Life Cycle Assessment and Cost Analysis of Distributed Mixed Wastewater & Graywater Treatment for Water Recycling in the Context of an Urban Case Study* (EPA/600/R-18/280). Retrieved from: https://cfpub.epa.gov/si/si_public_record_Report.cfm?dirEntryId=346484&Lab=NRMRL.

Chapter 14

Network Analysis

Chapter Concepts

- Networks are commonly found in nature, in industrial ecosystems, and in human interactions. The tools of network analysis apply to all
- Network analysis in industrial ecology is of particular value in understanding corporate and societal supply chains
- A second major use of network analysis in industrial ecology is in the study of corporate interactions in industrial symbiosis networks

14.1 Introduction to Food Web Analysis

In biological ecology (BE), a food web describes material and energy transfer interactions that occur among the organisms in a particular habitat. Organisms that provide similar functions can be grouped into trophic levels, each level representing a position within a food chain. An example food web, for a marine ecosystem, was shown in Figure 6.4. The web was mapped as a result of extensive observations and the conclusions were specific rather than general. Was this web typical or atypical of biological systems? To find out, the tools of network analysis are needed.

The network concept seems straightforward but is inherently limited by whether the appropriate data are accessible to the analyst. In BE, the constraints are related to the effectiveness and accuracy of counting and quantifying interactions among the organisms. In social ecology (SE), the limitations often involve the availability and recollections of the individuals whose interactions are the objects of study. In industrial ecology (IE), material flows may not have been measured or their availability may be restricted by the corporations involved. In all cases, the networks deduced by the analyst are likely to be incomplete so far as both species and interspecies interactions are concerned. Butts (2009) puts the challenge this way:

> To represent an empirical phenomenon as a network is a theoretical act. It commits on to assumptions about what is interacting, the nature of those interactions, and the time scale on which the interactions take place.

As it happens, one of the earliest industrial ecology publications involved a network, specifically of metal transfer transactions among firms in the U.S. state of Massachusetts. That network was constructed by asking each firm to which types of firms did it send or receive metals. The results were depicted in a "spaghetti diagram" (Figure 14.1) that shows commonly occurring (but unquantified) exchanges of metals, one line for each firm-to-firm relationship, the arrowheads indicating the direction of transfer. The diagram makes apparent the importance of scrap dealers, brokers, and agglomerators in transferring metal and of smelters, refiners, and alloyers in creating the forms of metals that are industrially useful.

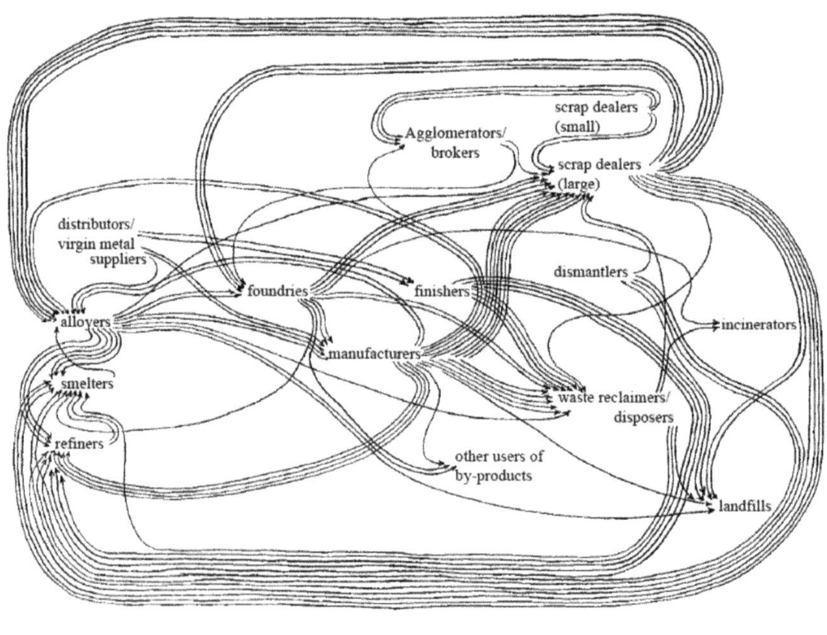

Figure 14.1. A "spaghetti diagram" of metal transfer relationships among firms in Massachusetts, US, in 1996 (Frosch *et al.*, 1997).

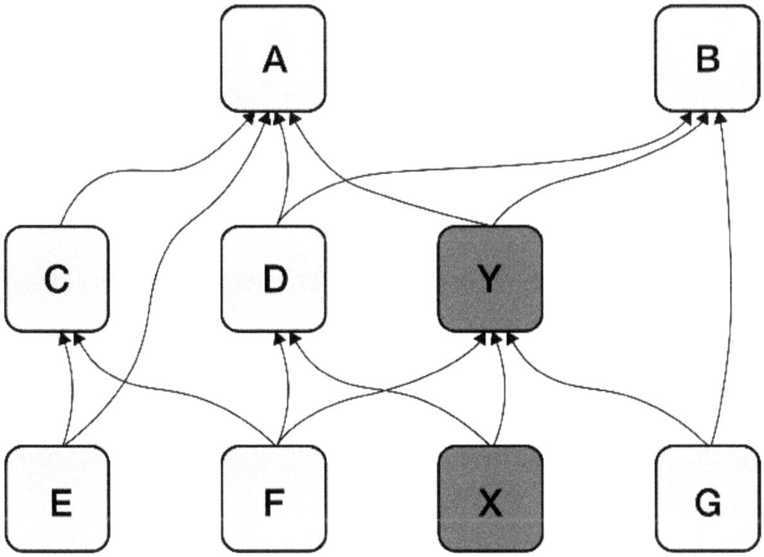

Figure 14.2. Example of a food web for nine organisms in three trophic levels (Hardy and Graedel, 2002). In such a diagram, the organisms are termed "nodes" and the links connecting predator and prey are termed "edges".

To inject more rigor into network studies, assume that a reliable assessment of links in an individual network (a biological one in this illustration) is given in Figure 14.2. The organisms in the diagram either eat (they are predators) or are eaten by (prey) other species. Two species are *linked* if there is an energy flow between them (e.g., Species X is eaten by Species Y). Quantitative descriptions of this network are then provided by computing a series of network metrics:

Species richness (N): where N is the total number of species (organisms or industrial entities) in the network under study. The network matrix size is $N \times N$.

Number of links (L):

$$L = \sum_{i=1}^{m} \sum_{j=1}^{n} f_{ij} \tag{14.1}$$

where the existence of links is being assessed $f = 0$ or 1.

Connectance (C): The number of actual interactions divided by the number of possible interactions, computed as

$$C = \frac{2L}{N(N-1)} \tag{14.2}$$

(Exercise for the student: Compute the network connectance of Figure 14.2).

Linkage density (Δ):

$$\Delta = L / N \tag{14.3}$$

(Note that a more connected, or linked network will utilize resources to a greater degree than one less connected, or linked.)

14.2 Network Analysis Metrics and Displays: The Resource User Perspective

Perhaps more than ecosystem food webs, the analyses of the complex networks often encountered in industrial ecology generally generate enhanced value by employing increasingly detailed analytical measures. Four of these are as follows:

- *Product complexity (in-degree)*: The number of incoming links that represent the number of species (elements, materials, parts, components) required to generate the final product.
- *Producer diversity (out-degree)*: The number of downstream material nodes and the number of production sites.
- *Average supply chain links (closeness centrality, undirected)*: The average number of steps from the node under investigation to any other node in the network.
- *Maximum number of physical transformation steps (eccentricity, directed)*: The greater the number of steps, the greater the potential for distortion of physical flows.

An example of a complex industrial supply chain is that for cadmium-telluride (CdTe) photovoltaic panels, by Nuss *et al.* (2016) shown in Figure 14.3(a). The complexity of this network is not unusual in industrial manufacturing. Rather, Figure 14.3(a) is probably simpler than the average supply chain. Cadmium (far left of diagram) does not occur in isolation in mineral deposits, but is instead a byproduct of zinc mining (bottom left), so the detailed connections of both must be evaluated. Similarly, tellurium does not exist in isolation but is available as a byproduct of copper mining (lower center and right). The resulting cadmium and tellurium are then employed in manufacturing the photovoltaic panel.

The complex network statistics for the CdTe panels are given in the top row of Table 14.1, followed by those for four other industrial products. (The details

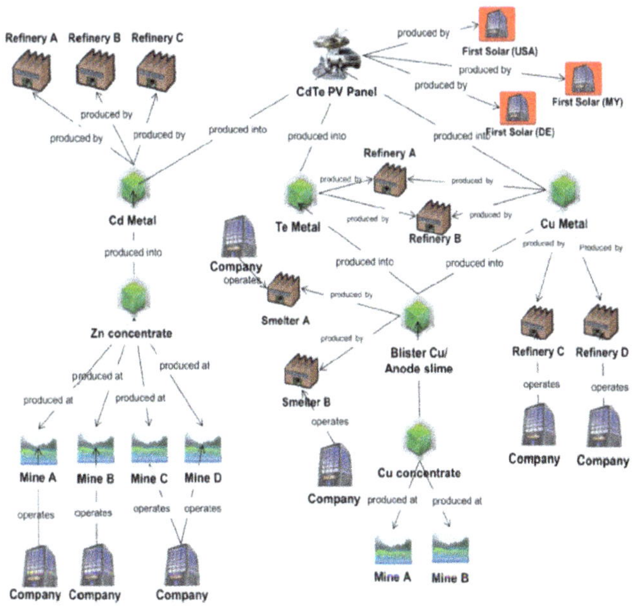

(a)

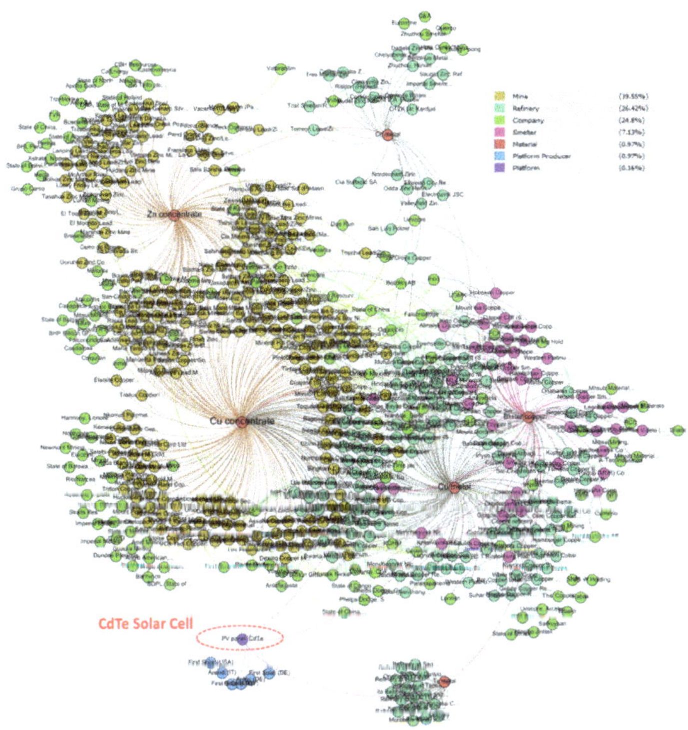

(b)

Figure 14.3. (a) Schematic diagram of the supply chain for cadmium telluride (CdTe) solar cells. (b) Visual representation of the nodes and links of all participants in the CdTe solar cell network (Nuss *et al.*, 2016).

Table 14.1. Comparison of Network Metrics for Five Product Platforms (Nuss *et al.*, 2016).

Product	Nodes	Links	Product complexity (in-degree)	Producer diversity (out-degree)	Supply chain length (closeness)	Physical transformation steps (eccentricity)
CdTe solar cell	617	999	3	6	3.1	3
Ge solar cell	549	826	2	4	3.4	5
Turbine blade	672	981	13	5	2.5	2
Pb-acid battery	250	321	2	35	2.7	2
High-density magnet	336	428	5	28	2.3	2

of the analysis are available in the Further Reading paper.) In the five product platforms, the number of nodes ranges from 250 for the lead acid battery supply chain to 672 for the turbine blade. In other comparisons, a hard drive magnet is an average 2.3 steps away from any other node in its network, while the Ge solar cell on average requires one additional step (closeness = 3.4) to reach the other nodes in the network. Analyses such as this can inform manufacturers what might be expected if they are contemplating producing a new product or can reveal opportunities for supply chain control in the product under manufacture.

The network map for the entire CdTe solar cell supply chain appears in Figure 14.3(b). It is obvious from this figure that the complexity of industrial network can be very great and that network analysis that can discover and express opportunities for significant complexity reduction has high potential value.

14.3 Network Analysis Metrics and Displays: The Resource Supplier Perspective

Network analyses that are complementary to those of resource users are those of resource suppliers. For such industrial firms, the network metrics inform those producers about the strength and diversity of their markets. Table 14.2 provides a set of metrics for 11 metals in the US economy in 2007 as derived by converting the economic input–output table into its physical counterpart and then deriving appropriate network metrics: nodes, links, connectivity, and supply chain length. The results can be tabulated (Table 14.2) or, perhaps better yet, plotted to illustrate markets for the resource in question (Figure 14.4 does so for the 2007 US tungsten market.)

The earlier examples have focused on metals, but network analysis tools are being applied to all manner of industrial ecology questions. IE is fundamentally

Table 14.2. Network Metrics for 11 Metal Networks in the 2007 US Economy.

Element	Network nodes	Network edges	Directed density (%)	Average degree	Network diameter	Avg. path length
Aluminum	390	64,575	43	165.6	3	1.55
Vanadium	389	18,589	12	47.8	8	2.14
Chromium	391	52,603	35	134.5	4	1.65
Manganese	391	54,624	36	139.7	3	1.63
Iron	400	72,063	45	180.2	4	1.52
Cobalt	389	16,614	11	42.7	6	2.13
Nickel	392	48,148	31	122.8	4	1.69
Copper	390	60,128	40	154.2	3	1.58
Niobium	390	22,313	15	57.2	7	2.04
Molybdenum	390	30,446	20	78.1	5	1.0
Tungsten	389	21,003	14	5.0	7	2.04

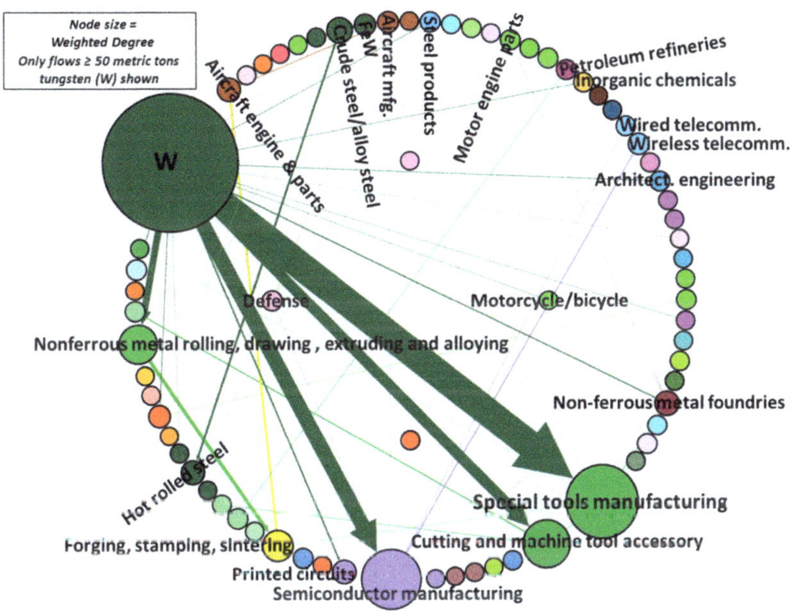

Figure 14.4. The tungsten network of the US economy, 2007 (Nuss *et al.*, 2019).

about entities interacting in a socio-technical system, so many of the IE tools you have learned about in the preceding chapters can be expressed as networks. For example, input–output tables express links between each pair of sectors in an economy, which can be modeled as a fully connected network where all links are bi-directional, or an *undirected graph.* The technology matrix of a life cycle inventory database is sparser, as not all processes are connected via physical flows and flows are often one-way; this is called a *directed graph.* Network science has developed numerous measures for each kind of graph that can be used in IE analyses. For example, measuring propagation through directed graphs can reveal important quantities like shipping times (in a supply chain graph), exergy destruction (in an energy graph), or dissipation of metals (in an SFA graph).

14.4 Enhancing Industrial Ecosystems by Network Analysis

It is obviously interesting to analyze and map the networks of natural and industrial ecosystems, but a more interesting and useful result of such a comparison would be if nature can serve as a teacher so as to upgrade and/or enhance industrial ecosystems. A potentially useful example in this regard is the study of Morris *et al.* (2021). Those researchers reviewed the scientific literature and recovered information on three sets of networks: (1) ecological food webs (31 examples), (2) urban-industrial ecosystems (UIE) (29 examples), and (3) eco-industrial parks (EIP) (48 examples). They then proceeded to compute network statistics for each group. A sampling of the results is shown in Figure 14.5. For these two structural

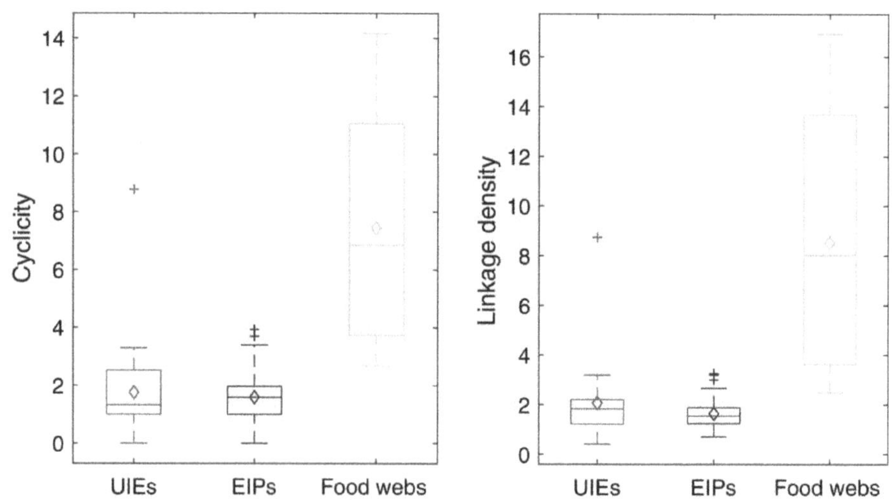

Figure 14.5. Comparisons of cyclicity (left) and linkage density (right) among groups of urban industrial ecosystems (UIEs), eco-industrial parks (EIPs), and ecological food webs (Morris *et al.*, 2021). These diagrams are box plots, in which the upper quartile of the data is indicated by the top of the box, the median by the line in the box center, the lower quartile by the bottom of the box, the average value by the diamond, and the extremes in the data by the vertical lines.

metrics, it is clear that the food webs show much greater cyclicity and linkage density than is the case for the human-centered systems. This result may well be due in part to the significantly larger number of nodes (actors) in natural ecosystems, thereby providing many opportunities for linkage. For the human networks to approach the performance of the natural networks, the former will need to strongly enhance their linkage densities and cyclicities.

Further Reading

Ashton, W. (2008). Understanding the organization of industrial ecosystems: A social network approach. "Uncovering" industrial symbiosis. *Journal of Industrial Ecology*, *12* (1), 34–51.

Brehm, C., and Layton, A. (2021). Nestedness of eco-industrial networks. *Journal of Industrial Ecology*, *25*, 205–218.

Butts, C. T. (2009). Revisiting the foundations of network analysis. *Science*, *325*, 414–416.

Frosch, R. A., *et al.* (1997). The industrial ecology of metals: A reconnaissance. *Philosophical Transactions of the Royal Society A*, *355*, 1335–1347.

Hardy, C., and Graedel, T. E. (2002). Industrial ecosystems as food webs. *Journal of Industrial Ecology*, *6* (1), 29–38.

Layton, A., Bras, B., and Weissburg, M. (2016). Industrial ecosystems and food webs: An expansion and update of existing data for eco-industrial parks and understanding the ecological food webs they wish to mimic. *Journal of Industrial Ecology*, *20* (1), 85–98.

Morris, Z. B., Weissburg, M., and Bras, B. (2021). Ecological network analysis of urban-industrial ecosystems. *Journal of Industrial Ecology*, *25*, 193–204.

Nuss, P., Graedel, T. E., Alsonso, E., and Carroll, A. (2016). Mapping supply chain risk by network analysis of product platforms. *Sustainable Materials and Technologies*, *10*, 14–22.

Nuss, P., Ohno, H., Chen, W.-Q., and Graedel, T. E. (2019). Comparative analysis of metals use in the United States economy. *Resources, Conservation, and Recycling*, *145*, 448–456.

Veleva, V., Todorova, S., Lowitt, P., Angus, N., and Neely, D. (2015). Understanding and addressing business needs and sustainability challenges: Lessons from Devens eco-industrial park. *Journal of Cleaner Production*, *87*, 375–384.

Chapter 15

Data Science in Industrial Ecology

Chapter Concepts

- Data science techniques are offering new capabilities to industrial ecologists for recognizing trends, gathering data, and creating predictive models
- Innovative industrial ecology work has applied data science techniques to sustainability problems in transportation, the built environment, and manufacturing processes
- Open-source code and data platforms can enable collaboration and transparency, especially around common data sets

15.1 The Data Science Revolution

As in so many areas of research, advances in computing power and techniques are revolutionizing the field of industrial ecology. Old problems are being re-examined with a level of detail or comprehensiveness not available previously, and new data streams and algorithms are opening entirely new topics to inquiry. A full treatment of data science methods is of course outside the scope of this textbook; instead, we will use examples of cutting-edge industrial ecology (IE) research to demonstrate the different types of data science techniques that are being used and the sorts of insights that they can enable.

Data analysis is familiar territory for many industrial ecologists who are used to working with large data sets to discover patterns in material and energy flows. In some ways, data science is a natural extension of statistical techniques that are commonly used, but in which the data sets are bigger. In other ways, however, a fundamental advancement in the tools at our disposal is occurring.

First, consider our increasingly digitized, interconnected, and sensed world. At this moment, you may be wearing a watch that monitors your physiology, using a phone that transmits your location, or one in a room that senses your occupancy and behavior (and maybe even your metabolism!) Each day, you may use credit cards or digital payment platforms that track what you purchase, use a car that records your driving behavior, or be recorded by cameras on roads and in buildings. These data streams, while having a primary business or safety purpose, also present an enormous potential resource in understanding behavior and decision-making at the individual level. Now, instead of relying on expensive, manual surveys in order to collect data on mobility or consumption for a small sample of people, one can use some of these new data streams to track an entire population. This capability affords unprecedented coverage and accuracy, and also allows us to examine outliers or differences among groups for which data have been difficult to find.

Second, computers can now be trained to process images and text that allow researchers to routinely access data that used to be laborious and time-consuming to extract. An early advance was the development of optical character recognition (OCR) that allowed computers to recognize printed text. Combined with the commercial introduction of scanners, OCR led to widespread digitization of paper records. Especially for industrial ecologists doing dynamic or historical material flow analysis (MFA), where many of the records are on paper, sometimes going back centuries, this advance has allowed them to process more records from more locations than the old-fashioned academic method of personally visiting the archives and transcribing by hand.

What this new wave of data science brings is the capability for computers not just to *recognize* text but to *process* it. Now, rather than reading the digitized records and extracting relevant data themselves, IE researchers can program an algorithm to comb through millions of documents that would have taken a lifetime to analyze by hand. Not just text, but images can also be digitized and processed. Published graphs, building diagrams, aerial images, personal photographs, and Google Street View shots can all now be analyzed *en masse* to answer important and unexplored questions about how we interact with the natural and built environments. 'Computer vision' techniques have been developed to analyze video feeds as well, so that we can now conduct highly accurate surveys of people or traffic or use computers for detecting anomalies across a wide range of systems for which we have video surveillance.

Finally, new analytical techniques from data science are allowing IE researchers to build models of socio-technical systems, models that have better performance or accuracy than has been possible with traditional statistical methods. Many of the systems that IE researchers study are non-linear and complex, and might involve hundreds or even thousands of variables. Using a small number of observations to model such a large system suffers from what is called the 'curse of dimensionality' (Bellman, 1961), meaning that very large data sets are needed to properly calibrate and validate predictive models. A separate paradigm is to consider the behavior of the system overall rather than its constituent parts. With large enough data sets, data-driven or machine learning techniques can be trained to discern how the system works based on experiences rather than on functional or causal relationships among variables. Both approaches have the potential to improve the breadth and performance of predictive models used throughout IE.

15.2 Location Data and Mobility

Transportation and mobility have been important topics for data science, in part because it was previously impossible to track the travel behavior of large numbers of people and understand the utility of various analytical options based on a wide range of individual travel patterns. Industrial ecologists are interested because transportation is a key factor in determining urban emissions, air quality, and connectivity. One important question is which transportation options will have the greatest benefits. Cai and Xu (2013) examined vehicle electrification by mining a large data set of more than 10,000 GPS-enabled taxis in Beijing. They used this data set to create a distribution of energy requirements and thus find an optimal vehicle type, battery size, and configuration of charging stations for the fleet. At the individual level, researchers have used accelerometer and positional data from cellphones to infer transportation modes, enabling such useful capabilities as real-time greenhouse gas (GHG) emissions tracking. Geotags created by social media and camera apps on smartphones also present rich potential data streams. By examining 150,000 photographs of various natural sites around Europe, Lenormand *et al.* (2018) were able to determine how far people were traveling for ecological tourism, informing studies of consumption and environmental pressures.

15.3 Remotely Sensed Data and the Built Environment

The built environment is another IE research area that is benefiting enormously from data science tools. Whereas cities used to be studied in terms of aggregate top-down statistics or a small subset of buildings, we can now build urban models with detailed physics-based representations of every single building, road, bridge, or green space. For example, Heeren and Hellweg combined data from a federal registry of building characteristics with 3D remote sensing data from airborne

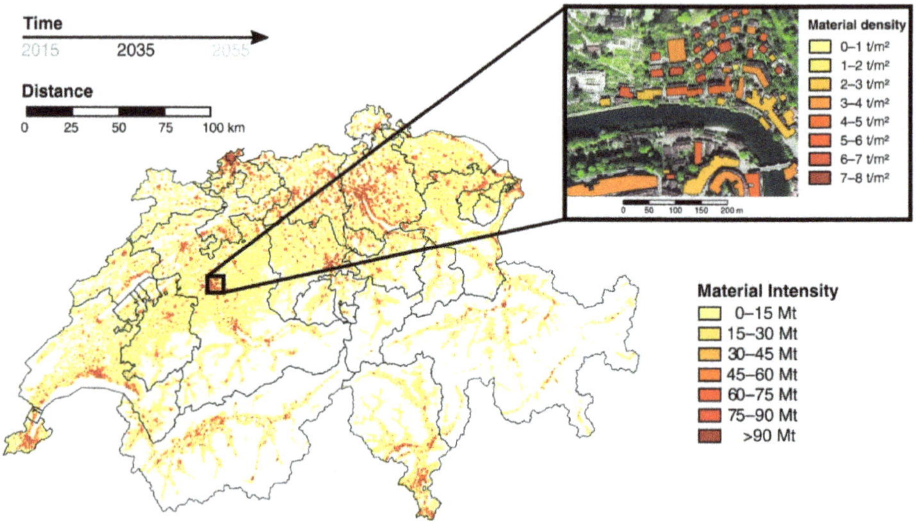

Figure 15.1. Material stock intensity map of Switzerland (Heeran and Hellweg, 2018).

laser scanning to create a spatial database of every building in Switzerland, a database that can be used for understanding in-use stocks (Figure 15.1), testing housing scenarios using dynamic MFA, or understanding life cycle resource requirements or emissions of buildings in unprecedented detail.

Similar high-resolution mapping efforts across the world have been used to analyze important urban design questions such as the overall potential for rooftop solar panels or urban agriculture, or the response and vulnerability of urban buildings to natural hazards. Coastal cities are especially at risk. Saha and Eckelman (2014) used building tax assessor information combined with remote sensing data to build a dynamic map of which concrete buildings across the city of Boston, U.S., are most at risk of corrosion, a study that informs inspection efforts and design codes.

15.4 Textual Process and Life Cycle Inventory Data

Oftentimes, IE research involves extracting information from text records. This is especially common in LCA, where process details might be contained in published articles or patents. The scientific literature is vast and it can be very difficult to obtain a comprehensive view of a process or material, so often LCI data are extrapolated from a single study. To tackle this challenge of scale, Kim *et al.* (2017) developed a text mining technique that combed through more than 10,000 scientific manuscripts to extract process data and build a robust predictive model for synthesis parameters for various materials.

Missing LCI data is another common issue, where there is simply no information provided on a specific emission and so it is omitted. Data mining provides one solution. In the U.S., many industries are required to report emissions to the Environmental Protection Agency (EPA), which has constructed a database query tool that allows researchers to mine these records to fill in LCI gaps or improve the accuracy of certain flows (Cashman *et al.*, 2016). Another LCI data approach is based on looking at similarity to existing data sets. 'Read-across' or the use of proxies is common in LCA and toxicology, but the process of choosing the best proxy can be subjective. Instead, data science researchers convert textual and numerical information on chemicals into a mathematical high-dimensional space and use algorithms to find the chemical nearest to one for which there is missing data. Hou *et al.* (2018) took this approach to estimate missing data in the popular ecoinvent LCI database. For data such as toxicity that are based on inherent chemical properties, missing data can be estimated by predictive models. In a separate study, Hou *et al.* (2020) train neural network models with data for nearly 2000 substances in the USEtox model to predict chemical ecotoxicity for chemicals for which data are missing.

Industry has also seized on machine learning models and artificial intelligence to aid all sorts of business innovations, including in advancing environmental management efforts. It used to be expensive for manufacturers to conduct or commission an LCA, such that only a few products were ever assessed. Using data science approaches that combine information on material sourcing, processing, and packaging, companies are moving to a paradigm where they can automatically generate LCA results or environmental product declarations (EPDs) across their entire product portfolio.

15.5 Data Standardization and Open Data Efforts

A final note in this chapter is that effective data science requires that researchers be able to share their data and algorithms, not just their results. IE researchers are increasingly coming together to develop standards and protocols for data and metadata (information that describes data). LCA has long used standard data formats like ecospold2 so that unit process data can be read by LCA software. New efforts like the USEPA's LCA data harmonization tool (Ingwersen *et al.*, 2015) provide a framework for managing and standardizing LCA metadata, including such core considerations as how flows are named and recognized.

The widespread adoption of open-source software and fully documented model codes means that IE research is becoming more collaborative and transparent, making it easier to compare work and leading to more robust science. For example, the python-based tools Brightway2 for LCA, pySUT for I-O modeling, and dynamic_stock_model for MFA (Pauliuk *et al.*, 2015) comprise an IE software 'toolbox' that python users can use and manipulate. IE researchers are also publishing entire data sets so that the rest of the IE community can make use of them, such as the multi-decade Yale Stocks and Flows Database (Myers *et al.*, 2019).

As the field of industrial ecology continues to advance, the data science trends and approaches described in this chapter will play an increasingly important role and may come to dominate entire areas of inquiry. As future scientists, training in coding, data management, and data science should be encouraged for all budding industrial ecologists.

Further Reading

Bellman, R. E. (1961). *Adaptive control processes: A guided tour*. Princeton, NJ: Princeton University Press.

Cai, H. and Xu, M. (2013). Greenhouse gas implications of fleet electrification based on big data-informed individual travel patterns. *Environmental Science & Technology*, *47* (16), 9035–9043.

Cashman, S. A., *et al.* (2016). Mining available data from the United States Environmental Protection Agency to support rapid life cycle inventory modeling of chemical manufacturing. *Environmental Science & Technology*, *50* (17), 9013–9025.

Davis, C., Nikolic, I., and Dijkema, G. P. (2010). Industrial ecology 2.0. *Journal of Industrial Ecology*, *14* (5), 707–726.

Font Vivanco, D. *et al.* (2019). Interactive visualization and industrial ecology: Applications, challenges, and opportunities. *Journal of Industrial Ecology*, *23* (3), 520–531.

Heeren, N. and Hellweg, S. (2019). Tracking construction material over space and time: Prospective and geo-referenced modeling of building stocks and construction material flows. *Journal of Industrial Ecology*, *23* (1), 253–267.

Hertwich, E., *et al.* (2018). Nullius in Verba 1: Advancing data transparency in industrial ecology. *Journal of Industrial Ecology*, *22* (1), 6–17.

Hou, P., *et al.* (2018). Estimating missing unit process data in life cycle assessment using a similarity-based approach. *Environmental Science & Technology*, *52* (9), 5259–5267.

Hou, P., *et al.* (2020). Rapid prediction of chemical ecotoxicity through genetic algorithm optimized neural network models. *ACS Sustainable Chemistry & Engineering*, *8* (32), 12168–12176.

Ingwersen, W. W., *et al.* (2015). A new data architecture for advancing life cycle assessment. *The International Journal of Life Cycle Assessment*, *20* (4), 520–526.

Kim, E., *et al.* (2017). Materials synthesis insights from scientific literature via text extraction and machine learning. *Chemistry of Materials*, *29* (21), 9436–9444.

Lenormand, M., *et al.* (2018). Multiscale socio-ecological networks in the age of information. *PLoS One*, *13* (11), e0206672.

Myers, R. J., Reck, B. K., and Graedel, T. E. (2019). YSTAFDB, a unified database of material stocks and flows for sustainability science. *Scientific Data*, *6* (1), 1–13.

Pauliuk, S., *et al.* (2015). Lifting industrial ecology modeling to a new level of quality and transparency: A call for more transparent publications and a collaborative open-source software framework. *Journal of Industrial Ecology*, *19* (6), 937–949.

Saha, M., and Eckelman, M. J. (2014). Urban scale mapping of concrete degradation from projected climate change. *Urban Climate*, *9*, 101–114.

Xu, M., Cai, H., and Liang, S. (2015). Big data and industrial ecology. *Journal of Industrial Ecology*, *19* (2), 205–210.

The Resources of Modern Technology

Chapter 16

The Material Resources of Modern Technology

Chapter Concepts

- The abundances of the elements vary widely, and the use of the more abundant elements recognizes the potential for more dependable supplies and lower costs
- Materials choice depends on anticipated uses and the associated material properties that will be needed
- The growth of urban regions worldwide has stimulated a corresponding growth in resource demand

16.1 The World of Materials

Late science-fiction author Arthur Clarke's observation about modern technology, that "any sufficiently advanced technology is indistinguishable from magic", resonates with almost everyone in this day of technological wonders: buildings powered by solar panels or by radiation through their windows, jet aircraft capable of flying almost anywhere on Earth in less than a day, the smartphone (incredibly smart!), orbiting satellites that map our planet and monitor its atmosphere, surface temperatures, forest fires, and much more. This 'magic' is made possible by a wide variety of materials, natural and engineered.

The use of materials, of course, began in order to respond to the requirements of the whims of humans in positions that enable them to implement what they wish to do. All humans have basic needs (such as protection) and desires (such as status), as will be discussed later in this book. Needs and desires in the modern world generally require (or at least are benefited by) the use of materials. Examples are quite obvious: self-protection generally is achieved through materials deployment in housing, while realizing desires often involves the acquisition of possessions such as communications technology or upscale vehicles.

However, the relationship of most humans to the materials on which many facets of their lives depend, or benefit therefrom, is very tenuous. Unless trained as scientists or engineers, most people could not name more than a handful of the elements, or what utility is provided by their existence. Such individuals are also generally not aware that few materials are used in pure form — this is because material performance can usually be improved in one way or another by combining two or more materials: increased strength, improved resistance to corrosion, minimization of toxicity, etc. Product designers work hard to select the optimum mix of a product's physical properties.

No matter what materials a designer might wish to use, those materials must be reasonably abundant and routinely available. The starting point for the designer is the element makeup of our planet, that is, what is the potential list of elements that can be considered. Figure 16.1 shows the relative concentrations of the

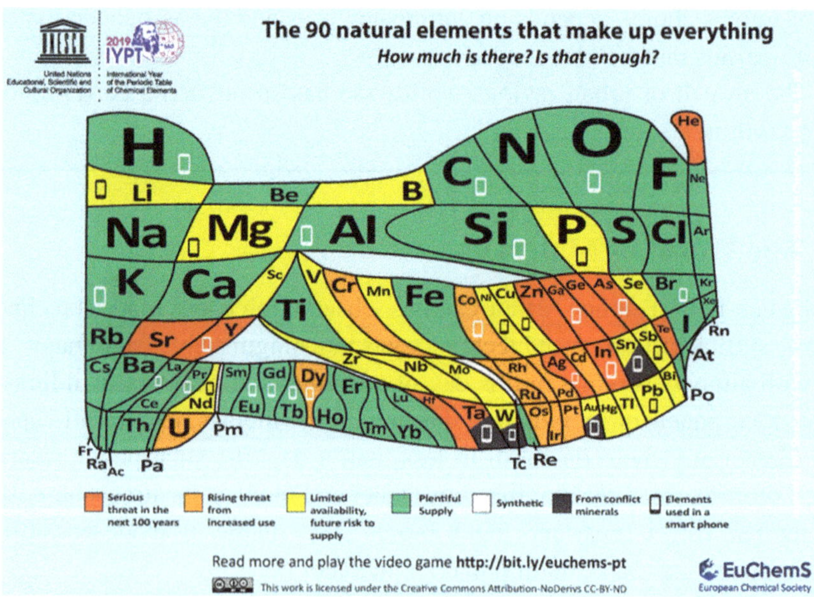

Figure 16.1. Abundances of the elements (European Chemicals Society). This diagram is a diffusion cartogram, in which the relative sizes of "regions" are shown in proportion to some property they possess — in this case, the relative abundances of elements in Earth's crust.
Source: https://www.euchems.eu/euchems-periodic-table/.

elements in Earth's upper crust — the section of the planet to which we have access. One readily sees the dominance of hydrogen, silicon, and oxygen, and to a lesser degree, carbon, nitrogen, magnesium, aluminum, calcium, and iron, and then all the rest. Many other details are of course important: extent and depth of mineral deposits, geographical location of those deposits, regulatory constraints, etc., but the point to be made here is that it will likely be much more technically challenging and much more expensive to design a product around the use of a scarce element rather than an abundant one.

16.2 The Materials Diversity of Modern Technology

Why does modern technology need a wide variety of materials? This is because we ask them to be so many different things: strong, light, corrosion-resistant, visually attractive, electrically conductive (or not), stable in different temperature regimes, and so forth. No single material can meet all these challenges and, in many cases, does not need to. Indeed, if a specific level of strength is needed, a wide variety of materials can meet the challenge. If, at the same time, the designer wishes to minimize the embodied energy of the chosen material (the energy required to extract and process the material), the choices may narrow greatly. For example, at a strength requirement of 10 MPa and an embodied energy of 10^4 MJ/m^3, only wood and bamboo appear likely to be suitable. Tools for evaluating material sustainability in the context of design are discussed further in Chapter 24.

In specific cases where a particular set of properties in a material is needed, the solution may be to invent or adopt a mixture of materials. An example of such a case is a fiber metal laminate that combines the properties of several materials to produce a lightweight material with superior mechanical properties (Figure 16.2). The result is a material type that is widely used for aerospace and

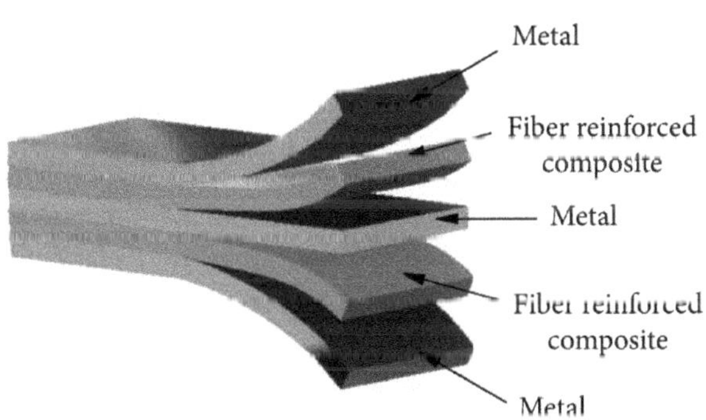

Figure 16.2. An example of a modern composite material — a fiber metal laminate (Chen *et al.*, 2020).

transport applications. It fits no particular category: it is not completely metal, alloy, or fiber but is perfect for specific applications.

Notwithstanding the above, some material choices are generally not appropriate, at least not in all applications. This is because materials not only need to have the desired physical properties, but they need to be suitable for other reasons as well. For example, the properties of materials generally need to be stable over time — resisting corrosion in a particular application, resisting fracture under unanticipated stresses, maintaining their physical appearance, etc. They also must not be potentially harmful in the application for which they are being considered, i.e., not poisonous (as with arsenic or lead), or radioactive (as with radium or uranium). Materials selection is thus a complex specialty for which theory, laboratory studies, and computer simulation under in-use conditions are commonly utilized.

As a consequence of their education and training, most scientists and engineers are "metals people", "plastics people", "composite material specialists", and the like. Almost no one is an expert across the very wide range of materials in widespread use, however. This book, while not a textbook on materials science, nonetheless provides an introduction to the principal materials of modern technology — the sort of raw materials that are employed, the implications for use and reuse that obtain when materials are combined, and so forth. We do so because industrial ecologists commonly evaluate aspects of the manufacture, use, and loss of diverse groups of physical and chemical resources and thus a general introduction to material groups and their properties is relevant. More detailed information, if desired, can be obtained from specialized journal articles and texts, examples of which are included in the References sections of each of the next several chapters.

16.3 The Pencil: A Perspective on Modern Industry

The spectrum of modern industrial materials can be divided into several groups: metals, plastics, construction minerals, and fibers, and composite materials. Each of those groups is addressed in the chapters that follow. Industrial ecologists are often told to "follow the materials" in order to understand resource acquisition, material processing, manufacturing, use, and recycling or loss. As will be seen, the very large and diverse materials of modern technology provide plenty of opportunities to follow the materials.

Is it possible, however, for industrial ecologists to follow the materials of the modern technological society? They are inevitably presented with complex constructions of materials of all kinds, and who could expect to be expert on them all?

More than 60 years ago, this issue was addressed by Leonard Read, an economist. I, Pencil, written from the perspective of a common pencil, presented a simple but important fact: that no single individual possesses all the knowledge and skills needed to produce a product so comparatively unpretentious as a pencil (Read, 1958). Nonetheless, analyzing the resource consumption, environmental impacts, and societal implications of such products is a major responsibility of industrial ecologists. Reading *I, Pencil* is an activity of 10 minutes or so, but it exemplifies the challenges faced by those reading this book. Industrial ecologists cannot be experts on everything with which they interact, but comprehending that challenge and addressing it directly in all its linkages is a useful and necessary goal.

Further Reading

Ali, M. M. (2008). Energy efficient architecture and building systems to address global warming. *Leadership and Management in Engineering*, *8* (3), 273–123.

Chen, Y., Wang, Y., and Wang, H. (2020). Research progress on interlaminar failure behavior of fiber metal laminates. *Advances in Polymer Technology*, 3097839. https://www.hindawi.com/journals/apt/2020/3097839/.

Ciacci, L. *et al.* (2013). Historical evolution of anthropogenic aluminum stocks and flows in Italy. *Resources, Conservation and Recycling*, *72*, 1–8.

Greenfield, A., and Graedel, T. E. (2013). The omnivorous diet of modern technology. *Resources, Conservation and Recycling*, *74*, 1–7.

Read, L. E. (December 1958). *My Family Tree as Told to Leonard E. Read*. Vol. 8. The Freeman, pp. 32–37.

Chapter 17

The Industrial Ecology of Metals

Chapter Concepts

- Metals are the foundation of all materials science and technology and have been for millennia
- Metals are predominantly used in alloy form in order to provide materials of enhanced physical and chemical properties
- The functional rates of recycling of metals tend not to be high, especially for elements and alloys used in small quantities in complex products

17.1 Why Use Metals?

Metals have a very long history as preferred materials for construction, artistic objects, and practical devices. Bronze provides an example; it is an alloy (a material mixture) of copper and tin and was probably discovered when stones rich in each of those metals were heated in cooking fires and the alloy was generated from the melted minerals. The variety of uses to which bronze could be put dominated metallurgy for roughly the period 3300–700 BC: the "Bronze Age". A bit later, methods for utilizing iron were developed: the "Iron Age", roughly marked at 500–300 BC. As early metallurgists learned to add small amounts of manganese to the iron, steels were created; they had the useful ability to make the iron harder without making it more brittle (modern steel typically contains 0.2–0.8% manganese).

Metals continue to dominate much of technology because of their material properties, which include high fracture toughness (resistance to crack propagation), electrical conductivity, similar strength in both tension and compression,

Table 17.1. Global 2018 Production of Selected Metals.

Metal	Symbol	2018 Global production (Tg)	Principal uses and industrial sectors
Iron	Fe	1810	Construction, automotive
Aluminum	Al	64	Automotive, electricity
Copper	Cu	20	Electricity, brass
Manganese	Mn	19	Iron alloy metal
Zinc	Zn	13	Iron corrosion protection, brass
Silicon	Si	7.4	Iron alloy metal
Lead	Pb	4.6	Batteries
Nickel	Ni	2.7	Iron alloy metal, superalloys
Magnesium	Mg	1.1	Aluminum alloy metal
Chromium	Cr	0.40	Iron alloy metal
Arsenic	As	0.33	Electronics
Tin	Sn	0.32	Electronics, bronze
Beryllium	Be	0.24	Electronics
Rare earths	RE	0.19	Magnets
Antimony	Sb	0.15	Batteries
Cobalt	Co	0.14	Batteries, superalloys
Tungsten	W	0.09	Iron alloy cutting tools
Vanadium	V	0.07	Iron alloy metal
Silver	Ag	0.03	Electronics
Cadmium	Cd	0.03	Solar power

Note: Data are from U.S. Geological Survey, Mineral Commodity Summaries.

mechanical properties that are independent of direction, and stability of those properties up to several hundred degrees Celsius. In addition, the long history of metal use in evolving technologies has provided a wealth of experience in mining and processing, as well as extensive experience with many combinations of compositions and applications.

More recently, as reuse has entered the picture, metals have proven much more recyclable and reusable than materials such as plastics and composites. Experience with re-melting and re-refining has enabled this effort, while reuse has encountered major challenges (as with many other materials). Metals have also proved adaptable to additive manufacture ("3-D printing") and thus to the relatively facile production of intricately shaped parts with minimal material loss. Given these unique and useful properties, metals will remain the backbone of materials use for a long time to come. Table 17.1 shows the global production volumes of the major metals.

17.2. The Metal Resource

The periodic table of the elements is dominated by metals. About half of all the elements are either "transition metals" (Columns 3–12, including the lanthanides) or "other metals" (aluminum, lead, and several more). More generically, metals are materials with high electrical and thermal conductivity, good ductility, malleability, high density, and high melting point. They are crystalline when solid. The metalloids (boron, silicon, germanium, arsenic, antimony, and tellurium) see much of their use in a wide variety of metal alloys and so form part of the metal use picture.

The periodic table can give the impression that metals are found in nature one by one, but that is certainly not the case. The properties of one metal are often sufficiently similar to those of others that two, three, or more commonly occur in the same mineral deposits. Therefore, they must be mined together and subsequently separated if they are to be employed for the uses to which they are uniquely suited. Nassar *et al.* (2015) conducted a careful analysis of the degree of this "companionality"; their results are shown in Figure 17.1(a). It turns out that some 30 different elements are available almost entirely as companions, and a number of others are recovered either from deposits in which they are the principal target of opportunity or from deposits where they are quite minor fractions. This complex mineralogy has proven important in establishing the degree to which companion metals are likely to be available in periods of increasing demand for companion metals but static demand for their metal hosts.

Another important feature of metal use is that it very often occurs in alloy form. For example, some three-quarters of cobalt and nickel use occurs in a variety of alloys such as stainless steel, and nearly all of vanadium and chromium use is in alloys. The alloy fraction for the elements is pictured in Figure 17.1(b), which shows that the alloy fraction for some 20–25 elements is greater than 50%, thereby providing another potential recovery and reuse challenge. Other metals, including platinum, tantalum, and silver, have alloy fractions of around 30%, and the uses of those alloys are quite important to modern technology.

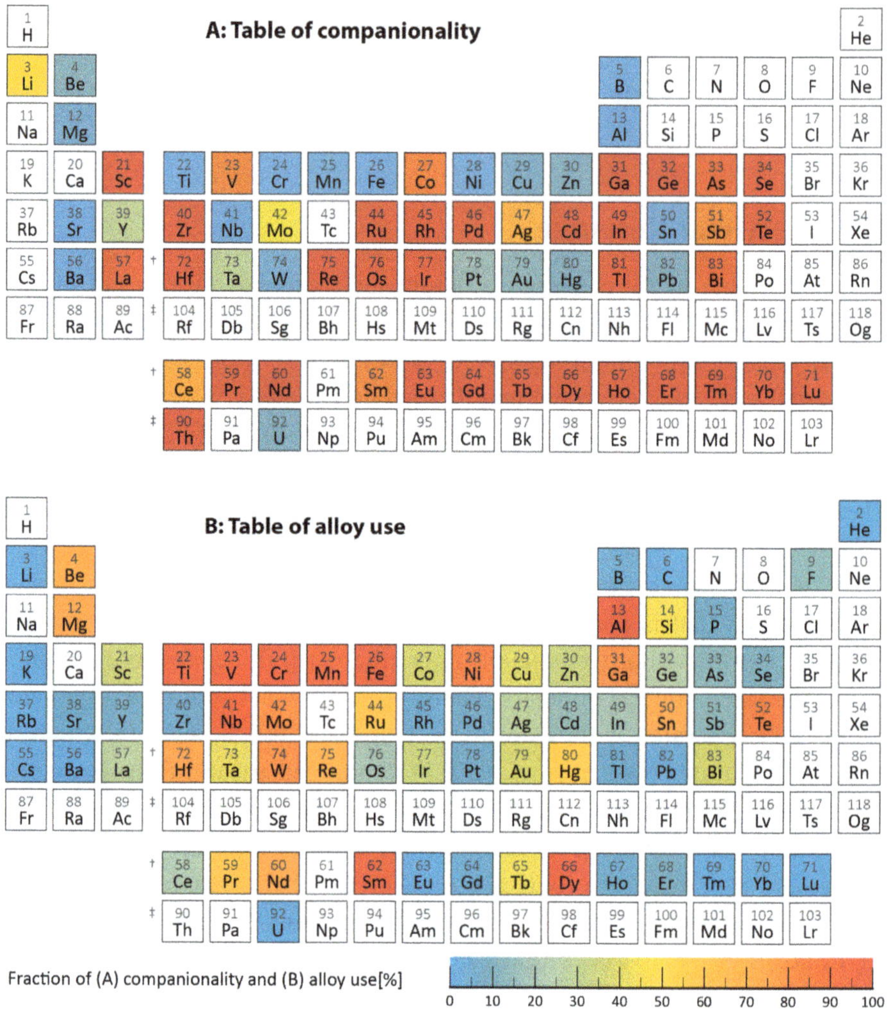

Figure 17.1. (a) Metal companionality; (b) fractional use of metals in alloy form (Graedel *et al.*, 2022).

17.3 Metal Ore Processing

Figure 17.2 shows the processing sequence for metal ores. The figure is for iron, but most major metal ores follow variations of this sequence which begins with the extraction of raw materials. Then, in combination with coal and limestone, the iron ore is smelted and refined into raw steel. This material is cast into "semis" (intermediate forms) which are then rolled and processed to generate desired final

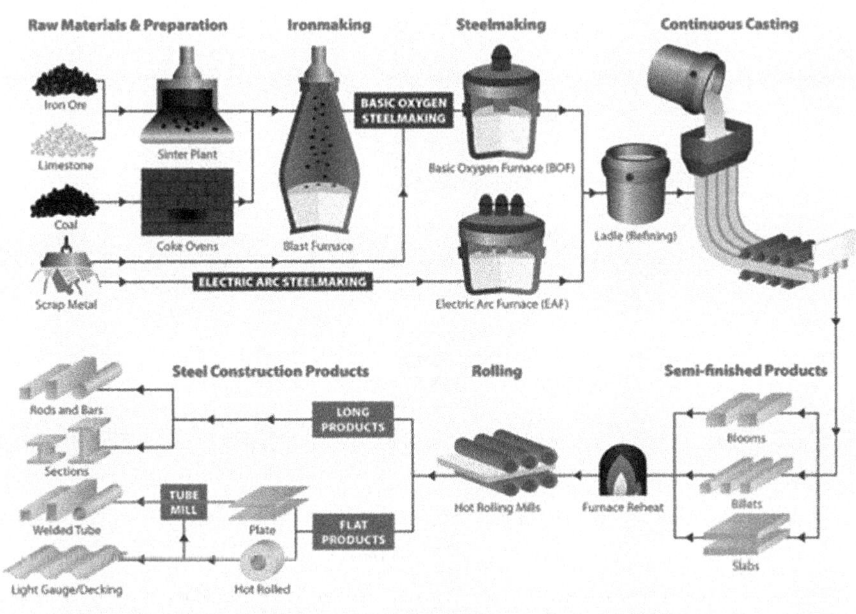

Figure 17.2. The processing sequence for iron ore (Nieto, 2019).

products such as bars, tubes, and sections. An alternative approach, shown at the bottom of the diagram, uses an electric arc furnace to process mixtures of iron ore and recycled steel.

For elements present in ore at low concentrations, additional steps are needed to produce materials used in small but vitally important quantities in such specialty products as modern electronics, aviation, renewable energy technologies, and many more.

17.4 Metal MFAs

Material flow analyses (MFA) of metals, when carefully and accurately done, provide a great deal of information throughout the metal life cycle. This is best illustrated by example, using the 2010 Asian cycle of zinc shown previously as Figure 8.4. (This evocative display, developed by Barbara Reck of Yale University, is a combination of a process flow diagram and a Sankey diagram.) In this circular depiction of the cycle, the starting point is at the lower left, which shows the mass flow of zinc extracted on the Asian continent. More than 5 Tg Zn (more than 5000

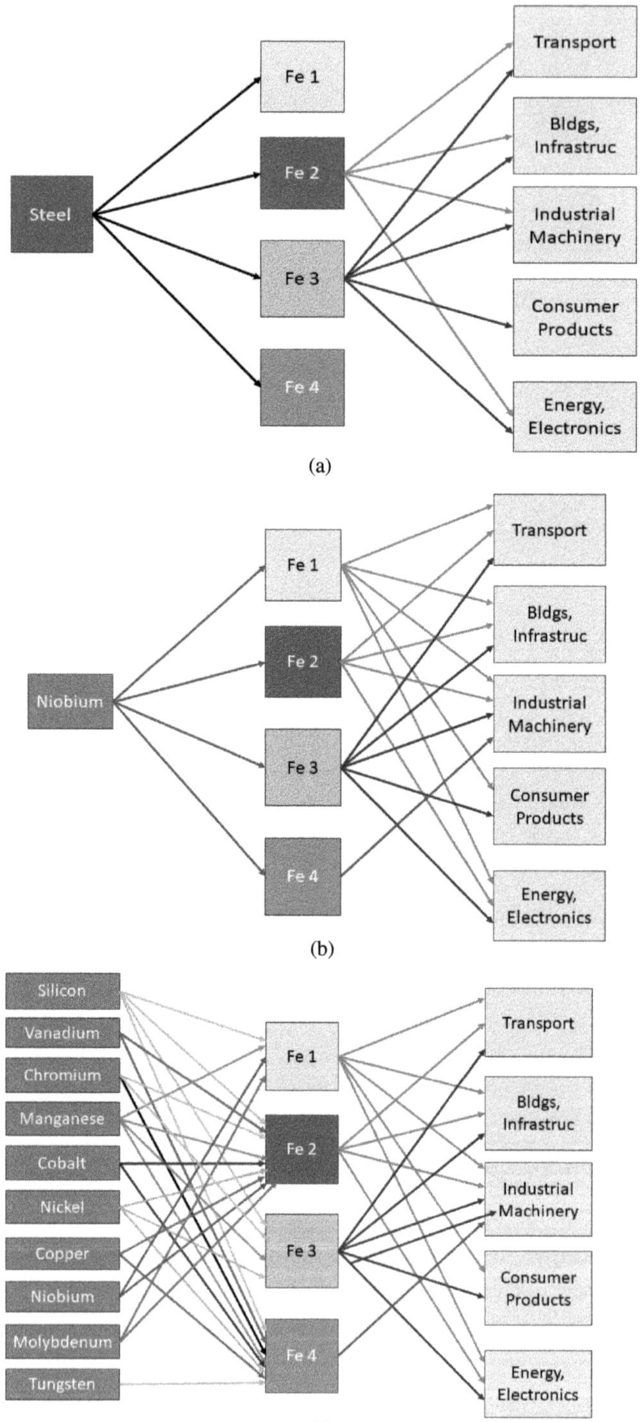
Steel
Fe 1
Fe 2
Fe 3
Fe 4
Transport
Bldgs, Infrastruc
Industrial Machinery
Consumer Products
Energy, Electronics
(a)
Niobium
Fe 1
Fe 2
Fe 3
Fe 4
Transport
Bldgs, Infrastruc
Industrial Machinery
Consumer Products
Energy, Electronics
(b)
Silicon
Vanadium
Chromium
Manganese
Cobalt
Nickel
Copper
Niobium
Molybdenum
Tungsten
Fe 1
Fe 2
Fe 3
Fe 4
Transport
Bldgs, Infrastruc
Industrial Machinery
Consumer Products
Energy, Electronics
(c)

Gg Zn) was mined (Min), and about 85% sent on. The dark circles represent "markets" at which zinc may be imported or exported. In this case, zinc exports and additions from recycling are roughly in balance, so the amount sent for processing at the smelter (S) is about the same as that emerging from the mining stage. The smelter/refinery output and a small import flow at the second market station proceed to fabrication (F). The production efficiency at the refinery stage is high in Asia (and elsewhere, incidentally). It is worth noting, however, that refinery losses of zinc to the scrap stage are about double the amount returned from that stage. Market stages 3 and 4 each show small zinc export flows, the result being that some 6.2 Tg Zn entered use in Asia in 2010. Of that amount, some three-quarters were added to in-use stock. Flows to waste management (W) could not therefore be great, however, about ¾ was recovered and sent back to the scrap market for reprocessing and reuse.

In Chapter 8, the introduction to MFA, the paucity of reliable information at many stages of the life cycles was discussed. Information challenges exist as well with the Asian zinc cycle. The best evidence for this is the dotted line from the scrap stage to the dotted line circle at the right, which shows a flow of some 2 Tg Zn that cannot be reliably placed in any of the normal flow channels. Illustrating this challenge openly reveals the discrepancy, thereby providing an impetus for similar studies to mark the flows that contributed to that imbalance for special attention.

17.5 Implications for Recycling Statistics of the Iron Family Analysis

Because many metals are used in alloys (combinations of metals), we can consider the metals utilized in iron alloys as forming an "iron alloy family". The first level subdivision of that family is generated by examining steel's four major divisions, as shown in Figure 17.3(a), together with the use of each of the alloy families in five principal end-use sectors. The four groupings (carbon steels, alloy steels, stainless steels, and tool steels) differ as a consequence of moderate to small amounts of other elements that are added to the carbon steel starting material. As an example, Figure 17.3(b) reveals that niobium is part of the compositions of the

Figure 17.3. (*Continued on facing page*) (a) The flow of iron and alloying metals through the U.S. economy, 2015. Because the rate of use in Fe 1 alloy is so much greater than those of the other Fe alloy families, logarithms of flows are plotted here. (b) The material flow of niobium in iron alloys in the U.S. economy, 2015. (c) The material flows of all iron alloying metals in the U.S. economy, 2015.

alloy steels and the stainless steels but not of carbon steels or tool steels. To recover the niobium contained in steel alloys following use, the carbon steels and the tool steels can be neglected, but the alloy steels and stainless steels must be identified, separated, and processed in such a way that the niobium is not uselessly combined into large recycled flows of carbon steels (and perhaps the much smaller amount of tool steels).

It is also worthwhile to visualize the flows of *all* elements involved in the steel cycle (Figure 17.3(c)). This diagram enables us to see that each of the steel families is composed of iron and between four and nine additional elements (generally in small to modest amounts). Depending on the end-use market, the steel recovered after use will (as mentioned earlier) need to be identified, separated, and processed in such a way that the alloying elements are not discarded without identification but rather are reprocessed for a reuse that takes advantage of the special material properties that the elements add to the steel. This situation is quite important for iron, the most recycled of all the elements, but applies as well to alloys of all other metals.

Steel is widely advertised as "the most recycled metal". This is certainly true on a mass basis, but that statement is not particularly impressive when the full panoply of iron's discard potential and discard flows are concerned. In the case of steel in general (almost entirely carbon steel by mass), it has been estimated that some 30% of obsolete ferrous scrap is not recoverable, largely because of dissipation or low rates of collection. For the specific situation of iron alloy families, functional end of life recycling rate statistics are available only for the stainless steels. Those rates, at more than 60%, are quite respectable and result in the functional recovery of chromium and nickel. In the case of alloy steels, however, detection and sorting can be quite challenging; it appears that little or no functional recycling of alloy steels occurs. A different situation exists for tool steels: they are largely used industrially and are thus easier to identify and recover. However, tool steel production is a tiny fraction of total steel production, so tool steel recycling plays little role in "circular economy" efforts.

The failure to functionally recycle the alloy steels (and probably a significant fraction of the tool steels) gains significance when one realizes that some of the alloying elements in those families are scarce (e.g., molybdenum) or (even more concerning) are designated as U.S. critical (i.e., vanadium, cobalt, niobium, and tungsten). It is relevant, therefore, to discuss three iron alloy elements that are perhaps of most concern. Vanadium in iron alloys deploys about ⅔ to tool steels and ⅓ to alloy steels. About 80% of the vanadium in iron alloy families is then deposited in the Industrial Machinery sector, which is thus the obvious place from which to capture recyclable vanadium in iron alloys.

In the case of cobalt in iron alloys, about ⅔ flows to alloy steels and most of the remainder to tool steels. About 60% is eventually employed in the Transportation end-use sector and most of the remainder in the Industrial Machinery sector. Recyclables from the other end-use sectors can be ignored so far as cobalt is concerned.

For niobium, whose patterns of flow in iron alloys are shown in Figure 17.3(b), about 60% is employed in stainless steels and the remainder in alloy steels. From those families, it is distributed widely to the five end-use sectors, at fractions varying from about 30% (Transportation) to about 13% (Buildings and Construction).

As Figure 17.3(c) demonstrates, the iron alloying elements are used predominantly in alloy steels and tool steels. The total metal flow into tool steels is comparatively quite small, however, so the first recourse in iron alloy metal recovery and recycling might well be to focus on the alloy steels in specific end-use sectors. Nickel and chromium are exceptions, flowing largely to stainless steels from which they are rather efficiently recycled. This conclusion suggests that a more general guideline for the reuse of metal employed primarily in alloy form is to identify the end-use sectors in which the alloys largely reside, and emphasize alloy recovery, identification, and reuse.

17.6 Challenges to Metal Use

Metals are so widely used and so familiar to us that it may not occur to the non-expert that they can have problematic properties, but such is indeed the case. The first is that natural processes have not distributed them equally over the planet. Some countries (Australia, for example) have rich deposits of many of the common metals. Others (France, for example) do not. This situation has been dealt with for decades by relatively unconstrained international trade in metal ores, but there have been occasions such as wars or political disputes when trade restrictions have been imposed, with industrial manufacturing challenges as a result. At present, Brazil is by far the largest global source of beryllium and China of tungsten, rendering products that depend on those materials a bit riskier than those containing copper or aluminum, which are much more widely available.

A second challenge to the unfettered mining, processing, and use of metals relates to their environmental implications. The existence and operation of mining sites are widely known, and they are often feared for the potential of their air emissions and breached tailings dams to cause widespread harm. A second challenge is that several metals, including arsenic, cadmium, mercury, and lead have high toxicities, and their uses must be closely controlled. Finally, a number of metal uses are inherently dissipative, such as zinc in fertilizers or vehicle tires, or strontium in fireworks. For each dissipative use, the case must be made that the use outweighs the potential environmental implications, thereby providing an additional hurdle for metals' applications.

The third challenge to metal use relates to the energy required to mine and process them (expressed by the embodied energy, generally in units of MJ/kg). Most metals have melting points in the range of 300–3000°C, and the amount of metal used is such that metals are responsible for at least 8% of global greenhouse gas (GHG) emissions. The more apparent the impact of climate change becomes, the more the pressure to utilize alternative materials will grow. Further, ore deposits have grown poorer over time, with new discoveries typically being lower in metal content than has historically been the case. As a consequence, the amount of energy used per weight of final product is increasing rapidly. There is no obvious way around this geological challenge.

Notwithstanding these challenges, metals possess unique and valuable properties not shared by other materials. As is the case with many challenges of all kinds, well-informed decision-making tends to use materials of all kinds in ways that optimize their benefits while minimizing their liabilities. Given their importance in modern technology, metals will continue to provide rich benefits for a very long time to come.

Further Reading

Ali, S. *et al.* (2018). Mineral supply for sustainable development requires resource governance. *Nature*, *543*, 367–372.

Calvo, G., Mudd, G. M., Valero, A., and Valero, A. (2016). Decreasing ore grades in global metallic mining: A theoretical issue or a global reality? *Resources*, *5* (4), 36. https://www.mdpi.com/2079-9276/5/4/36.

Graedel, T. E. (2019). Material flow analysis from origin to evolution. *Environmental Science & Technol*ogy. *53*, 12188–12196.

Graedel, T. E., Reck, B. K., and Miatto, A. (2022). Alloy information helps prioritize material criticality lists. *Nature Communications*, *13*, 150.

Lu, K. (2010). The future of metals. *Science*, *328*, 319–320.

Nassar, N. T., Graedel, T. E., and Harper, E. M. (2015). By-product metals are technologically essential but have problematic supply. *Science Advances*, *1*, e1400180.

Nieto, V. (2019). Steel production: from iron ore to functional industrial products. *Vepica News*. Retrieved from: https://www.vepica.com/blog/steel-production-from-iron-ore-to-functional-industrial-products.

Nuss, P., and Eckelman, M. J. (2014). Life cycle assessment of metals: A scientific synthesis. *PLoS One*, *9*(7), 0101298.

Chapter 18

The Industrial Ecology of Plastics

Chapter Concepts

- Almost all plastics are currently made from petrochemicals, but bio-based substitutes are possible in nearly every end use
- Packaging dominates the use of plastics, has short lifetimes, and has very low recycling rates
- It is generally not possible to recycle plastics so that they can be reused in modern commercial products

18.1 Introduction

Plastics are synthesized from organic molecules (monomers) and can be molded into shape while soft and then hardened into rigid or semi-rigid forms. Most are relatively strong and lightweight. In order to satisfy a variety of demands during processing and use, polymers contain a multiplicity of additives. The pellets that are manufactured for input to injection molding or other processes are thus very complex mixtures, and the components are incapable of being separated after use. As a consequence, the typical fate of plastics at the end of their initial use is either to be downcycled into less valuable and less demanding applications, such as agglomeration into park benches, or to be incinerated to recover some of the energy used in their manufacture.

18.2 Plastics Manufacture

Most plastics are products of oil and natural gas and can be made efficiently from either of those sources. Polymers can also be made from biological materials, although the quantities are, at present, significantly less than those from petrochemicals. The manufacturing processes begin by generating the monomers, the basic building blocks of plastics, which have structures that can be chemically linked together to form long chains. The resulting polymers, generally formed into pellets, are of two types: (1) the *thermoplastics*, which can be re-melted and reused if desired, and (2) the *thermosets*, which form cross-linkages that cannot be reversed without the destruction of the basic monomeric structures. A third form, which occurs naturally or can be manufactured from appropriate monomers, is the *elastomers*, loosely cross-linked polymers that mimic natural rubber in their properties.

Thermoplastic or thermoset products are formed by a variety of industrial techniques (Figure 18.1). The thermoplastics are made either by injection molding, in which a screw machine heats and mixes the pellets before injecting them into a heated mold and allowing them to harden, or by blowing air up through the center of a die to form a polymer film. The film cools as it expands, and is

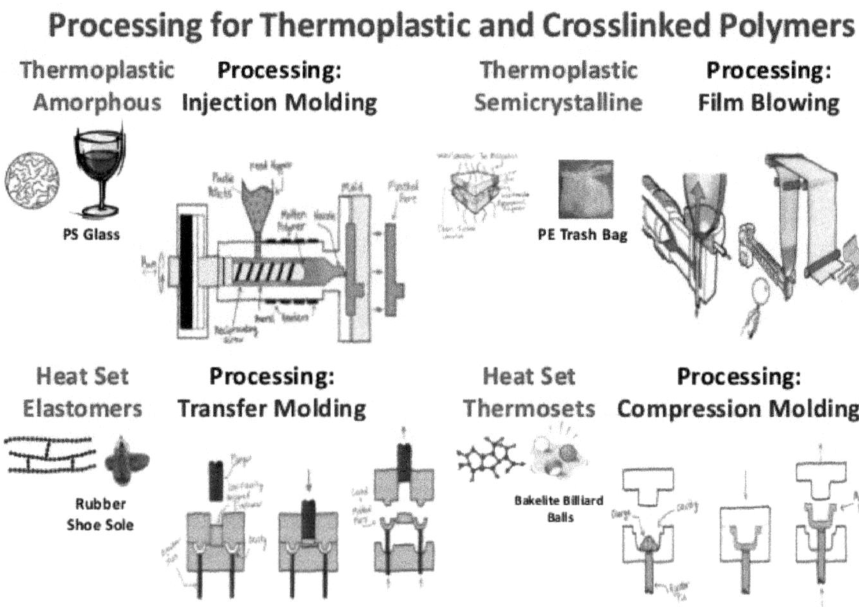

Figure 18.1. Processing sequences for plastics and thermoplastic polymers. (Courtesy of Materials Science & Engineering Department, Arizona State University, Tempe, AZ).

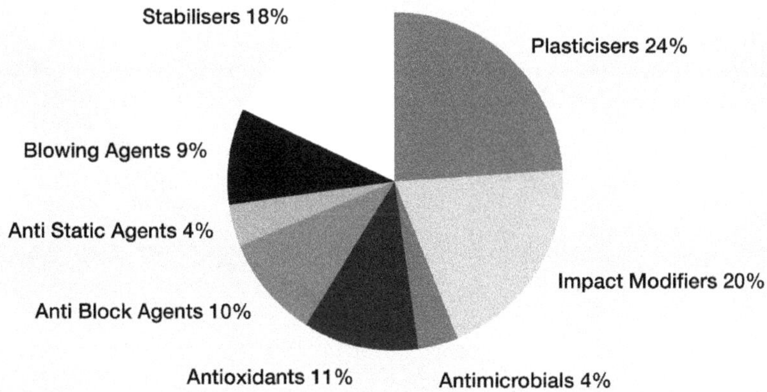

Figure 18.2. Major additives added to plastics (Plastics Europe).

collected and sealed by rollers atop the tubular die. The film is then converted into products such as the ubiquitous plastic bag.

Elastomers are converted into products by transfer molding, in which the pellets are forced by plunger into a mold through a sprue and allowed to harden. This approach allows for placing a metal component within the mold prior to pellet injection, if desired.

Most thermoset products are made by the process of compression molding, which places monomer pellets in the mold cavity, closes the cavity with a cover closed by plunger, and then employs both pressure and temperature to form and cure the product.

In all cases and for most polymers, a wide variety of additives are included in the pellets. The additives provide mechanical reinforcement, thermal stability, fire suppression, color, and a variety of other properties. The degree to which additives and fillers are used in the major plastics is indicated in Figure 18.2.

This variety of added materials constitutes a significant challenge to reuse or recycling when films and sheets of different polymers are comingled on collection because the mixed additives no longer have uniform properties.

18.3 Plastics Types and Applications

The use of plastics is about a century old, but plastics use expanded dramatically toward the end of the 20th century and is predicted to continue its very rapid rise (Figure 18.3), thanks to the wide array of physical properties that plastics possess (e.g., low density, flexibility, transparency, chemical resistance, stability). Packaging is the largest plastics use sector, with textiles, building and construction, and

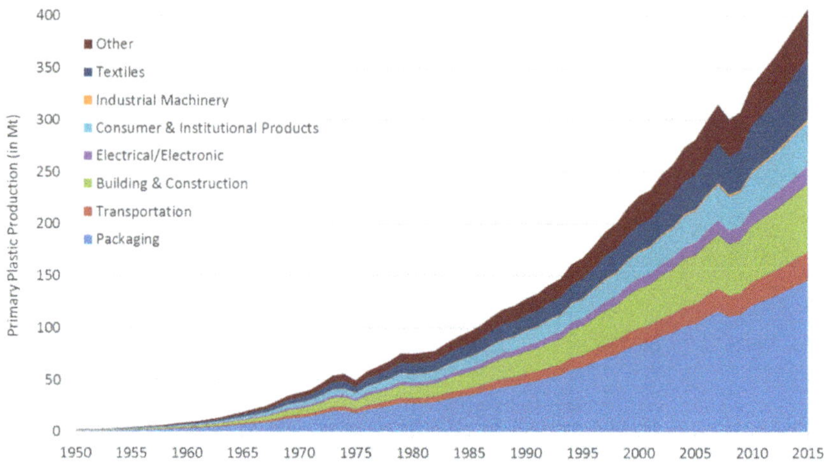

Figure 18.3. Global plastics primary production by industrial use sector (Geyer *et al.*, 2017).

Table 18.1. Nomenclature, uses, and use rates of common plastics.

Plastic name	Typical uses	EU use rate, Tg, 2014*	US use rate, Tg, 2015#
Low density polyethylene (LDPE)	Packaging	7680	7370
High density polyethylene (HDPE)	Bottles, piping	5770	6620
Polypropylene (PP)	Packaging, auto parts	9130	9550
Polyethylene terephthalate (PET)	Bottles, clothing	6400	5350
Polyvinyl chloride (PVC)	Piping, insulation	5080	4630
Polystyrene (PS)	Boxes, housings	1680	2350
Acrylonitrile butadiene styrene (ABS)	Home appliances, auto parts	0.3	0.3

Note: *Kawecki *et al.* (2018); #Di *et al.* (2021).

consumer and institutional products also being major uses. Thanks to the wide array of physical properties that plastics possess, plastics have been broadly utilized throughout industry and consumer products ever since the mid-20th century.

There are seven major types of plastics, as summarized in Table 18.1. Some of the most common uses are indicated in the table, as are fairly recent estimates of the roughly similar rates of plastics production in Europe and the U.S. Many other plastics are manufactured in smaller quantities for specialized uses; the table includes ABS, a rugged and useful polymer, as an example of the wide variety of plastics used in modern technology.

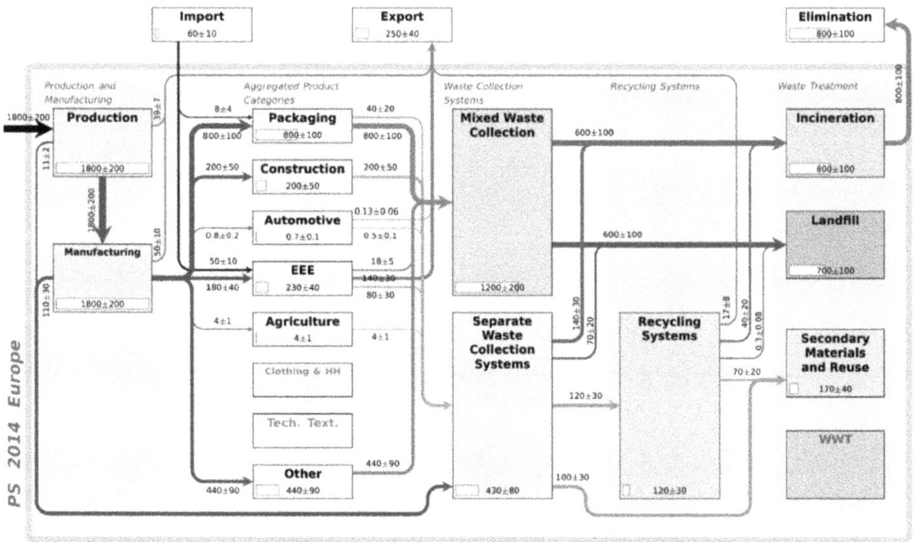

Figure 18.4. The estimated 2014 material flow analysis for polystyrene in Europe (Kawecki *et al.*, 2018). The units are Gg.

18.4 Plastics MFAs

Material flow analyses (MFA) of plastics, when carefully and accurately done, provide a great deal of information throughout the life cycle. This can be illustrated by example, using the 2014 European cycle of polystyrene (PS) shown in Figure 18.4. Half of all PS used in Europe in that year was for packaging, and about a sixth for electrical and electronic equipment, 10% for construction, and the remainder for miscellaneous applications.

A more comprehensive look at plastics used in a particular geographical region can be provided by combining the flows from feedstock to final use, reuse, and loss for all the major plastics on a single Sankey diagram, as shown for the U.S. in 2015 in Figure 18.5. It needs but a quick glance to see that most of the polymers have significant export flows (upper left on the diagram), that inputs to different end-use sectors can be readily distinguished (center), and that there is near-total loss of all plastics at end of use (lower right).

Another way of examining the U.S. data is to look from the perspective of the end uses rather than from that of the polymer itself. The decision of using one plastic resin over another for a given end use is examined in Figure 18.6. The results show that packaging (43.2%), consumer and institutional products (28.7%), and building and construction (16.3%) were the 2015 top three application sectors for plastic resins. PET, LDPE or LLDPE, PS, and PP are mostly used in the

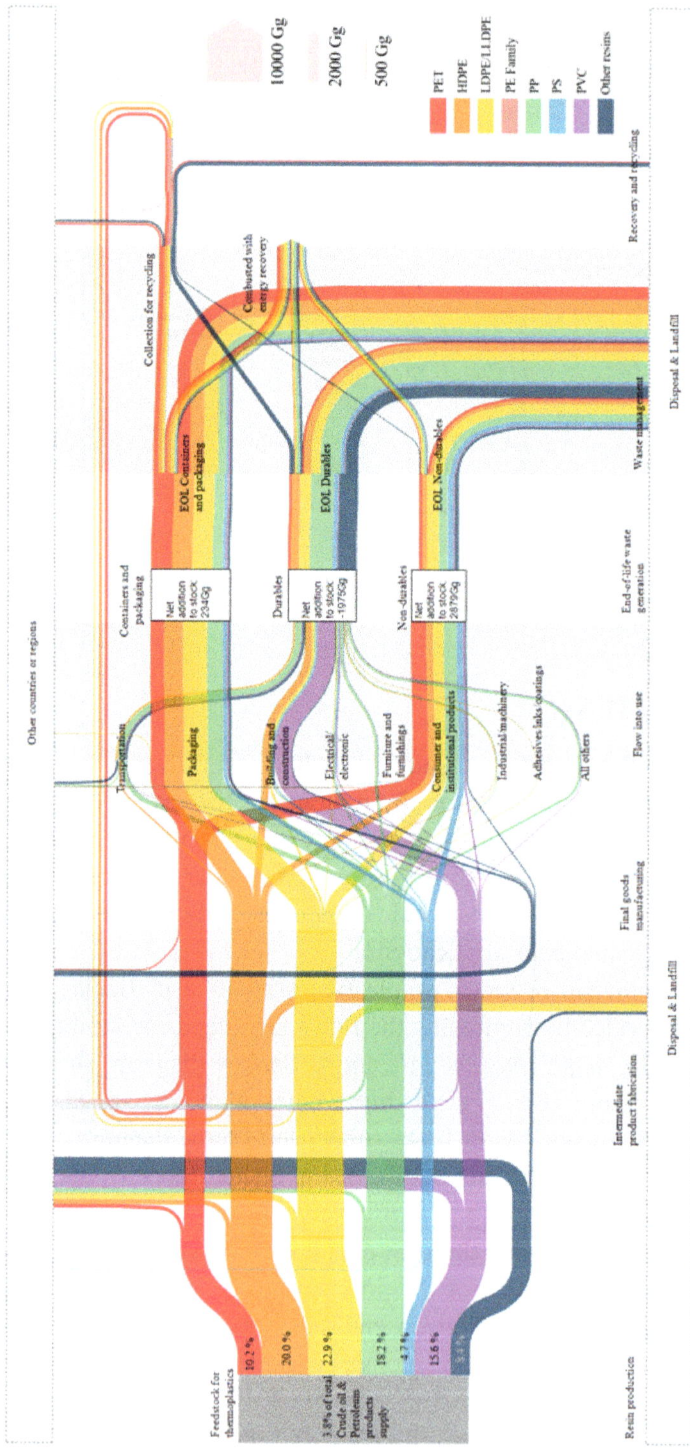

Figure 18.5. Material flow diagram for the predominant plastics manufactured in the U.S. in 2015 (Di *et al.*, 2021).

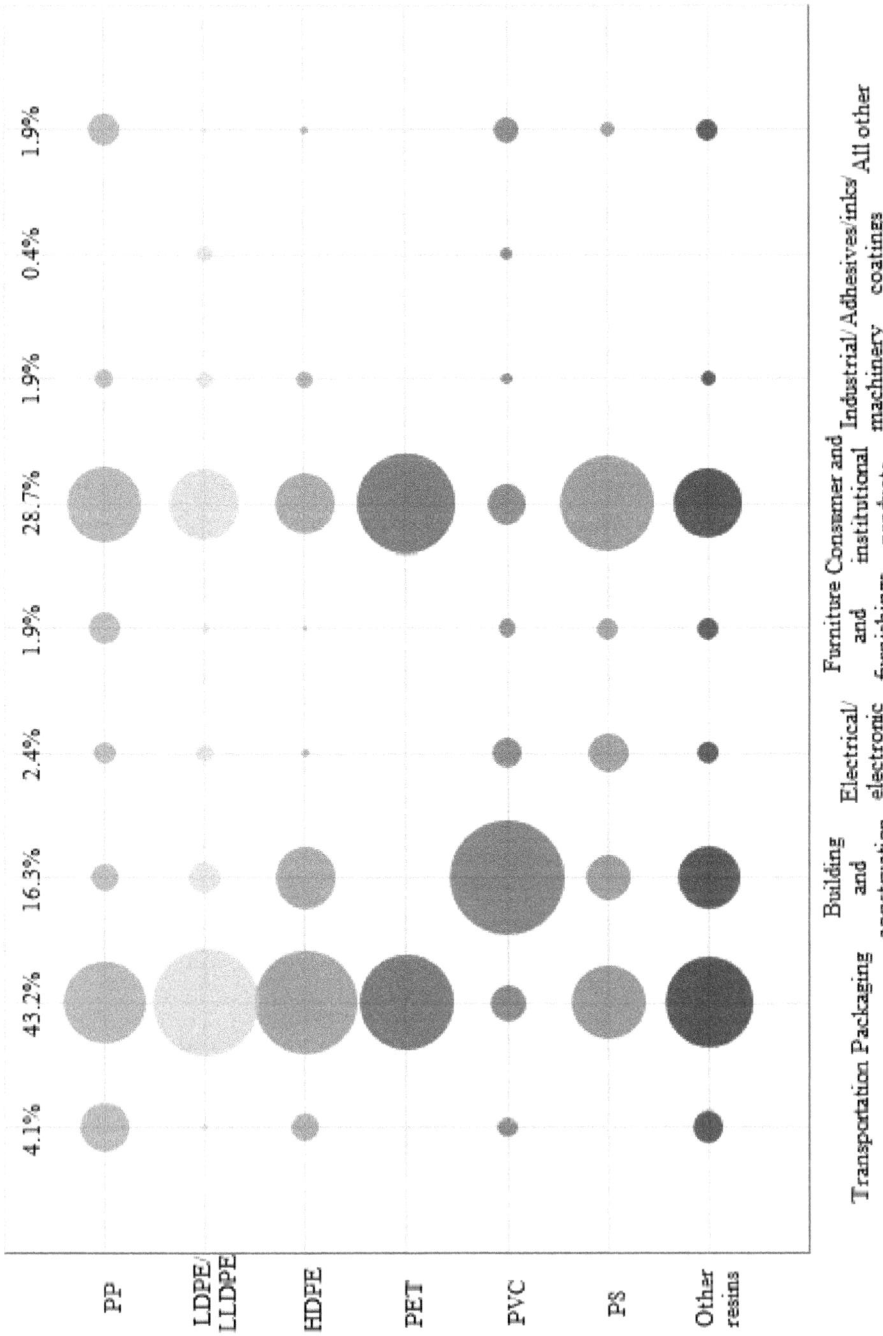

Figure 18.6. The uses of the major plastics in the end-use sectors of the U.S. economy in 2015 (Di *et al.*, 2021).

production of wrapping sheets and containers (packaging), and in mass-produced products (consumer and institutional products). HDPE, a specialized resin with high resistance to corrosion, finds extensive application both in the packaging sector (e.g., bottle caps) and in the building and construction sector (e.g., water piping). Whereas packaging is typically disposed of within a year of their production, building components can be in use for decades. The average lifespan of HDPE products is thus highly dependent on the nature of their application. PVC and Other Resins (including ABS) have extensive application in the building and construction sector and electrical/electronic sector.

18.5 The Plastics Resource

The vast majority of plastics are made from fossil-fuel feedstocks. To the degree that oil and natural gas continue to see widespread high-volume use as plastics starting materials, the potential for continued and substantial fossil fuel-based plastics production is high. However, polymers can also be made from biological materials. Figure 18.7 shows that all of the major plastics now in widespread use can also be produced from plant-based resources.

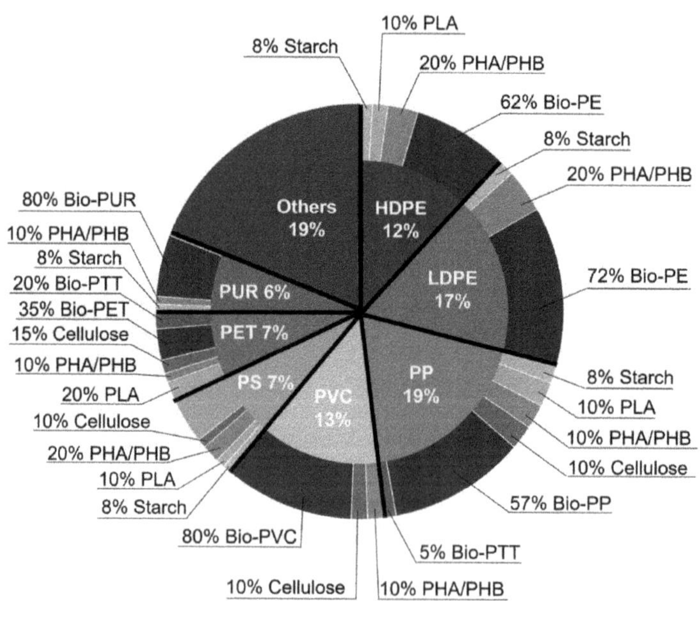

Figure 18.7. The potential for bioplastics to substitute for petroleum-based plastics, PUR = polyurethane, PHA = polyhydroxyalkanoate, PHB = polyhydroxybutyrate, PTT = polytrimethylene terephalate, PLA = polylactic acid, PUR = polyurethane (Zhao *et al.*, 2020).

As of 2021, bioplastics currently account for only a few percent of total plastics production, but those materials are used in all major final product sectors, flexible and rigid packaging being the largest. For bioplastics to come into widespread use, however, some significant challenges must be overcome. Because most research and development of these materials has taken place relatively recently, current bioplastics often have performance issues (mechanical properties, chemical stability, etc.). They tend also to be higher in cost than the fossil-based materials they aim to supplant. Considerations also exist with respect to biopolymer recycling, which tends to be more complicated than that of fossil-fuel plastics. Biodegradation, anaerobic recycling, and chemical recycling of these materials are all being actively investigated. To the degree that bioplastics can become suitable replacements in all performance and price aspects, or by regulatory requirements, the much lower climate change impacts of their use compared with those of fossil-based plastics for a strong argument for their replacement as the routine starting materials for plastics production.

18.6 Plastics as Environmentally Problematic

The enormous production of plastics comes at a cost: many plastics are short-lived products and thus the amount of plastic that is turned into waste has grown substantially over time, as shown in Figure 18.8. With minor variations, this figure

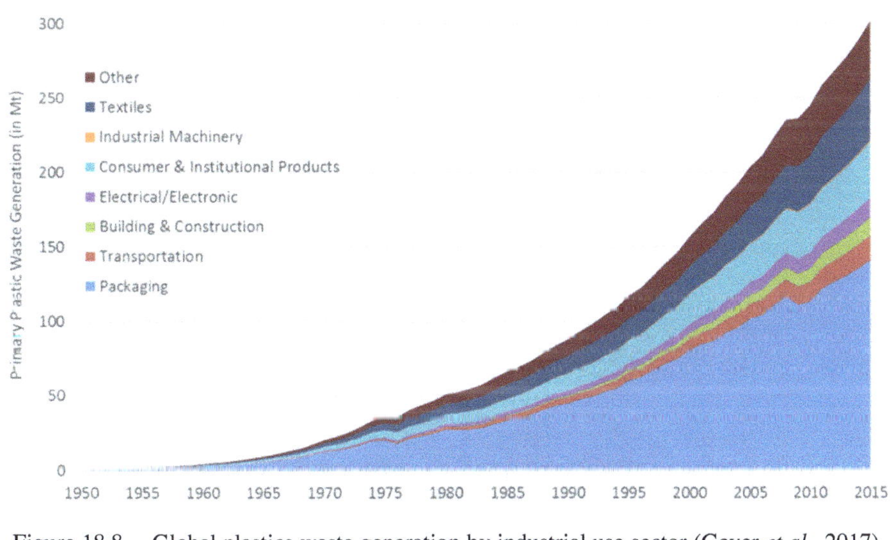

Figure 18.8. Global plastics waste generation by industrial use sector (Geyer *et al.*, 2017).

roughly matches that of Figure 18.3, indicating that little recycling of any of the major polymers is occurring.

A notable portion of waste plastics ends up in the oceans, causing harm to both marine species and, ultimately, to humans. The ocean plastics crisis calls for immediate action: two obvious steps are to replace plastics with alternative materials and/or to maximize recycling rates. Unfortunately, current substitutes such as bioplastics can have up to four times higher environmental costs than traditional plastic products, rendering improved recycling the preferred option. As noted earlier, however, the presence of fillers and other additives significantly complicates recycling and reuse.

Research that attempts to track the flows of plastics can be categorized by spatial scale and plastic type. At the global scale, researchers have accounted for the entire cycle for plastics (i.e., all types of plastics lumped into a single category) from fossil-fuel feedstocks to chemical products, plastic production, use, and waste management; and the impact of various policies on the global plastic waste trade. Regionally, research has been focused on Asian, North American, and European countries; this work has generated a detailed analysis of the material flows of combined plastics and disaggregated flows of the major plastic types.

Despite the general consensus that recycling is paramount for tackling the environmental problems associated with plastics, global recycling rates remain

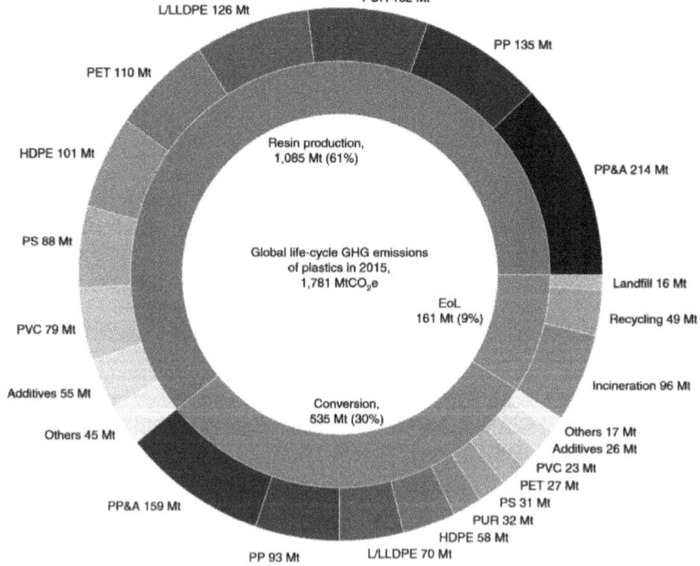

Figure 18.9. Global carbon emissions for plastics in 2015 (Zheng and Suh, 2019).

low. In 2015, the average global recycling rate was thought to be only 9%. Historically, the U.S., Europe, and other countries have relied on China as a recipient of a large fraction of its waste plastics. This practice was suddenly disrupted in 2017 when China issued a ban on the import of plastic waste. Plastic waste exporting countries thus had to look for alternative approaches.

The carbon emission situation for plastics is shown in Figure 18.9. The inner circle on the diagram represents the processing stages: 61% for resin production, 30% for conversion of resins to plastics, and 9% from the degradation of plastics following use. The segments in the outer circle indicate the product of the energy required for processing and the demand for the individual plastic.

It is a major challenge to industry to find ways to better close the plastics cycle or produce plastics that are environmentally benign.

Further Reading

Andrady, A. L., and Neal, M. A. (2009). Applications and societal benefits of plastics. *Philosophical Transactions of the Royal Society B*, *364*, 1977–1984.

Di, J., Reck, B. K., Miatto, A., and Graedel, T. E. (2021). United States plastics: Large flows, short lifetimes, and negligible recycling. *Resources, Conservation, and Recycling*, *167*, 105440.

Gavigan, J., Kefela, T., Macadam-Somer, I., Suh, S., and Geyer, R. (2020). Synthetic microfiber emissions to land rival those to waterbodies and are growing. *PLoS One*, *15* (9), e0237839.

Geyer, R., Jambeck, J. R., and Law, K. L. (2017). Production, use, and fate of all plastics ever made. *Science Advances*, *3* (7), e1700782. DOI: 10/1126/sciadv:1700782.

Hahlodakis, J. N., Velis, C. A., Weber, R., Jacovidou, E., and Purnell, P. (2018). An overview of chemical additives present in plastics: Migration, release, fate, and environmental impact during their use, disposal, and recycling. *Journal of Hazardous Materials*, *344*, 179–199.

Jambeck, J. R., Geyer, R., Wilcox, C., Siegler, T. R., Perryman, M., Andrady, A., Narayan, R., and Law, K. L. (2015). Plastic waste impacts from land into the ocean. *Science*, *347*, 768–771.

Karn, R., Mandon, P., and Prasad, E. (2022). Bioplastics market by type (biodegradable plastic and non-biodegradable plastic) and application (flexible packaging, rigid packaging, textiles, coating & adhesives, agriculture & horticulture, consumer goods, and others): global opportunity analysis and industry forecast 2022–2031, *Allied Market Research*. Retrieved from: https://www.alliedmarketresearch.com/bioplastics-market.

Kawecki, D., Schneeder, P. R. W., and Nowack, B. (2018). Probabilistic material flows analyses of seven commodity plastics in Europe. *Environmental Science & Technology*, *52*, 9872–9888.

Miller, S. A. (2020). Five misperceptions surrounding the environmental impacts of single-use plastic. *Environmental Science & Technology*, *54*, 14143–14151.

Plasteurope.com. (n.d.). Retrieved from: https://www.plasteurope.com/news/PLASTIC_PACKAGING_ADDITIVES_t214184/.

Zhao, X., Cornish, K., and Vodovotz, Y. (2020). Narrowing the gap for bioplastic use in food packaging. *Environmental Science & Technology*, *54*, 4712–4732.

Zheng, J. and Suh, S. (2019). Strategies to reduce the global carbon footprint of plastics. *Nature Climate Change*, *9*, 374–378.

Chapter 19

Industrial Ecology and Food

Chapter Concepts

- Life cycle assessment and material flow analysis of food systems provide much useful information, but constraints are often imposed by data limitations
- Food production, use, and loss, the use of water for food production, and the consumption and impacts of energy for food production and delivery form a nexus of interaction suitable for grouped analysis
- Food demand is proportional to population and per capita wealth, both of which are anticipated to increase markedly in the next several decades

19.1 Food, Demographics, and Wealth

David Tilman of the University of Minnesota is a specialist in human domination of Earth. As such he is vitally concerned with the human use of resources of all kinds, including those related to food. In a landmark publication in 2011 (see Further Reading), he and his colleagues dramatically demonstrated that the average caloric intake of humans is directly related to per capita income. A portion of the Tilman results is reproduced in Figure 19.1(a); it clearly shows that wealthier people have higher daily caloric demand than do those who are poorer.

Not all of that demand ends up being utilized, however. Worldwide, about one-third of all food produced for human consumption is thought to be lost or wasted. Lopez Barrera and Hertel (2021) show that the share of food waste is around 10–12% for low-income countries, 10–28% for middle-income countries, and 30–35% for higher-income countries (Figure 19.1(b)). Losses occur from the

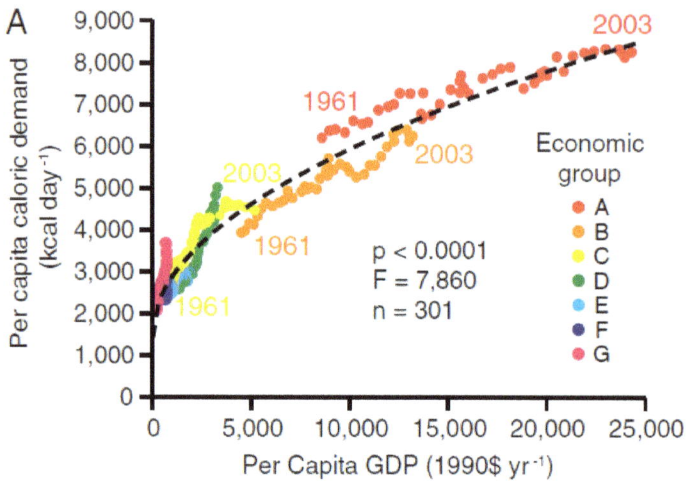

(a)

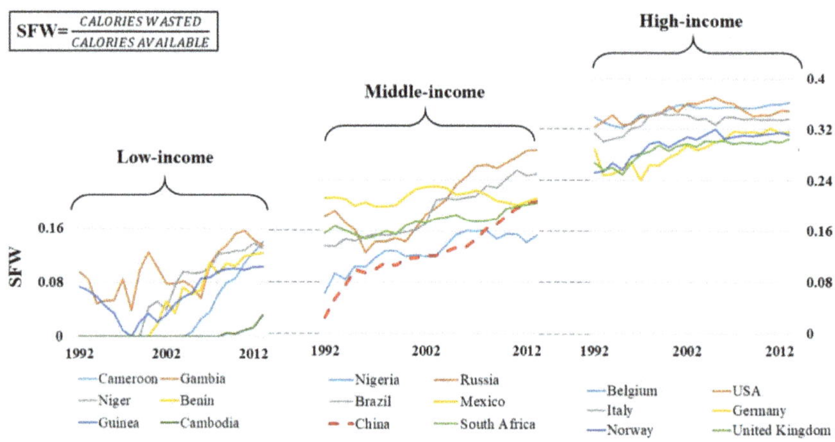

(b)

Figure 19.1. (a) The dependence of per capita caloric demand on per capita gross domestic product for seven populations in 100 nations in 1961–2007. Each color of point shows the trajectory for a particular economic group (Tilman *et al.*, 2011). This is a scatterplot. (b) Share of food waste for 18 countries at different income levels (Lopez Barrera and Hertel, 2021).

farm-to-retail stage, at the retail level, and the consumer stage; they are functions of farm and distribution practices, spoilage, and uneaten food. These losses are very high when expressed in monetary value terms, but reducing those losses is quite challenging as it would involve significant and probably quite costly transformations of agricultural practice, international trade, food marketing, and consumer behavior.

The food use and food loss pictures seem challenging enough for the present, but they may portend even more effort in the future given the growth in world population anticipated in the coming decades. As discussed in Chapter 2, it is anticipated that Earth's population will be some 9.5 billion people by 2050, with growth occurring especially in Africa and Asia. How wealthy might all these new people be? The experts are uncertain about wealth futures because differences in developmental and political trajectories can strongly influence personal wealth. Instead of providing a variety of opinions, they have developed scenarios that describe feasible futures for the planet and explore the implications of those futures. In all but one case, people look to become wealthier, although the relative changes are quite different.

More people, more wealth, and a demonstrated preference of the wealthier population for more calories — it all adds up to major implications for the future of food on Earth. Without significant changes in the average diet, in the ways in which food is produced, in the ways in which food travels from farm to person, and in the rates of food loss, a serious food crisis may be in the offing. Many parameters of this challenge are amenable to study by industrial ecologists, however, as will be discussed in the following sections.

19.2 Food and Water

Agriculture is, by a large margin, the biggest consumer of water worldwide, and in most countries as well. On a global basis, crops and animal products utilize about two-thirds of all water that is extracted. Another quarter of the water comprises the water footprint of animal feed. The final tenth of extracted water sees a variety of industrial uses such as biofuel and biofiber production. Because of the close water–food connection, water demand is anticipated to increase as populations rise, per capita incomes grow, and diets increasingly favor water-intensive foods. The water–food connection is well understood, and research on the topic is well advanced.

Water use in the production of food is customarily measured in three categories: (1) green, (2) blue, and (3) gray water. *Green water* is water that enters soil from precipitation and is stored there, thus making it available for plant uptake. *Blue water* is water that is added to crops by irrigation, following its extraction from nearby water bodies or from groundwater. *Gray water* is water that has been contaminated by agrichemicals. Typical data on the "water footprints" of different food products are shown in Figure 19.2. A few aspects of the data are easy to see

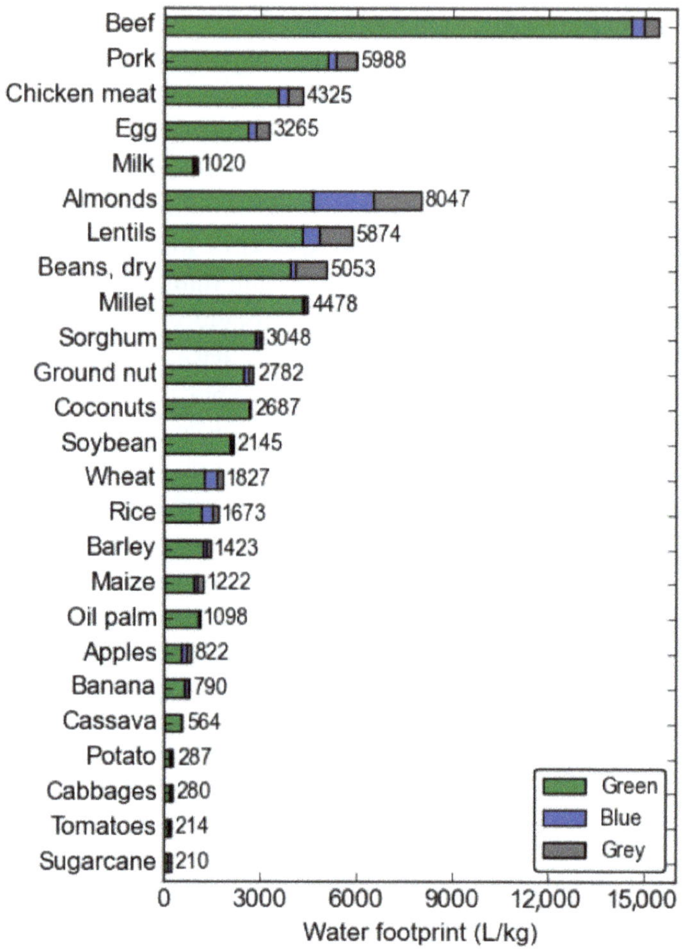

Figure 19.2. The water footprint of several crop and animal products (liters of water per kilogram of product (Mekonnen and Gerber-Leenes, 2020).

in the figure: the very large water footprint of beef (arising largely from the need to grow the crops to feed the cattle), the ranking of almonds as the second most water-hungry crop, and the dominance of green water for meeting most of the needs of all crop and animal products. If a particular diet is specified, the data also permit the calculation of dietary water footprints. It is obvious from Figure 19.3 that diets that emphasize meat and meat products are likely to have much larger water footprints than would be the case for vegetarian or vegan diets.

Countries, regions, and the planet contain some areas that are highly suitable for crop production, others that are of moderate suitability, and some that are largely unsuitable. As a result, food travels long distances from where it is grown to where it is used, and the water employed in its production rather than where the food is used is termed *virtual water*. This is sensible if the producing region has water in abundance and the consuming region does not.

The trade of virtual water is enabled if the products that are traded are properly quantified and if the water footprints of those products are well determined. This has become the case in recent years, and the "flows" of virtual water from one country to another have become determined with a fair degree of precision. A rather recent example is shown in Figure 19.3. Most countries are both importers and exporters of food products (and thus of virtual water). For example, China (at the bottom of the diagram) has major outflows to Japan and to the U.S., and major inflows from Russia and Indonesia.

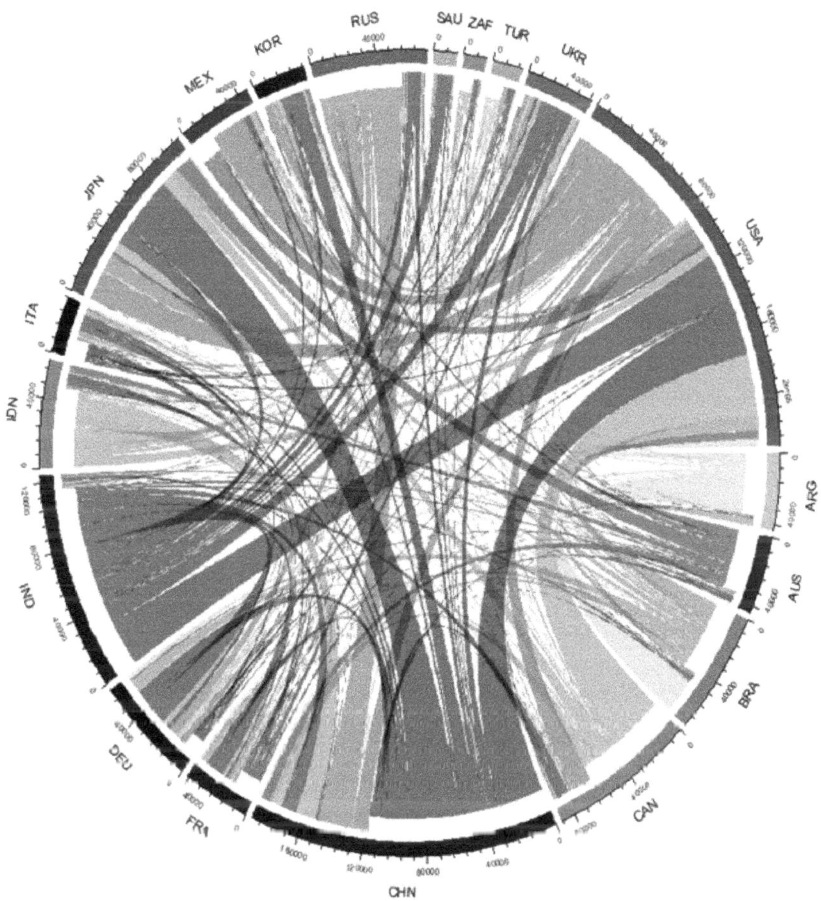

Figure 19.3. The summed virtual water trade among major countries. The flows *from* a particular country are shown by the chords farther from the outer circle, while the flows *to* a particular country are the chords that end *nearer* the receiving country. The units are in million cubic meters (Deng *et al.*, 2021).

Looking to the future, many countries realize that their use of blue water at current rates will eventually become unsustainable (i.e., more blue water is withdrawn from reservoirs than is replaced). As blue water shortages intensify, virtual water trade can be expected to increase. A final consideration in this regard is related to climate change because as the planet warms, the air can hold more water. World regions with water-producing weather are expected to see increased precipitation and more severe storms. In contrast, wind flows will render some regions to a future of decreased precipitation and increased drought frequency. The result is likely to be that virtual water shipments around the world will need to increase. Given the amount of food that is likely to be involved, ocean shipping could become even more intensive than is currently the case.

19.3 Food and Energy

Water is not the only resource required for food production. Energy is involved as well, being needed for pumping the water needed for irrigation, for operating harvesting machinery, for transporting the food to the market, and so forth. In an extensive study published in 2015 see Further Reading, Monforti-Ferrario and colleagues measured the total energy footprint of a number of common food products in the European Union (EU). They did this by statistically determining the average diet of the average European, and then deriving the energy use at the different stages of the food life cycle. Their results appear in Figure 19.4. Among the

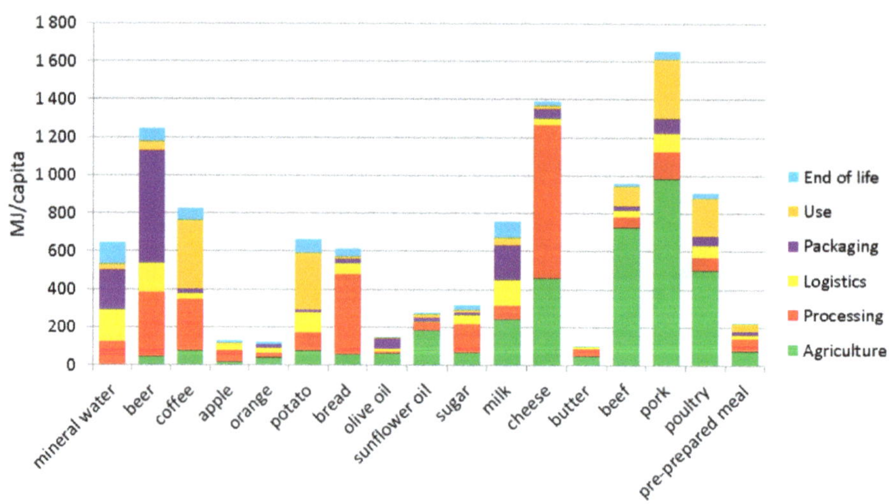

Figure 19.4. The energy footprint of several crop and animal products (Monfortti-Ferrario *et al.*, 2015).

features of the results are that the highest energy uses are for pork and cheese, two energy-intensive products widely consumed in Europe. In the case of pork, about 60% of the energy use goes to raising and processing the animal feed, about 20% for processing, packaging, and logistics, and the remainder for use-related energy for refrigeration, cooking, and so forth. The case for cheese is quite different: only 30% is agriculture-related, 55% of the energy use occurs in processing, and the remainder in other stages of the food sequence. Beer, the third-highest ranking product by the measure employed, sees the most energy use in processing and packaging.

If one contemplates food sourced through ocean shipping, substantial energy charges would then relate to transport energy, logistics before and after shipping occurs, and a variety of related activities.

An extension of analyses of this type occurs by realizing that food is closely coupled to water as well as to energy and, in fact, that the three resources are tightly bound, each needing some quantity of both of the others in order to exist. This is the food–energy–water (FEW) nexus, and any analysis is incomplete

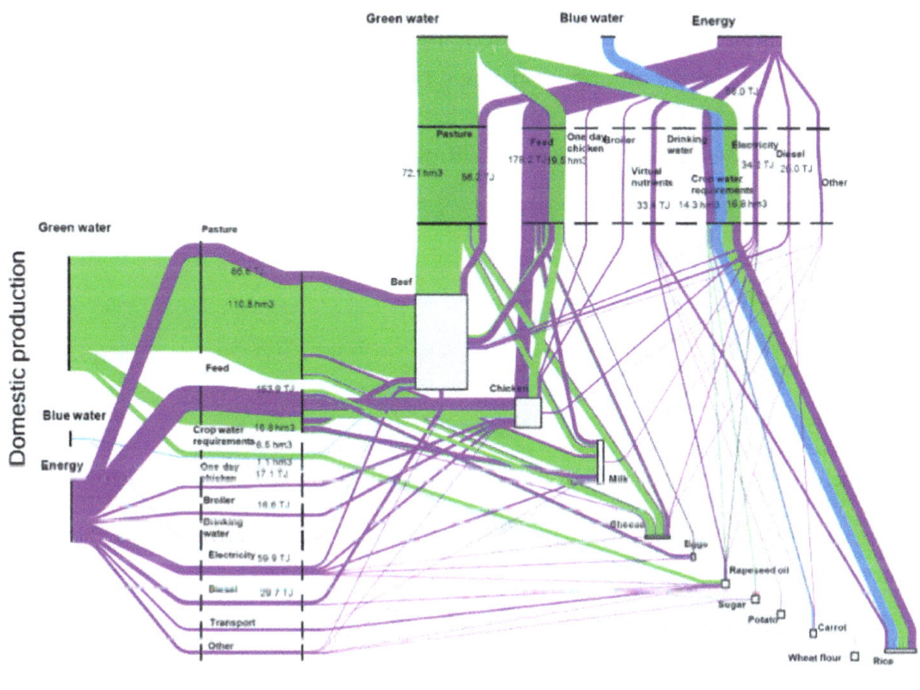

Figure 19.5. The embodied energy (TJ), blue virtual water (hm^3), and green virtual water (hm^3) needed to produce agricultural products for citizens of Tamar catchment, U.K. in 2013 (Salmoral and Yan, 2018). Note that domestic production and imports are distinguished, as are green and blue water. The energy units on the diagram are TJ for energy and cubic hectometers for water.

without realizing and analyzing the tight couplings among them. The concept is perhaps best illustrated by example, which we do in Figure 19.5 for food production in the Tamar Catchment in South-West England. The region itself provides much of the water and energy needed for the production of the food consumed domestically, but virtual water and virtual energy are also provided through imports of these food products. The three examples in Figure 19.5 illustrate the results. In the case of beef (the rectangle at the center of the figure), large amounts of water and moderate amounts of energy are acquired from domestic sources and somewhat smaller amounts of each resource by import. Cheese (the smaller rectangle to the lower right of the beef rectangle) consumed roughly equal measures of energy and water from domestic and imported sources. Rice (the rectangle at lower right) draws its resources entirely from outside the catchment.

19.4 Food and Life Cycle Assessment

One of the early studies that used life cycle assessment (LCA) to examine environmental impacts was that of Weber and Matthews (2008). The data that were utilized were for the U.S. population's use of various food groups and the related carbon intensities of transport fuels. Among the results that are illustrated in the publication are those for annual food group use per household (Figure 19.6(a)) and equivalent carbon dioxide emissions per household (Figure 19.6(b)).

Figure 19.6(a) shows the large quantities of red meat and cereals that were consumed. Domestic rail transport (red bar) is high for these commodities, as is

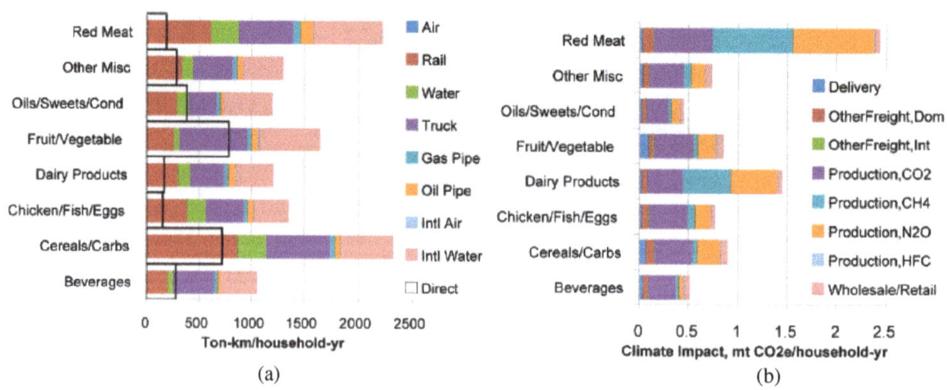

Figure 19.6. (a) Total ton-kilometers of food-related freight per year by household in the U.S. in 1997; (b) total greenhouse gas emissions associated with household food consumption in panel (a) (Weber and Matthews, 2008).

international marine shipping (pink bar). The related carbon emissions and thus the climate impacts are very high for red meat, reflecting the direct or indirect production of the greenhouse gases (GHG) CH_4, CO_2, and N_2O. The paper also addresses the topic of "food miles" — the idea that the environmental impacts of the food one eats are proportional to the distance that food travels from producer to consumer. In rebuttal, Weber and Matthews say "Although food is transported long distances in general... the GHG emissions associated with food are dominated by the production phase". A dramatic demonstration of that assessment was made by Saunders and Barber (2008) who compared the energy implications of raising lamb in the U.K. with raising lamb in New Zealand. In the latter locale, fuel, electricity use, fertilizer, and other energy-related emissions are so much lower than in the U.K. That the emissions from ocean transport are very modest by comparison.

A final note on LCA approaches to analyzing the environmental impacts of food: Notarnicola *et al.* (2017) make the point that the complexity of food systems is such that traditional LCA methods may be incomplete or inappropriate. For example, within different years, the effects of variable weather conditions may result in differences in LCA results for the same products grown in the same places. Additionally, LCA methodology may not capture issues such as soil fertility evolution, changes in soil erosion, or loss of biodiversity. It may be that increased attention needs to be given to the development of LCA approaches that are specific to the special properties of food systems.

19.5 Food and Material Flow Analysis

At first thought, one might imagine a material flow analysis (MFA) of food production to be of the same order of precision, magnitude, and challenge as the MFAs for iron or plastics that were discussed earlier in this book. However, data limitations and the complexities of food systems have traditionally combined to limit the detail of food-related MFAs. Two examples in the References section — Amicarelli *et al.* (2021) and Kytzia *et al.* (2004) — suffice to make the point; each was quite limited in scope.

An exception to this situation is a study of the food supply chain, food use, and food losses in Japan in 2008 based on detailed data from the Japanese Ministry of Agriculture, Forestry, and Fisheries (Ju *et al.*, 2017). The dataset enabled the authors to quantify flows and losses for up to 10 food categories in three home use categories (fresh food, processed food, and cooked food) and in restaurants. An example of the results is shown in Figure 19.7; it quantifies both food flow and food loss at each stage from input to the food system to consumption or loss.

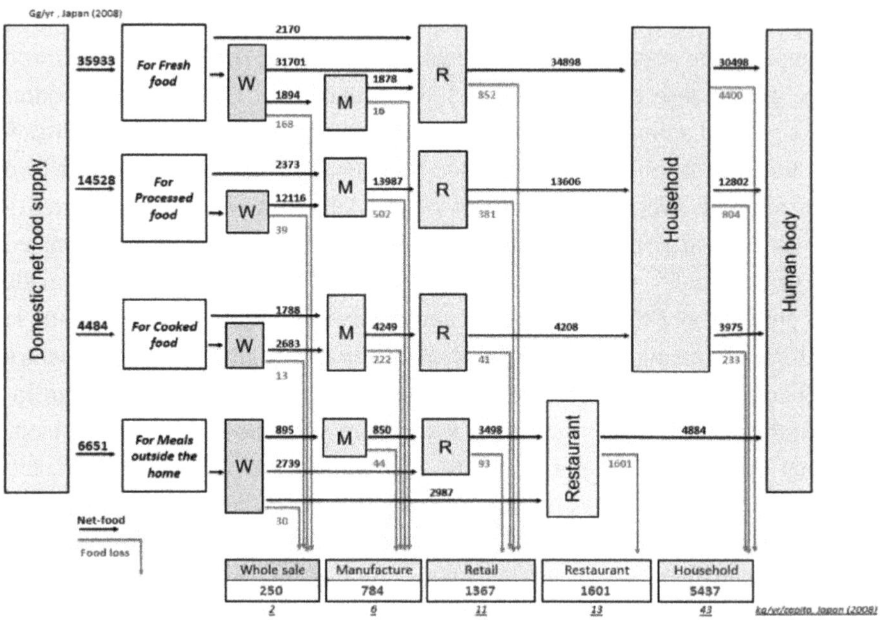

Figure 19.7. Food use and loss in Japan, 2008 (Ju *et al.*, 2017).

As is the case with other flows of resources, the more detail that is contained in an MFA the more useful the analysis is likely to be. As increasingly detailed information in food sourcing, distribution, use, and loss becomes available, future material flow analyses will make that information increasingly communicable and useful.

Further Reading

Amicarelli, V., Bux, C., and Lagioia, G. (2021). How to measure food loss and waste? A material flow analysis application. *British Food Journal*, *123* (1), 67–85.

Deng, G., Lu, F., Wu, L., and Xu, C. (2021). Social network analysis of virtual water trade among major countries in the world. *Science of the Total Environment*, *753*, 142043.

Dias, G. M., *et al.* (2017). Life cycle perspectives on the sustainability of Ontario greenhouse tomato production: Benchmarking and improvement opportunities. *Journal of Cleaner Production*, *140*, 831–839.

Ju, M., Osako, M., and Harashina, S. (2017). Food loss rate in food supply chain using material flow analysis. *Waste Management*, *61*, 443–454.

Keairns, D. L., Darton, R. C., and Irabien, A. (2016). The energy-water-food nexus. *Annual Review of Chemical and Biomolecular Engineering*, *7*, 239–262.

Kytzia, S., Faist, M., and Baccini, P. (2004). Economically extended — MFA: A material flow approach for a better understanding of food production chain. *Journal of Cleaner Production*, *12*, 877–889.

Ladha-Sabur, A., Bakalis, S., Fryer, P. J., and Lopez-Quiroga, E. (2019). Mapping energy consumption in food manufacturing. *Trends in Food Science & Technology*, *86*, 270–280.

Leimbach, M., Kriegler, E., Roming, N., and Schwanitz, J. (2017). Future growth patterns of world regions. *Global Environmental Change*, *42*, 215–225.

Lopez Barrera, E., and Hertel, T. (2021). Global food waste across the income spectrum: Implications for food prices, production, and resource use. *Food Policy*, *98*, 101874.

Mekonnen, M. M., and Gerber-Leenes, W. (2020). The water footprint of global food production. *Water*, *12*, 2696.

Monforti-Ferrario, F., *et al.* (2015). *Energy Use in the E.U. Food Sector: State of play and opportunities for improvement*, JRC Science and Policy Report, Luxembourg (Luxembourg): Publications Office of the European Union. http://dx.doi.org/10.2790/266295; http://dx.doi.org/10.2790/158316.

Notarnicola, B., *et al.* (2017). The role of life cycle assessment in supporting sustainable agri food systems: A review of the challenges. *Journal of Cleaner Production*, *140*, 399–409.

Salmoral, G., and Yan, X. (2018). Food-energy-water nexus: A life-cycle analysis on virtual water and embodied energy in food consumption in the Tamar catchment, U.K. *Resources, Conservation and Recycling*, *133*, 320–330.

Saunders, C., and Barber, A. (2008). Carbon footprints, life cycle analysis, food miles: Global trade trends and market issues. *Political Science*, *60*, 73–88.

Tilman, D., Balzer, C., Hill, J., and Befort, B. L. (2011). Global food demand and the sustainable intensification of agriculture. *Proceedings of the National Academy of Sciences of the U.S.A.*, *108*, 20260–20264.

Weber, C. L., and Matthews, H. S. (2008). Food-miles and the relative climate impacts of food choices in the United States. *Environmental Science & Technology*, *42*, 3508–3513.

Chapter 20

The Industrial Ecology of Biomass Materials

Chapter Concepts

- Natural and artificial fibers see widespread use in clothing, home furnishings, industrial applications, and floor coverings
- As of 2020 and beyond, material flow analyses for pulp, paper, and textiles are available globally and for a few countries
- Composite materials combine strength with lightweight and are used increasingly in transportation and other specialized applications

20.1 Fibers in the Modern World

Fibers are natural or manufactured threadlike or bundled structures found in animal and plant tissue or made industrially. The latter are generally derived from petroleum intermediates. Fibers can be extracted and/or manufactured to have a variety of properties. Those flows that are designated textile fibers (wool, cotton, etc.) are slender, flexible, and relatively strong. They have low density, are generally inexpensive, and are capable of being recycled. Their satisfactory performance and appearance in many applications make them attractive alternatives to synthetic fibers for selected applications in construction, packaging, transportation, and elsewhere (rope, for example). However, their usefulness is limited by generally poor adhesion between fibers, moisture adsorption, poor fire resistance, and low impact strength.

Wood fibers, in contrast, have great strength-to-weight ratios, are visually attractive, and are good insulators. They can be processed in a variety of ways and are easily workable. They are the classic building material of antiquity after stone.

Synthetic fibers are produced from oil and gas feedstocks. They are readily transformed into a great variety of shapes, forms, and colors, and have a wide range of properties. Some synthetic fibers such as the carbon fibers used in carbon fiber-reinforced polymers (CFRPs) have exceptional strength.

20.2 Textile and Fabric Fibers and Products

Textiles, formally defined, are fibrous materials suitable for being spun or woven into fabrics. Those fabrics, originating from animal (e.g., wool), vegetable (e.g., cotton), and industrial (e.g., polymer) starting materials are, as the name suggests, used in clothing, housewares, industrial applications, and elsewhere. Their processing has often been problematic in the past, largely because toxic dyes were common colorants; this problem is mostly old history at present.

These fibers can be processed into many forms — for example, cotton rope is often used decoratively (sash cord, etc.) as it is relatively soft and easy to handle. Additionally, natural textile fibers pose no environmental problems and can be freely used around animals and wherever loss to nature is a potential concern.

Cotton has traditionally been the dominant fiber used for fabrics. However, polymers have played an increasing role since the middle of the 20th century. Around the year 2000, polymer production, especially of polyester, eclipsed cotton production (Figure 20.1). The disparity has continued to increase, with polyester production about twice that of cotton as of 2019. Other fibers have much smaller roles, finding their markets in fabrics that take advantage of particular properties such as toughness and solvent resistance.

Numerical insight into the uses and losses of textiles appear in the global textile cycle for 2017 shown in Figure 20.2. Five different natural fibers are distinguished, including polyester, nylon, and cellulosics. The flows enter four principal end-use sectors: (1) apparel, (2) home furnishings, (3) industrial applications, and (4) floor coverings. About one-sixth of the apparel amount is added to in-use stock, one-fifth of home furnishings to its in-use stock, one-fourth of industrial input to industrial in-use stock, and one-third of floor coverings to floor covering in-use stock. About half of total flow is lost at the chemical pulping stage and the remainder fabricated into intermediate and final products. Most of the fabrics were

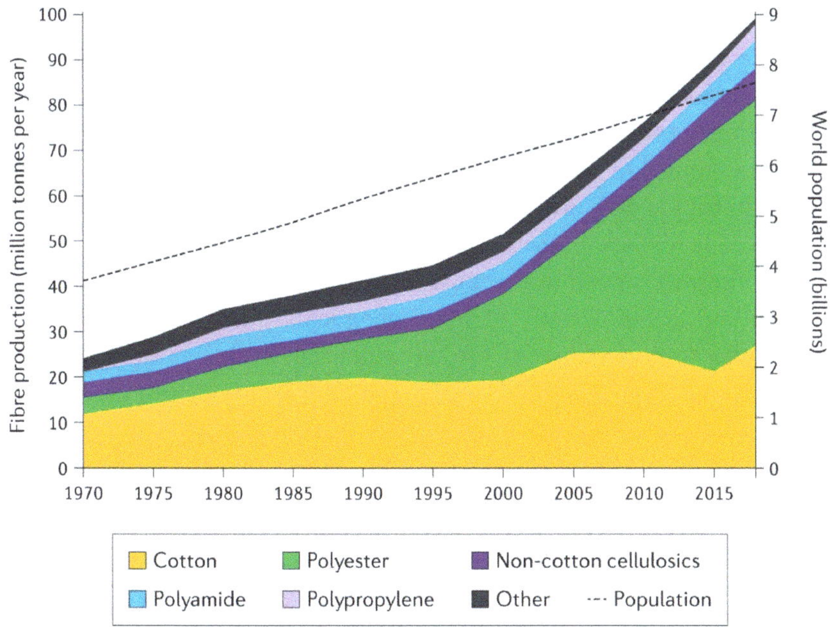

Figure 20.1. Global production of fibers, 1970–2019. "Other" indicates silk and wool combined. The dotted line indicates population (scale at right) (Niinimäki *et al.*, 2020).

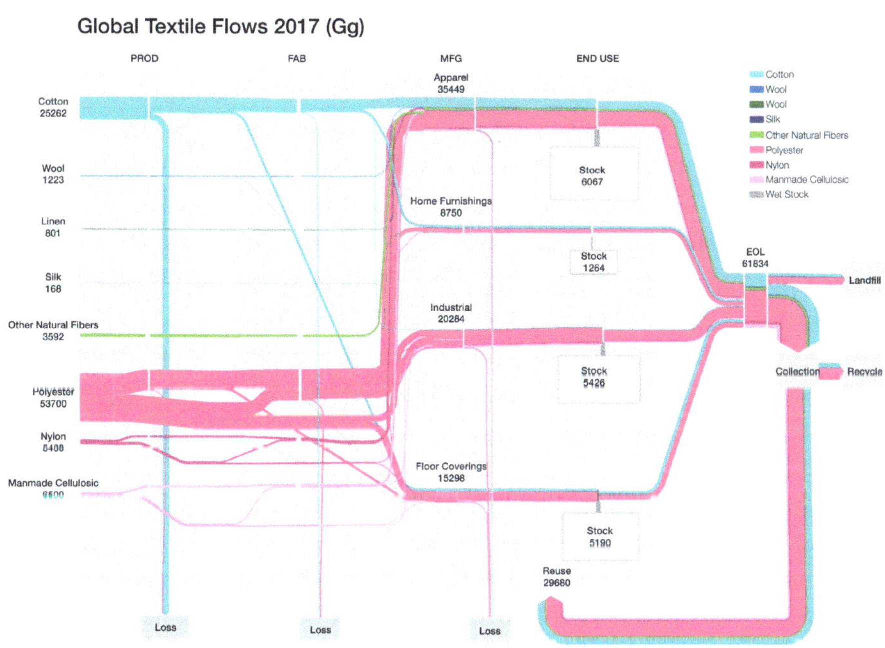

Figure 20.2. The global textile and fabric cycle for 2017 (Althaf and Reck, 2022a).

distributed among all the manufacturing sectors. Flows *into* end of life reflect large fractional flows *into* use for all but the home furnishings end-use sector.

As is the case with other materials employed in modern technology, the textile acquisition, processing, use, and end of life stages have inherent environmental interactions. For natural fibers such as cotton, depletion of water supplies and heavy pesticide use are potential issues of concern. In cases in which a petrochemical feedstock is the raw material for fabrics, significant greenhouse gas (GHG) emissions need to be considered. In the dyeing and finishing stages, water pollution and emissions of toxic chemicals must be addressed. A final concern relates to end of useful life, at which landfilling is the most common fate. Life cycle assessment (LCA) evaluations are increasingly used for assessment, but chemical coverage and methodology development remain works in progress.

20.3 Pulp and Paper Fibers and Products

Wood and bamboo have long histories of use in societies all over the world. They are renewable, are widely available, and are generally strong and visually attractive. The two major uses of wood are in construction and for paper and paper products. Bamboo is widely used, predominantly in laminated forms in construction, but very little quantitative information is available as to rates of use, recycling, or loss.

A typical processing sequence for forest products is shown in Figure 20.3. There are customarily three principal product segments: (1) wood chips burned for energy (especially in Europe), (2) "saw logs" that will be transformed into a variety of forms used in construction, furniture, and other products, and (3) "pulpwood" that is transformed into paper and paper products. In Figure 20.3, a diagram from Finland, import/export flows are shown at the right (mostly wood products), wood chips for combined heat and power (CHP) at the bottom center, and the pulp and paper industry at the left (typical of most forest product processing sequences). Many other details can be seen in this detailed diagram.

Figure 20.4 presents the global cycle for forest products in 2017. It is of interest that about half of the forest material is immediately discarded as unusable (but can be burned for energy recovery). The other half is treated with harsh chemicals to break down the lignin that binds the wood fibers together. The resulting pulp is processed into printing and writing paper (about one-third of the pulp input) and packaging materials (about two-thirds). About half of these products are recycled for further use.

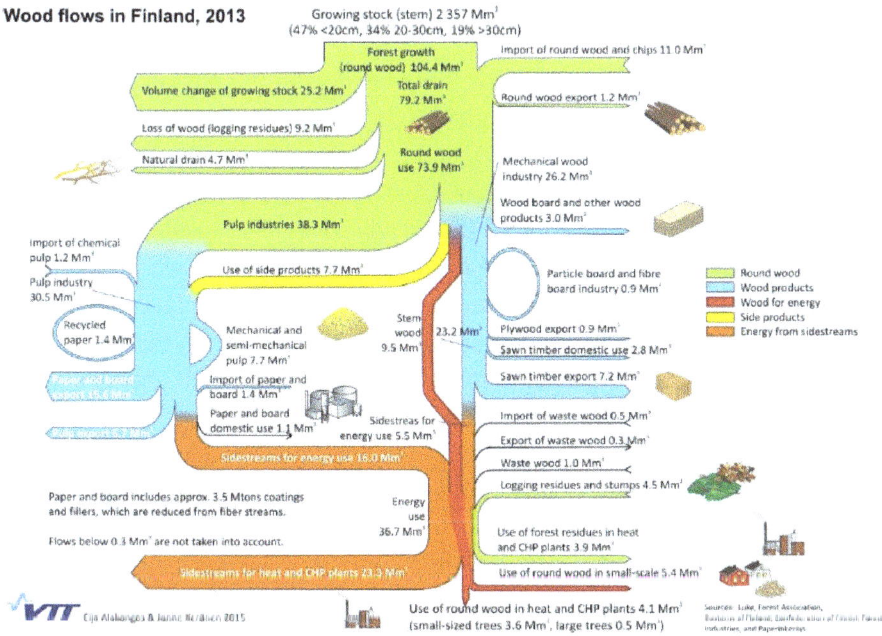

Figure 20.3. Processes in the manufacture of paper, chips, and other wood products (Koponen *et al.*, 2015).

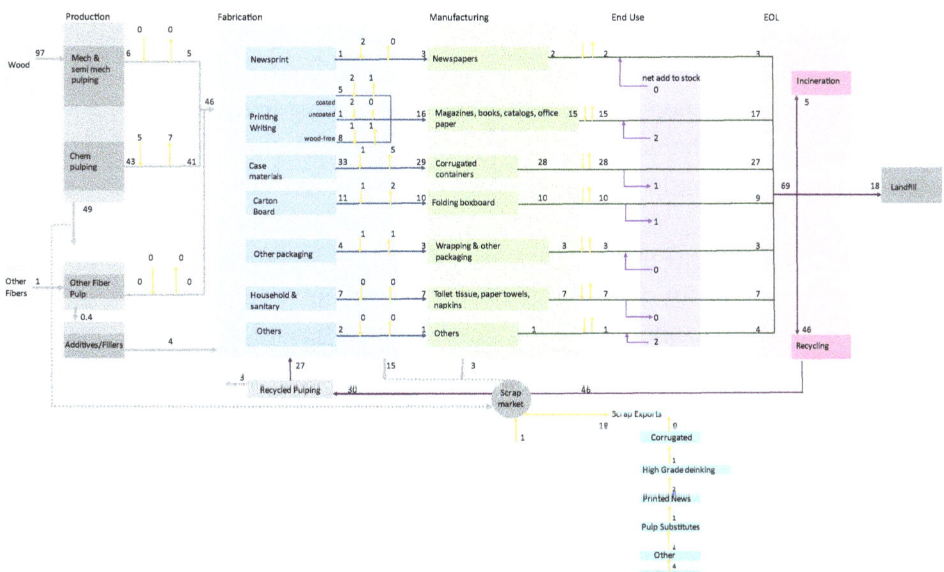

Figure 20.4. The detailed U.S. pulp and paper cycle for 2017 (Althaf and Reck, 2022b).

20.4 Wood and Bamboo Fibers and Products

A wide variety of wood products have long been used in construction, including laminated sheets of wood joined by glue ("glulam"), wood fiber insulation board (WFIB), oriented strand board (OSB), and medium-density fiberboard (MDF). A relatively recent addition to this list is cross-laminated timber (CLT), in which relatively thick boards are fused with strong adhesives. The result is a wood product eminently suitable for uses where high degrees of strength are necessary.

The substitution of highly engineered wood (CLT, etc.) in place of cement and steel turns out to have double benefits for climate stabilization. First, it can avoid the substantial greenhouse gas emissions resulting from cement and steel production. Second, it can turn buildings into a carbon sink because they store the CO_2 taken up from the air by trees when they are harvested and used as engineered timber. The storage lasts as long as the wood is in service, and even longer if discarded wood is stored in such a way that decay will be markedly slowed.

As a building material, timber comes with a number of interesting features detailed in the analysis. For instance, large structural timbers are comparatively fire resistant — their inner core becomes protected by a charring layer if the CLT is burnt, so it is hard for a fire to really destroy a CLT structure. Many national building codes already recognize these properties. However, while the required amount of timber harvest appears to be sufficient for widespread construction use (see the discussion by Ramage *et al.* (2017)), such an enhancement in rate of use would clearly need effective forest management and governance to avoid loss of environmentally vital ecosystems and the resulting ecosystem services.

A study by Hildebrandt *et al.* (2017) provides perspective on the potential for carbon storage in a variety of wood-based construction materials. In a growth-oriented scenario, they envision carbon storage in building construction in the European Union (EU) (Figure 20.5). Several types of wood products are involved in the storage, with the CLT fraction of the total increasing rapidly by 2030. Should the scenario results be realized over the next several decades, carbon storage in wood-based construction could be very important to the global carbon budget.

20.5 Composite Materials

Composite materials are materials made from two or more constituents that themselves have significantly different physical or chemical properties. One constituent (the *matrix*) provides the framework for the final product, while the *reinforcement* provides the strength. When the matrix and reinforcement are combined, they produce materials that have characteristics that improve upon those of the

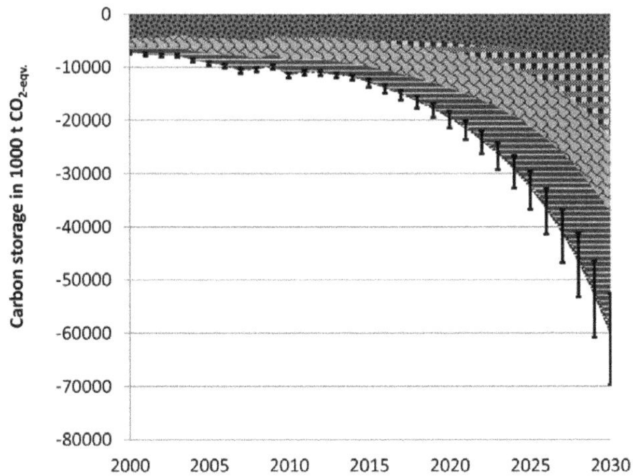

Figure 20.5. Carbon storage in wood product types in buildings (Hildebrandt *et al.*, 2017).

individual constituents. Common historical examples include mud bricks containing straw and steel-reinforced concrete (see Chapter 21, Construction Minerals). In the mid-to-late 20th century, fiberglass (an organic resin outer shell over glass fibers) became a widely used composite. Laminates (materials in which two or more layers are bonded together with an adhesive) have also become widely used in applications such as automobile windshields and outdoor clothing.

In recent decades, a number of specialized high-value products have utilized carbon fiber-reinforced polymers (CRFP) or other suitably reinforced fibers: composite materials comprised of an organic outer shell over exceptionally strong, lightweight carbon fibers. The manufacturing process is illustrated in Figure 20.6. To begin, a "prepreg", an epoxy-impregnated layered organic structure, is laid around the reinforcing fiber. The structure is then placed in a heated press and formed under pressure. The result is a material that combines lightweight with great strength.

Among the products manufactured by this approach are golf clubs, automobiles, and aircraft. Perhaps the most impressive products are the blades of wind turbines, which are composites of glass fiber, lightweight wood, and thermoset resin (or CFRPs or similar constituents). The longest blades being made (as of 2021) are nearly as long as the Eiffel Tower is tall.

At the other end of the spatial scale, carbon nanotubes embedded in organic matrices serve as reinforcement for a variety of small but important products, including auto parts and thin-film electronics.

Despite their advantages in numerous applications, composites at the end of life stage present major challenges to reuse and recycling. They are so well

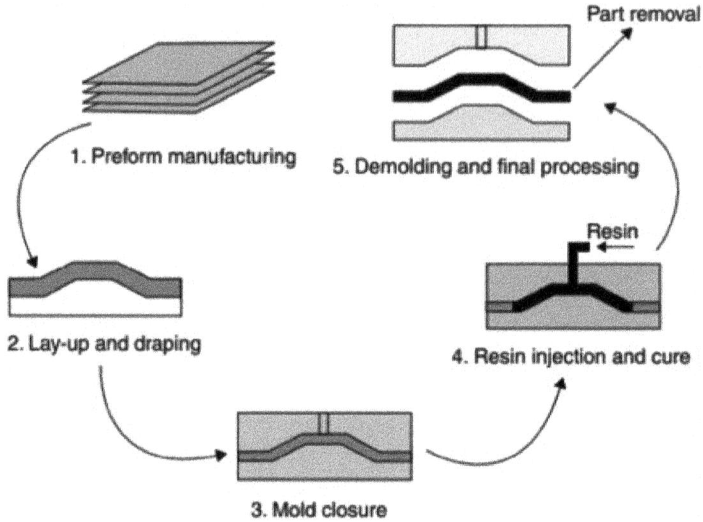

Figure 20.6. The process of manufacturing a carbon fiber-reinforced polymer component (adapted from "Molding", n.d.).

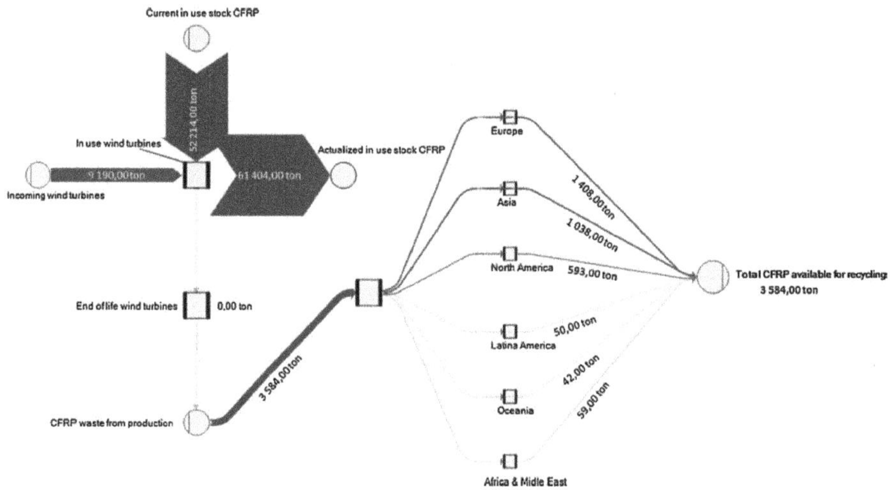

Figure 20.7. Carbon fiber-reinforced polymer stock and flows in wind turbine applications, global and regional, 2017 (Lefeuvre *et al.*, 2019).

bonded, one material to another, that they frustrate recycling almost completely. In a world striving for circularity in the use of materials (see Chapter 37), the lack of a capability to recycle composites calls for intensive efforts to surmount this inability for reuse. A perspective on this problem is provided by the wind turbine example shown in Figure 20.7, a material flow diagram for 2019. If the average lifetime of these turbines meets the anticipated 20–30 years, the CFRP discard flows in the 2040–2050 period will be substantial and need to be planned for.

Further Reading

Althaf, S., and Reck, B. K. (2022a). The global and U.S. fiber cycles for 2017. Manuscript in preparation.

Althaf, S., and Reck, B. K. (2022b). The U.S. pulp and paper cycle for 2017. Manuscript in preparation.

Arup, Inc. (2019). *Rethinking timber buildings: Seven perspectives on the use of timber in building design and construction*. Retrieved from: https://www.arup.com/-/media/arup/files/publications/r/rethinkingtimberbuildings.pdf.

Beton, A., *et al.* (2014). Environmental Improvements Potential of Textiles (IMPRO Textiles). *JRC Scientific and Policy Reports*. DOI: 10.2791/52624.

Churkina, G., *et al.* (2020). Buildings as a global carbon sink. *Nature Sustainability*, *3*(4), 269–276.

Gholampour, A., and Ozbakkaloglu, T. (2020). A review of natural fiber composites: Properties, modification and processing techniques, characterization, application. *Journal of Materials Science*, *55*, 829–892.

Hiken, A. (2019). The evolution of the composite fuselage: A manufacturing perspective. In Dekoulis, G. (Ed.), *Aerospace Engineering*. IntechOpen. Retrieved from: https://www.perlego.com/book/2026459/aerospace-engineering-pdf.

Hildebrandt, J., Hagemann, N., and Thrän, D. (2017). The contribution of wood-based construction materials for leveraging a low carbon building sector in Europe. *Sustainable Cities and Society*, *34*, 405–418.

Koponen, K., *et al.* (2015). *Sustainability of Forest Energy in Northern Europe*. VTT Technical Research Center of Finland Ltd. Retrieved from: https://www.vttresearch.com/sites/default/files/pdf/technology/2015/T237.pdf.

Lefeuvre, A., Garnier, S., Jacquemin, L., Pillain, B., and Sonnemann, G. (2019). Anticipating in-use stocks of carbon fibre reinforced polymers and related waste generated by the wind power sector until 2050. *Resources, Conservation and Recycling*, *141*, 30–39.

Molding (n.d.). Retrieved from: http://songwoltech.com/en/molding/.

Niinimäki, *et al.* (2020). The environmental price of fast fashion. *Nature Reviews Earth and Environment*, *1*, 189–200.

Pei, S., van der Lindt, J. W., and Popovski, M. (2013). Approximate R-factor for cross-laminated timber walls in multistory buildings. *Journal of Architectural Engineering*, *19* (4), 245–255.

Ramage, M. H., *et al.* (2017). The wood from the trees: The use of timber in construction. *Renewable and Sustainable Energy Reviews*, *68*, 333–359.

Roos, S., Posner, S., Jönsson, C., and Peters, G. M. (2015). Is unbleached cotton better than bleached? Exploring the limits of life cycle assessment in the textile sector. *Clothing and Textile Research Journal, 33* (4), 231–247.

Van Ewijk, S., Stegemann, J. A., and Ekins, P. (2017). Global life cycle paper flows, recycling metrics, and material efficiency. *Journal of Industrial Ecology*, *22*, 686–693.

Woodford, C. (2022). *Composite Materials*. Retrieved from: https://www.explainthatstuff.com/composites.html.

Wulfsberg, J., *et al.* (2014). Combination of carbon fibre sheet moulding compound and prepreg compression moulding in aerospace technology. *Procedia Engineering*, *81*, 1601–1607.

Chapter 21

The Industrial Ecology of Construction Minerals

Chapter Concepts

- Construction minerals have been used by humans since the dawn of history and continue to be widely used today
- Except for cement, which is capable of being exported at a profit over long distances, construction minerals are mostly used geographically near to where they are extracted
- The environmental impacts of the acquisition and processing of construction minerals are large and growing

21.1 Introduction

After food, the next goal of humans arriving at a new place has almost always been finding or constructing suitable shelter. The earliest approach to this need, at least where local conditions provided it, was a cave of suitable size and security. These were often superseded by structures of biomass (trees, branches, etc.) or stone. The weight of these materials generally made it challenging to move them very far, although the statues of Easter Island and the "standing stones" of the British Stonehenge are obvious exceptions. Over the course of human history, a variety of materials have been used in construction, from clay brick to wood to concrete, to steel, to reinforced concrete, to modern varieties of composite materials. The buildings and infrastructures that were fashioned from these materials generally form the largest part of in-use material stocks.

21.2 Construction Minerals — Historic and Contemporary

Stone has been a major construction mineral for millennia. Simple stone structures, if well fashioned, are strong and long-lasting. Some types of stone are capable of being worked to some degree with simple tools. Early stone buildings were roofed with small logs, branches, leaves, and a variety of other vegetation. Starting around 7,000 years ago, brick-making began in the Middle East, using mud that was air-dried in the sun. Several thousand years later, clay, sand, and straw were mixed together and fired in kilns. This type of brick may be seen today in the Great Wall of China and elsewhere, and diverse brick products continue to be used in paths and in small to moderate size constructions.

Larger, taller, and heavier constructions, and those subject to stresses such as earthquakes, require stronger materials. Such considerations led to the invention of cement in the 1800s, which is manufactured by mixing lime (CaO, about 63% of the total mixture), silica (SiO_2, 22%), alumina (Al_2O_3, 5%), calcium sulfate (Ca_2SO_4, 3%), and other natural constituents (7%), and firing the mixture under very high heat (Figure 21.1).

Cement production is a big business and occurs throughout the world. Its rates are directly proportional to the levels of industrialization and urbanization in different countries and regions (Figure 21.2). The dramatic expansion of industrial

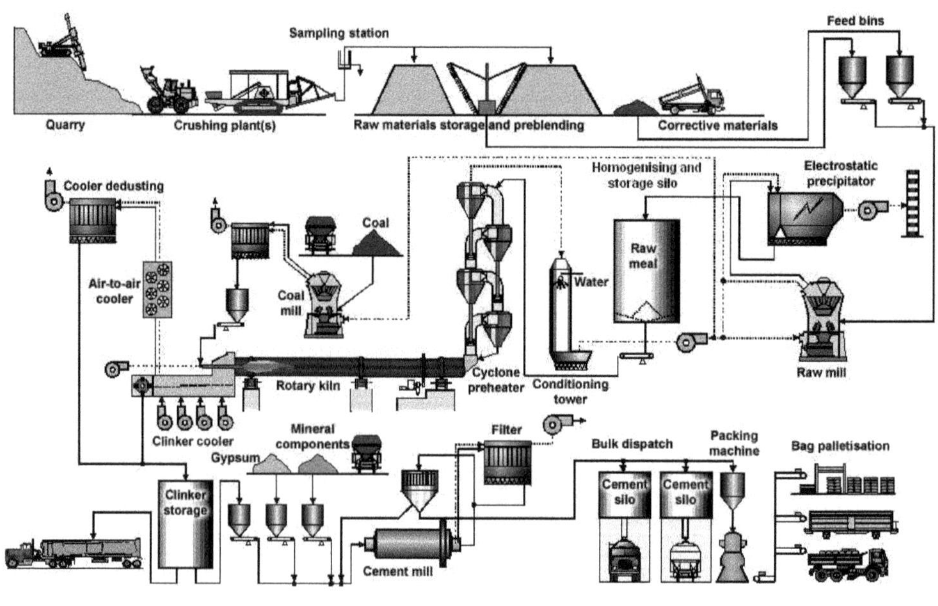

Figure 21.1. The manufacture of cement (CEMBUREAU, 2018).

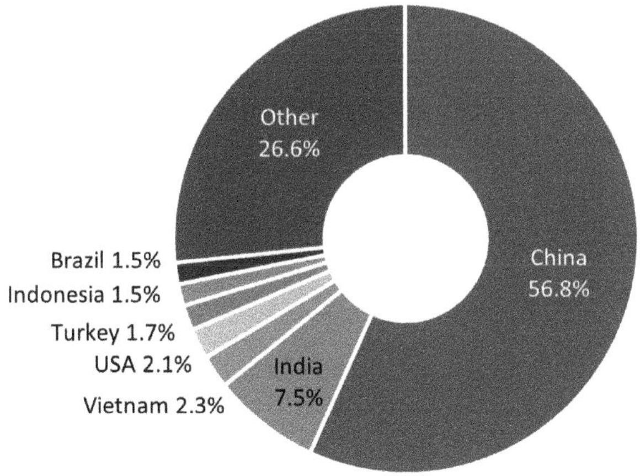

Figure 21.2. World cement production by country, 2021 (Matos, 2017).

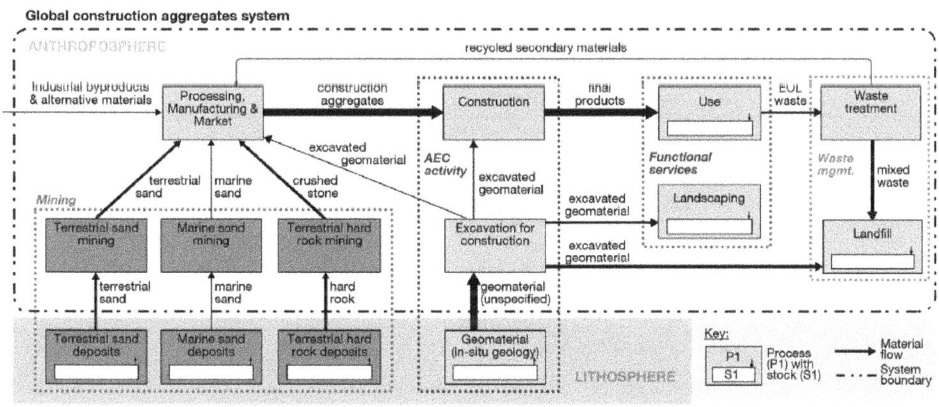

Figure 21.3. An approximation of the global construction aggregates system. Arrow width indicates the relative sizes of material flows (Torres *et al.*, 2021).

activity in China is obvious, but the rest of Asia ranked second worldwide, with India and Europe important as well.

Cement is used to manufacture concrete, by combining it with water, gravel, and sand, generally in proportions of about four parts gravel, two parts sand, one part cement, and added water. Together, these flows constitute the construction aggregates system shown in Figure 21.3. This figure is a Sankey diagram, but only an approximate one, because sand and gravel are generally procured near where the resulting concrete is to be used. Data on those flows are kept in haphazard fashion, if at all, so the arrow widths on Figure 21.3 indicate relative flows as derived from expert judgment. The figure is nonetheless a useful picture of the

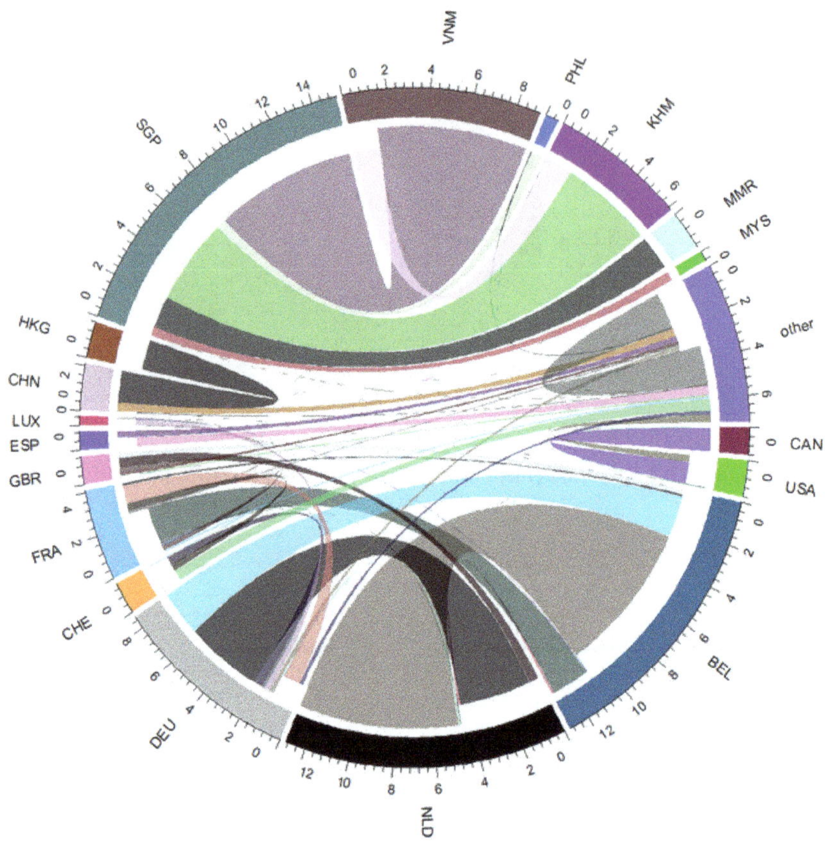

Figure 21.4. The trade in sand between countries in 2009–2010 (average of the two years) (Ioannidou *et al.*, 2020). For each arc, the end closest to the outer circle represents export, and the other end import.

processes and principal flows of material within the construction aggregates system.

Construction aggregates use has its challenges, a major one of which is that the minerals are heavy and large amounts are required, thus making it expensive to move them very far from their sources. Nonetheless, there are major exchanges of sand among countries geographically close to one another. Figure 21.4 illustrates this sand trade. Major exchanges occurred from Cambodia to Singapore, from Belgium to the Netherlands, and from China to Hong Kong. Poor countries without large available deposits of gravel and sand may not be able to afford to import the materials, however, thereby constraining their development potential. Fly ash from fossil fuel combustion can be an acceptable sand substitute in moderate quantities, but this option may disappear over time as energy production moves to renewable technologies.

Another aspect of the use of concrete that is worth noting is that the material is very strong in compression but not in extension. To ameliorate this deficiency, concrete is often reinforced with steel bars or other interlaced structures. This approach provides a major market for recycled steel, as careful sorting of the steel is not needed because most discarded steel is sufficiently strong to serve as reinforcing bars. A second problem with concrete is that it can decay over time as the reinforcing steel is corroded, as mechanical stresses degrade it, and as freeze-thaw cycles induce cracking. Several new innovations are currently under development to try to address this issue, such as including microcapsules containing biological healing agents to repair cracks, embedding shape-memory polymers to resist structural deformation, or forming the concrete with flow networks into which adhesives can be pumped if necessary. At the time of writing, targeted applications are in development, and some are in place, but widespread use of crack prevention agents or approaches is not yet common.

A less well-studied material used in buildings is flat glass, an approximate global cycle of which is shown in Figure 21.5. Glass is a product created by melting sand, soda ash, and limestone (again bringing the construction aggregate system into play in the construction industry). Because data on glass production by country, region, or planet are not routinely kept, the data shown here were generated by applying appropriate glass to steel ratios to the well-characterized steel data. Figure 21.5 makes it apparent that about half of all glass that is manufactured is incorporated in buildings, where it remains for perhaps 75 years. The other half of produced glass is container glass for beverages and food, and this glass typically has lifetimes of less than a year.

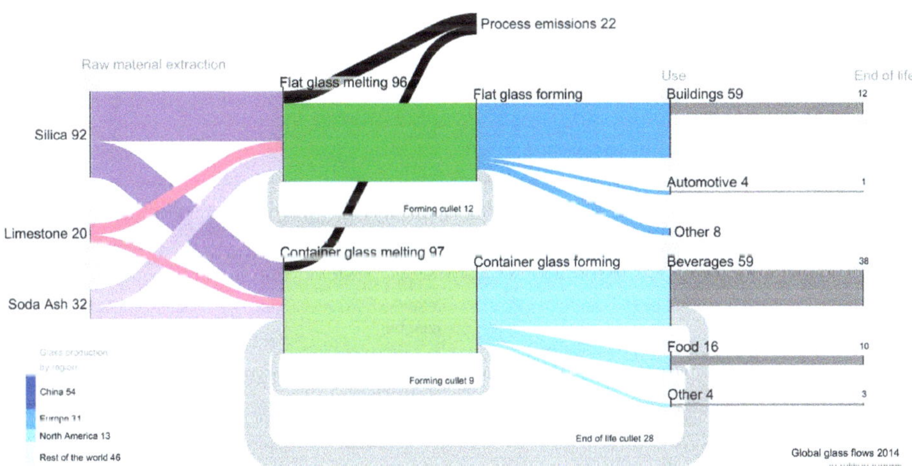

Figure 21.5. Material flow analysis for the global 2014 production, use, and loss of glass (Westbroek *et al.*, 2021).

21.3 Magnitudes of Construction Mineral Use

The rates of use of a variety of materials in the U.S. over the past 110 years are shown in Figure 21.6. It can immediately be seen that construction minerals completely dominate all other material uses. "Construction aggregate" is the catchall term that includes sand, gravel, crushed stone, and sometimes slag and recycled concrete. Of the amount reflected in Figure 21.6, roughly two-thirds is gravel ("aggregate") and nearly one-third is sand. It is also apparent from Figure 21.6 that construction mineral demand is highly sensitive to economic disruption — note the strong decrease during the 2009 financial crisis, for example. Data from other global regions differ in magnitude but not in message.

Construction material use tends to follow population. With the increasing movement of people into cities, urban construction of buildings and infrastructure has provided the principal market for construction materials, which are acquired locally and by import from other regions and countries. An example is shown in Figure 21.7 for Taipei, a metropolitan area in Taiwan. As a relatively wealthy

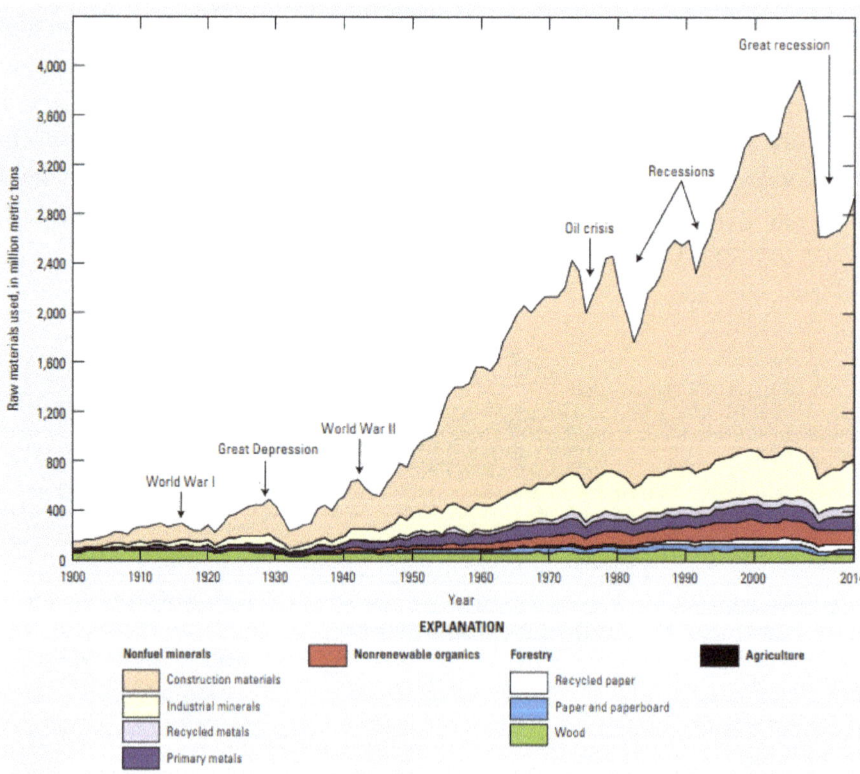

Figure 21.6. U.S. use of construction minerals (1900–2010) (Matos, 2017).

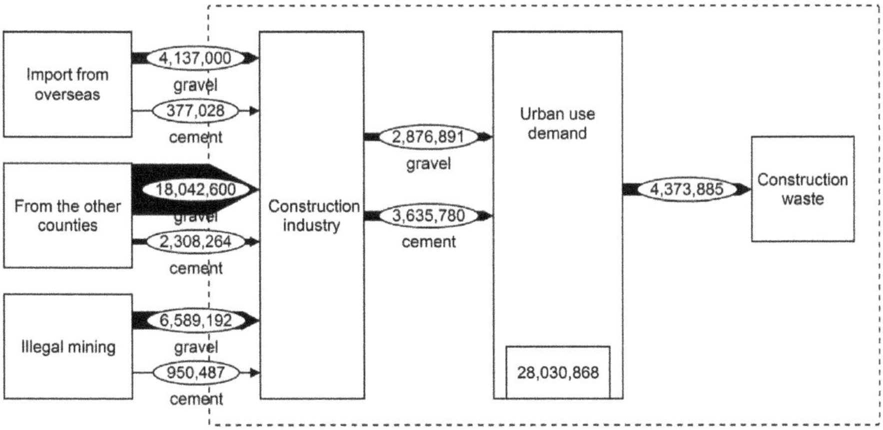

Figure 21.7. The flow of construction materials in Taipei, 2013 (Wang *et al.*, 2018).

country, Taiwan can afford to import a significant amount of its raw materials, and a substantial fraction of those imports are used in Taipei, Taiwan's largest city. Other regions in Taiwan (counties) supply the majority of the construction materials, however. It is interesting that nearly 25% of all construction material flows can be traced to illegal mining throughout Taiwan. Finally, because construction is generally preceded by the demolition of previously existing structures, flows of construction waste are nearly 70% of all new material flow. The construction waste goes primarily to landfills.

21.4 Limitations to the Use of Construction Materials

Construction mineral use has its limits. Another constraint on the use of construction minerals from natural deposits is that their acquisition can be environmentally problematic. The best gravel, for example, is from riverbeds where it is readily extractable, but erosion and pollution occur as a result. In the case of sand, the best tends to come from beaches where ocean currents and tides make it readily available, but again with significant environmental damage when it is extracted. Finally, cement and glass production are major factors in climate change, as will be discussed in a following chapter.

Further Reading

Ali, M. M. (2008). Energy efficient architecture and building systems to address global warming. *Leadership and Management in Engineering*, *8* (3), 113–123.

Bendixen, M., Hackney, C., and Iverson, L. L. (2019). Time is running out for sand. *Nature, 571*, 29–31.

CEMBUREAU — The European Cement Association (2019). Activity Report 2018. Retrieved from: https://cembureau.eu/media/sn3fohjd/actvity-report-2018.pdf.

Davies, R., *et al.* (2018). Large scale application of self-healing concrete: Design, construction, and testing. *Frontiers in Materials*, https://doi.org/10.3389/fmats.2018.00051.

Ioannidou, D., Sonnemann, G., and Suh, S. (2020). Do we have enough sand for low-carbon infrastructure? *Journal of Industrial Ecology*, *24* (5), 1004–1015.

Kohler, N., and Hassler, U. (2002). The building stock as a research object. *Building Research and Information*, *30*, 226–236.

Matos, G. (2017). Use of raw materials in the United States from 1900 through 2014, Fact Sheet 2017–3062. *U.S. Geological Survey* (USGS). https://doi.org/10.3133/fs20173062. Retrieved from: https://pubs.er.usgs.gov/publication/fs20173062.

Miatto, A., *et al.* (2021). Italian clay bricks manufacture: Shrinking production and substantial waste flows. *Journal of Industrial Ecology*. https://doi.org/10.1111/jiec.13208.

Pamenter, S., and Myers, R. (2021). Decarbonizing the cementitious materials cycle. *Journal of Industrial Ecology*, *25* (2), 359–376.

Torres, A., *et al.* (2021). Sustainability of the global sand system in the Anthropocene. *One Earth*, 4, 639–650.

Wang, Y., Chen, P. C., Ma, H.-W., Cheng, K.-L., and Chang, C. Y. (2018). Socio-economic metabolism of urban construction materials: A case study of the Taipei metropolitan area. *Resources, Conservation and Recycling, 128*, 563–571.

Westbroek, C. D., Bitting, J., Craglia, M., Azevedo, J. M. C., and Cullen, J. M. (2021). Global material flow analysis of glass: From raw materials to end of life. *Journal of Industrial Ecology, 25* (2), 333–343.

Chapter 22

Industrial Ecology and Energy

Chapter Concepts

- Energy use is a prerequisite for extracting and processing materials, which enables societal improvement
- The embodied energy of metals, plastics, and fibers are all highly significant to greenhouse gas emissions, thus suggesting the benefit of rapid transition to renewable energy sources for material production
- Complex and globally integrated supply chains for major technological products complicate attempts to minimize the embodied energy for many manufactured products

22.1 Energy and Societal Development

A transition of a material or a product from one state to another cannot happen without a source of energy. Indeed, it is not an exaggeration to say that a transition from one source of energy to an improved source has empowered humans to make great leaps in their own comfort, capabilities, and accomplishments, over and over again in history. The first energy source was fire, which along with the cooking of food also enabled the clearing of land for crop production and, later, the processing of small amounts of metal ores into tools and weapons. Waterpower, when harnessed from nearby rivers, made possible mills for the production of flour and also many early industrial processes, such as the manufacture of fabrics. Crude iron, made by burning wood and collecting the residues (i.e., charcoal), was created by heating ores to generate impure iron capable of being worked into tools. The next big jump was the discovery of coal deposits and their use in enabling the growth of embryonic industries in the 17th century CE. With the advent of oil drilling two centuries later, the Industrial Revolution transformed largely agricultural societies into the industrialized world we know today.

Energy use falls into two categories: (1) it can be used in a way that is productive over the long term, such as the manufacture of tools or housing, or (2) it can be productive in the short term as in the heating of buildings for the comfort of occupants. The latter dissipative energy use is obviously beneficial in the short term but potentially problematic in the longer term. As is well known, the widespread use of coal and oil for both productive and dissipative purposes resulted in environmental problems as it enabled technological and societal advances. Earth is now undergoing a massive shift in climate, with resulting changes in temperature, seawater level increases, ecosystem transformations, and numerous other transitions. The question now is whether ecosystem impacts can be minimized, and a more stable and sustainable society can emerge in a time of rapid planetary evolution.

22.2 Global Energy Use by Final Use Sector

As with most issues in industrial ecology, quantification related to the use and loss of materials is essential for detailed understanding. We have already made the case that some of the resources that the planet provides are scarcer than others. This fact suggests that more energy is required to access scarcer materials than those materials present in higher concentrations or more convenient repositories. The embodied energy of producing different metals is shown in Figure 22.1(a). The concept of embodied energy or cumulative energy demand should be familiar from the introduction to life cycle assessment (LCA) in Chapters 11 and 12; life cycle impact (LCI) databases contain embodied energy values for a wide range of materials. It turns out that rhodium (Rh) has the largest embodied energy (more than 35 MJ/kg). The other "platinum group" metals (Pt, Ir, Os, Pd, Ru) have very high embodied energies as well, as does gold (Au). In contrast, most of the commonly used metals (third row, center) have quite low embodied energies. Most other elements are between these two extremes.

An overview of the energy use attributable to the acquisition and processing of materials is that of Figure 22.1(b), in which energy use per unit of material is multiplied by the societal rates of use of those materials. It turns out that the rate of production of steel is so high that some 35% of all material-related energy use can be attributed to it. Cement and paper/cardboard production call for around 15% each. All other materials — metals, plastics, glass, rubber, etc. — constitute the rest.

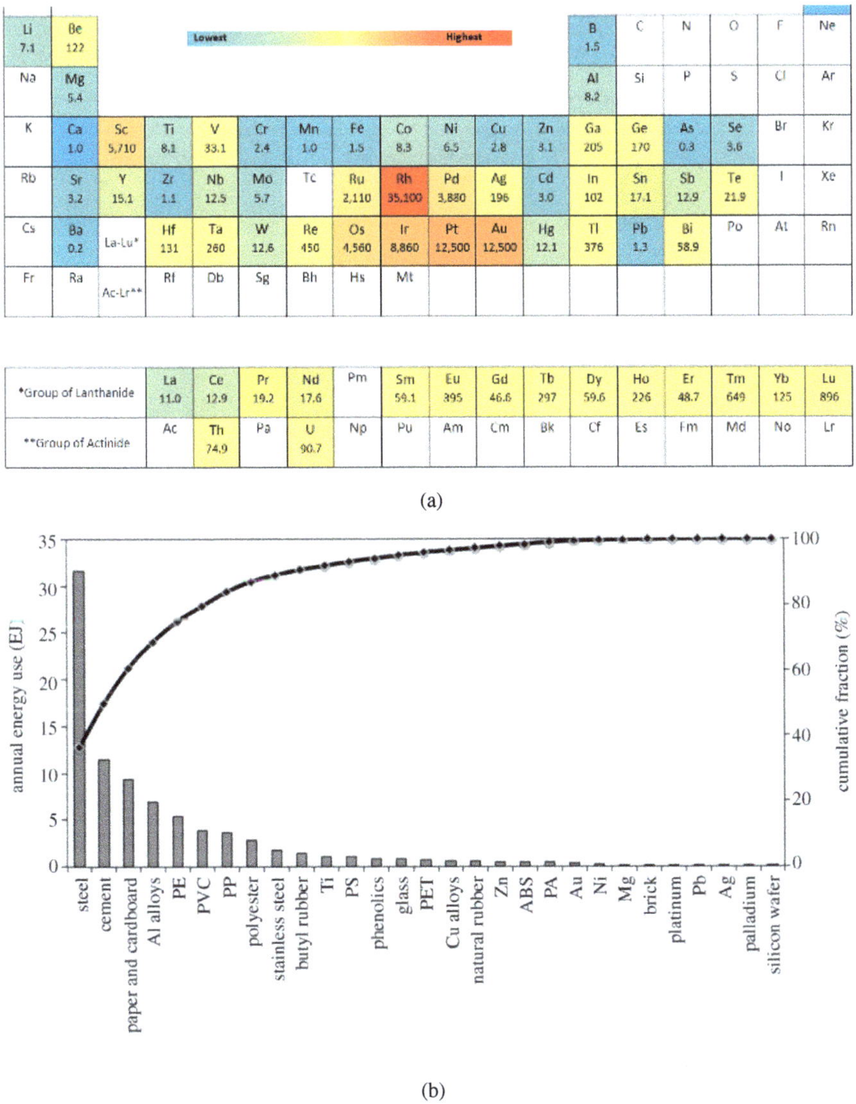

Figure 22.1. (a) Embodied energies in the elements of the periodic table (Nuss and Eckelman, 2014). (b) Annual energy use required by the demand of different widely used materials (Gutowski *et al.*, 2013).

22.3 Energy Use by Major Industrial Sector

The major industrial sectors can be distinguished from an energy perspective by tracking their energy use over time, as shown in Figure 22.2. As of 2010, metal acquisition and processing consumed nearly a quarter of all industrial energy. Cement at about 13% was next, with chemicals (including plastics) at 8%. "Other

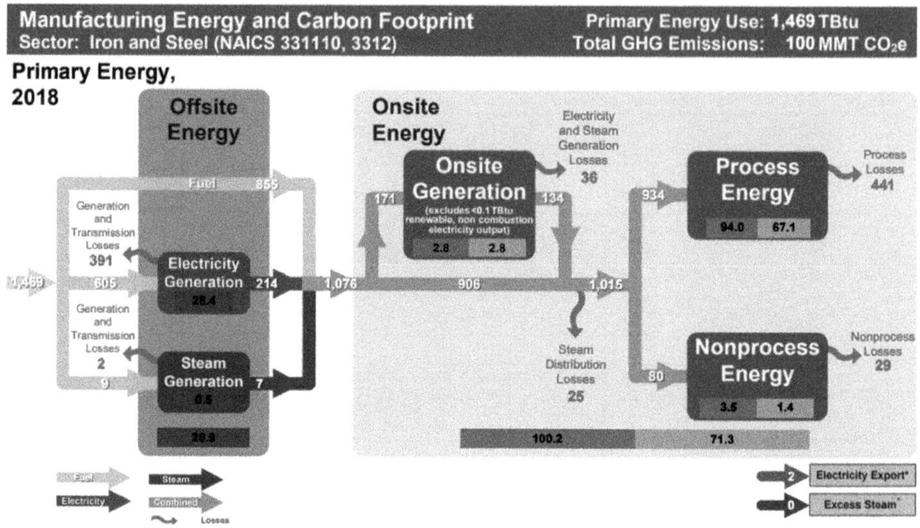

Figure 22.2. Process energy flows in the iron and steel sector of the U.S. (U.S. Department of Energy, 2021).

industries" constituted more than a third of all industrial energy use in activities such as transforming materials, machining and processing them, and transporting the final products to customers. Wastewater treatment and landfills, industries that deal with discards and methane emissions from those discards, consume a surprising 15% of industrial energy use.

It is possible, of course, to dig deeper into the energy use illustrated in Figure 22.2.

Once given a quantitative perspective on energy emissions from a particular process or energy sector, an analysis then can demonstrate opportunities for reducing energy demand. This type of study can be done in a variety of ways, and to greater or lesser degrees. An example of such an exercise by the U.S. Department of Energy (DOE) is shown in Figure 22.3, which addresses energy use in the U.S. iron and steel industries. The DOE has also completed similar studies for other industrial sectors, including aluminum, plastics, and carbon fiber-reinforced polymers. As shown in Figure 22.3, the gray, green, and blue areas represent the ranges of energy use resulting from process and operational differences throughout the U.S. iron and steel industrial sectors. Analyses done to this level of depth have the potential to stimulate investments in process improvement at the industry and factory levels.

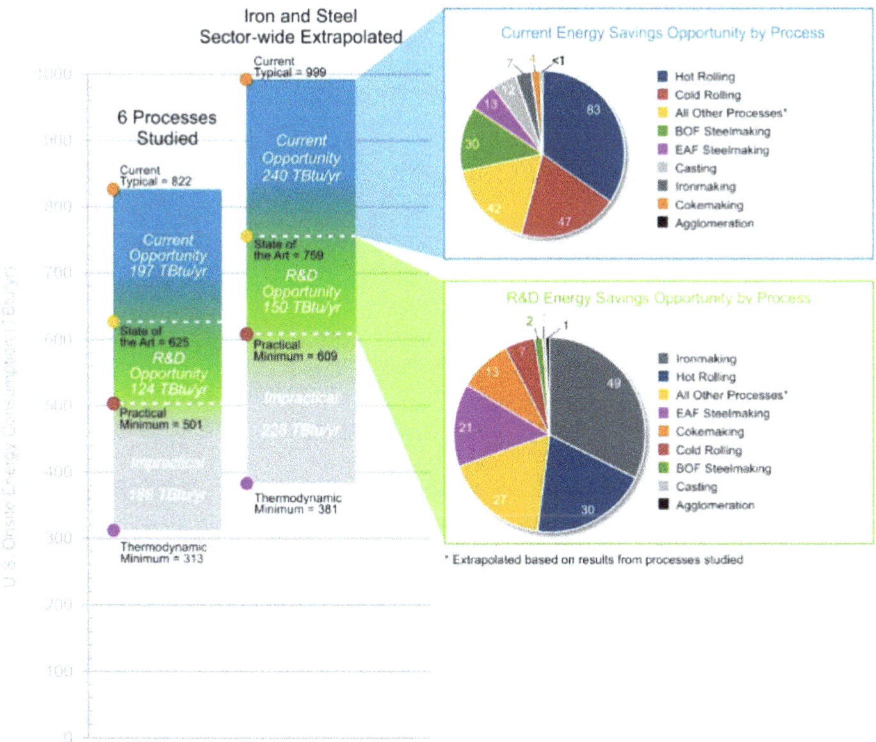

Figure 22.3. Distributions of energy in the iron and steel sector of the U.S. Energy saving opportunities in the iron and steel sector of the U.S. Savings could be realized are shown in blue, potential energy saving opportunities that require research and development to realize appear in green, minimum energies needed for the processes are shown in gray, thermodynamic minima are noted in purple (U.S. Department of Energy, 2015).

22.4 Future Prospects for the Materials–Energy Linkage

The relationships between materials and the energy needed for their acquisition and processing are not static but evolve with the sources and types of materials and with technology evolution. For example, we may wish to produce all plastics from biological sources rather than from fossil oil and gas. If doing so requires more energy per unit of product, there may be an energy penalty involved. Similar concerns apply to other materials. A well-studied example is that of the energy required for the extraction and processing of metal ores. As will be discussed in

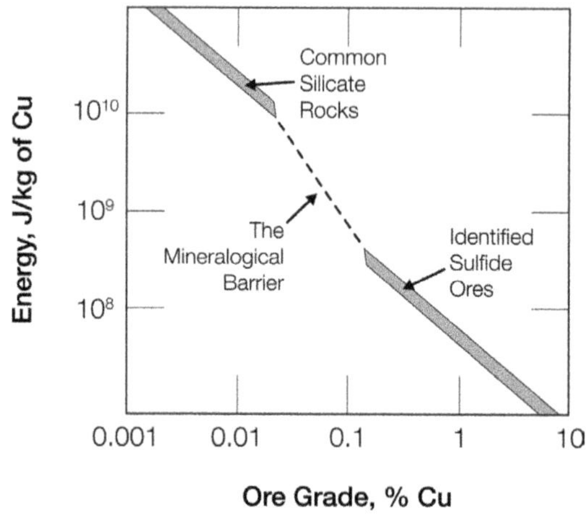

Figure 22.4. The energy necessary to recover copper from its ores and by atomic substitution from common silicate materials (adapted from Gordon *et al.*, 1987).

more detail in Chapter 36, the ore grade (percentage of the metal of interest in a typical ore deposit) is clearly decreasing over time, as richer deposits are mined out and poorer deposits are then exploited. It appears that a logarithmic relationship exists between ore grade and required energy (Figure 22.4), and thus the amount of energy required to access the same quantity of extracted metal is very likely to increase dramatically. Over time, new approaches to material use, vastly improved technologies for materials processing, and/or new energy/materials relationships may be required to simultaneously provide the material resources for humanity as well as to maintain a sustainable global environment in which to enjoy the fruits of those resources.

22.5 Energy–Materials–Society Challenges

Although the relationship between the employment of materials in various ways and for various purposes, together with the energy thereby required, is quite likely to constitute a challenge in that the environmental implications are directly related to bettering the human condition in one way or another. An obvious approach to this challenge is to provide the energy needed to serve humanity from renewable resources, or perhaps with some use of carbon capture and storage. Energy is clearly needed if society is to function, and thoughtful provisioning of that energy will therefore be clearly needed as well.

Further Reading

Gordon, R. B., Koopmans, T. C., Nordhaus, W. D., and Skinner, B. J. (1987). *Toward a New Iron Age? Quantitative Modeling of Resource Exhaustion.* Cambridge, MA: Harvard University Press.

Gutowski, T. G., Sahni, S., Allwood, J. M., Ashby, M. F., and Worrell, E. (2013). The energy required to produce materials: Constraints on energy-intensity improvements, parameters of demand. *Philosophical Transactions of the Royal Society A*, *371*, 20120003. https://doi.org/10.1098/rsta.2012.0003.

Nuss, P., and Eckelman, M. J. (2014). Life cycle assessment of metals: A scientific synthesis. *PLoS One*, *9*(7), e101298. https://doi.org/10.1371/journal.pone.0101298.

Oberle, B., *et al.* (2019). Global Resources Outlook 2019: *Natural Resources for the Future We Want*. Paris: United Nations Environment. Retrieved from: https://www.resourcepanel.org/reports/global-resources-outlook.

Rötzer, N., and Schmidt, M. (2019). Historical, current, and future energy demand from global copper production and its impact on climate change. *Resources, 9*, 44.

U.S. Department of Energy (2015). *Bandwidth Study on Energy and Potential Energy Savings Opportunities in U.S. Iron and Steel Manufacturing*. Retrieved from: https://www.energy.gov/sites/default/files/2015/08/f26/iron_and_steel_bandwidth_report_0.pdf.

U.S. Department of Energy (2021). *Manufacturing Energy and Carbon Footprints*. Retrieved from: https://www.energy.gov/sites/default/files/2021-12/2018_mecs_iron_steel_energy_carbon_footprint.pdf.

Chapter 23

Material Criticality

Chapter Concepts

- Critical materials are those with limited or restricted supply that are nonetheless embodied in products deemed to be of vital importance to industries or governments
- Depending on the corporate or government entity conducting the evaluation, as many as half of the elements employed by modern technology can be designated as critical
- Given the broad palette of materials in many of today's commercial products, avoiding the use of critical materials is increasingly difficult

23.1 The Criticality Concept

In 2008, a committee of the U.S. National Academies published an analysis of which non-fuel minerals might be termed "critical" to the national economy. In that work, the study identified 11 minerals or mineral groups as having at least some degree of criticality (U.S. National Research Council, 2008). Perhaps more important, the committee devised a two-axis rating system (shown in Figure 23.1) based on "supply risk" and "impact of supply restriction".

The idea was that if a country, a region, or the world was unable to procure a specific material that was needed for an important technology or application, either because the supply was shrinking for some reason or the demand was rapidly growing as new applications emerged, the inadequacy could hamper

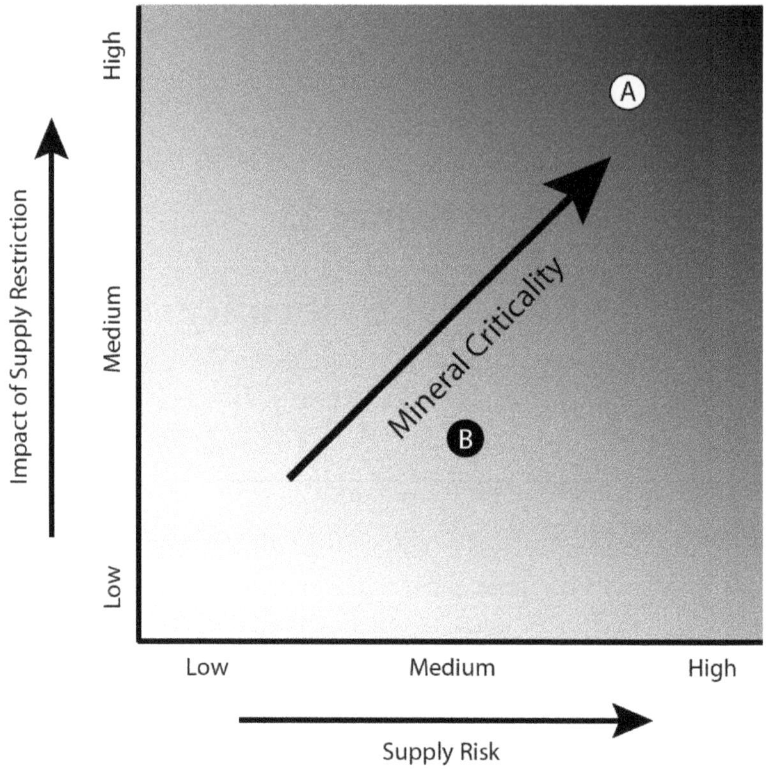

Figure 23.1. The criticality matrix originated by the U.S. National Research Council. The degree of criticality increases as one moves from the lower left to the upper right corner of the figure. In this example, Mineral A is more critical than Mineral B (U.S. National Research Council, 2008).

industrial progress. The second axis in Figure 23.1 was designed to evaluate the degree to which that supply–demand imbalance would cause disruption in industrial activities.

The National Research Council report generated short lists of "dimensions of primary availability" (for virgin resources) and "dimensions of secondary availability" (for previously used resources), as follows:

Dimensions of Primary Availability

- Geologic — Does the mineral resource exist?
- Technical — Can we extract and process it?
- Environmental and social — Can we produce it in environmentally and socially accepted ways?

- Political — How do governments influence availability through their policies and actions?
- Economic — Can we produce it at a cost users are willing or able to pay?

Dimensions of Secondary Availability

- Technical — Can we recover and reprocess it from old scrap?
- Environmental and social — Can we recover and reprocess it in environmentally and socially accepted ways?
- Political — Can we assign high political importance to the recovery and reuse of old scrap?
- Economic — Can we recover and reprocess it at a cost users are willing or able to pay?

These key questions formed the basis for criticality methodologies and discussions in 2008 and still do so today, but any program that deals with identifying critical materials and addressing the challenges that emerge is only effective to the degree that the questions are addressed in a structured manner and that effective responses result.

Two years after this initial criticality concept was formulated, it was utilized (with some revisions) in determinations of materials criticality for the European Commission and the U.S. Department of Energy. Since that time, a number of different approaches to critical materials assessment and responses have been devised. Topics that have been emphasized have addressed the importance of scale (national, regional, global), the economic aspects of supply shortages, the treatment of technology in the assessment, or changes in the environmental implications of material acquisition. Regardless of the precise methodology employed or the time scale addressed, the results have tended to single out for concern materials with low abundance in Earth's crust, those available largely as byproducts, and (for other than global assessments) those with little or no availability in domestic mineral deposits.

23.2 Criticality Methodology

As a representative example of criticality methodology, we illustrate and discuss a comprehensive approach developed at Yale University (Graedel *et al.*, 2012). The basic concept was to adopt the two axes suggested by the National Academies committee (Figure 23.1) and to add a third axis to address issues related to

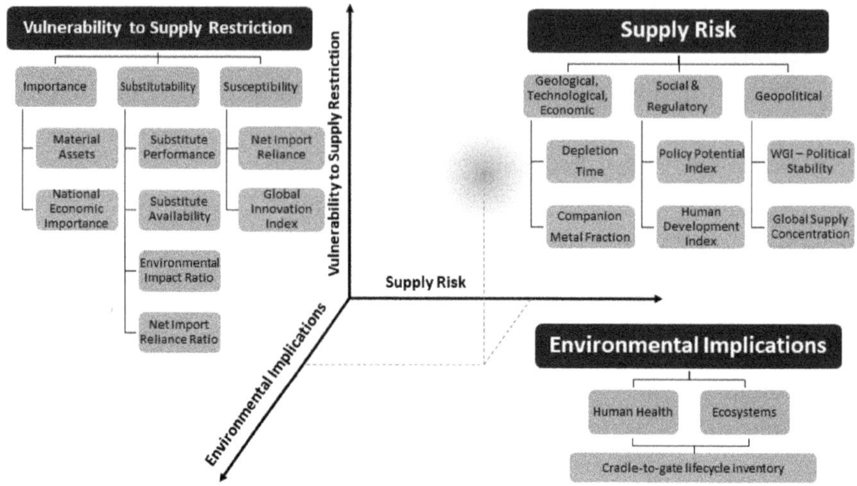

Figure 23.2. The Yale University criticality analysis methodology across axes of Supply Risk, Vulnerability to Supply Restriction (VSR), and Environmental Implications.

environmental impacts. The goal was to develop an approach that could be used (probably with variations) at the corporate, national, and global levels. The challenge was to select the metric or metrics that incorporated information from many different sectors and data sources. We will not provide a comprehensive description of the approach (which is available in Graedel *et al.*, 2012), but rather illustrate some of the final choices to give an example of the scope of a criticality analysis.

The Yale methodology is shown in Figure 23.2. For the Supply Risk axis (upper right) three components are chosen, each of which has two indicators, the values of which are quantitatively derived. At the long-term level, only the first of the three components is retained. The indicators are quite diverse methodologically. For the Vulnerability to Supply Restriction (VSR) axis, the components and indicators are shown at the upper left of Figure 23.2. The third axis, Environmental Implications (lower right), is evaluated using life cycle assessment (LCA) software.

An example of the results from applying this methodology is given in Figure 23.3, where criticality at the U.S. level (top diagram) and the global level (bottom level) is shown for the five elements commonly found in iron mineral assemblages. The estimated uncertainties around each axis arise from a Monte Carlo analysis, the results of which form an "uncertainty cloud" for each metal as indicated by the breadth of the plotted points on each axis. For niobium (Nb, green) the uncertainty level is greatest for supply risk (SR), while vulnerability is highest for manganese (Mn, purple). At the global level, the supply risk is

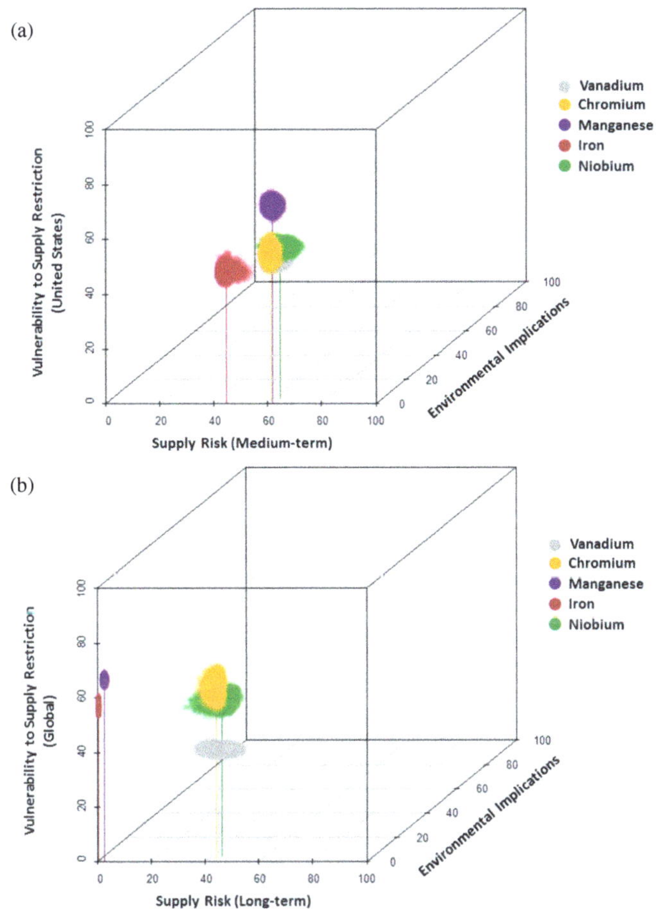

Figure 23.3. Locations of the geological iron family of elements in criticality space (Nuss *et al.*, 2014): (a) National level for the U.S., 2008. The highest level of criticality is at (100,100,100) (back right top). (b) Global level, 2008.

generally lower because mineral deposits around the world provide lower risk, but the vulnerability to supply restriction (VSR) ratings are roughly equivalent to the U.S. evaluations.

23.3 Alternative Determinations of Material Criticality

The methodology of the European Commission (2017) employs somewhat different metrics. That approach is distinguished by the imposition of firm lines that establish criticality boundaries for both the Supply Risk and the Economic Importance to the European Union (EU) axes (Figure 23.4).

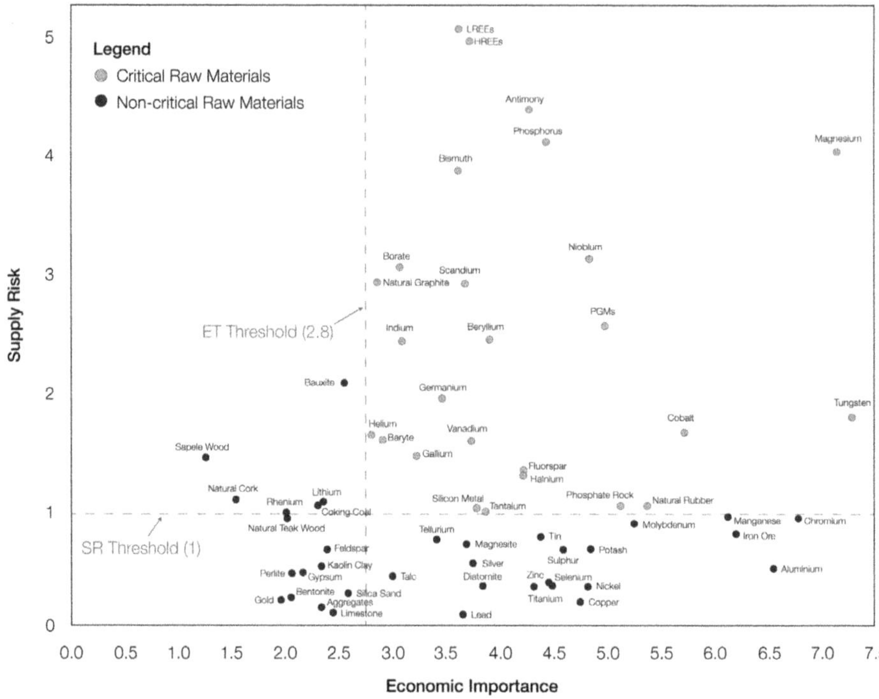

Figure 23.4. Elements deemed critical by the EU using its methodology, 2017 (adapted from European Commission, 2017).

Determinations of material criticality have major importance in the policy realm because country and regional governments are very interested in whether supply/demand imbalances might have impact upon their economies. As a result, there are a number of "criticality lists" that guide governmental policy, including Australia, Canada, the European Commission, Japan, and the U.S. Figure 23.5 graphically illustrates the elements selected by those entities. Together, the surveys select a total of 63 elements — not too far from selecting everything. To be optimally useful, it would seem that criticality methodologies should target the extreme end of the criticality spectrum, perhaps 15 or 20 elements at most. At the time of writing, it appears that criticality methodologies need to be revised to be more sharply focused.

23.4 Addressing the Challenges of Critical Materials

Chapters 17–21 discussed the properties and benefits of several different classes of materials. Depending on the use and the material characteristics that the use requires, materials are selected based on the degree to which those properties are met. Material criticality, however, places an additional constraint on the product

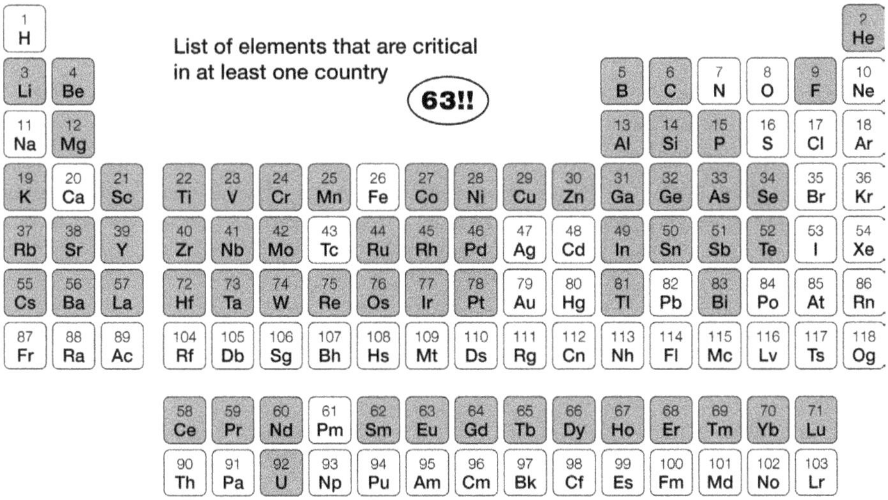

Figure 23.5. A composite periodic table of criticality using the assessment approaches of Australia, Canada, the European Union, Japan, and the U.S. as of early 2021.

designer. If a material that is desired is designated as critical by governmental agencies, its supply over the immediate or longer term may be uncertain, and political considerations may also play a role depending on the stability or cooperation of the principal producing countries. These considerations may lead the product designer to seriously consider materials with somewhat inferior mechanical or chemical properties but with much greater security of supply.

From a political perspective, a short-term response to a determination of criticality for a specific material is to establish a stockpile to compensate for the scarcity at a particular point in time. Over the longer term, governments can consider increasing supply by entering into new mining projects. Another alternative is to embrace enhanced recycling, which can be helpful for elements easily recovered and reused (such as platinum-group metals from motor vehicle catalytic converters). In other cases, such as the germanium used to optimize infrared radiation transmission in fiber-optic cables, recovery is problematic. If concern for germanium supply is high enough, therefore, it may be necessary to redesign the cables in order to minimize germanium use in that application. Similar considerations apply to many other materials identified in Figure 23.5.

Further Reading

Australian Government. (2020). Australian Critical Minerals Prospectus, 2020. Retrieved from: https://www.austrade.gov.au/ArticleDocuments/1358/Australian_Critical_Minerals_Prospectus_2020.pdf.aspx.

Government of Canada (2021). Canada's Critical Minerals List 2021. Retrieved from: https://www.nrcan.gc.ca/sites/nrcan/files/mineralsmetals/pdf/Critical_Minerals_List_2021-EN.pdf.

European Commission (2017). *Study on the Review of the List of Critical Raw Materials: Criticality Assessments*. Retrieved from: https://data.europa.eu/doi/10.2873/876644.

Graedel, T. E., *et al.* (2012). Methodology of metal criticality determination. *Environmental Science & Technology, 46*, 1063–1070.

Graedel, T. E., *et al.* (2015). Criticality of metals and metalloids. *Proceedings of the National Academy of Sciences*, *112* (14), 4257–4262.

McCullough, E., and Nassar, N. T. (2017). Assessment of critical minerals: Updated application of an early-warning screening methodology. *Mineral Economics, 30*, 257–272.

Nakano, J. (2021). *The Geopolitics of Critical Minerals Supply Chains*. Retrieved from: https://usea.org/sites/default/files/event-/USEA_April2021_presentation%20Jane%20Nakano.pdf.

Nuss, P., Harper, E. M., Nassar, N. T., Reck, B. K., and Graedel, T. E. (2014). Criticality of iron and its principal alloying elements. *Environmental Science & Technology*, *48* (7), 4171–4177.

U.S. Department of the Interior (2018). *Final List of Critical Materials 2018*, Federal Register, 23295–23296.

U.S. National Research Council (2008). *Minerals, Critical Minerals, and the U.S. Economy*, Committee on Critical Mineral Impacts on the U.S. Economy. Washington, DC: The National Academies Press.

Industrial Ecology in Industry

Chapter 24

Sustainable Design Frameworks: Design for Environment, Cradle-to-Cradle, and the Principles of Green Chemistry and Green Engineering

Chapter Concepts

- Sustainable design frameworks provide general guidance for creating sustainable products and processes
- Design software used by designers, engineers, and architects is increasingly incorporating environmental considerations to allow for rapid assessment and iteration to meet environmental targets
- Heuristics and 'rules-of-thumb' allow for widespread implementation of IE-informed recommendations but should be updated based on changing circumstances

24.1 Sustainable Design Frameworks

Industrial ecology (IE) tools allow us to quantify industry's use of resources, identify environmental hotspots in the supply chain of a product, or compare product options over their life cycles. This is critical information in understanding *what* and *where* improvements can be made in a manufacturing process or product, but not *how* to make them. Designers and engineers need guidelines in order to innovate on the next generation of products and materials. IE has long been allied with various sustainable design frameworks that provide general heuristics or principles to guide practitioners. In this chapter, we will review three important frameworks that have broad applicability: (1) Design for Environment (DfE), (2) Cradle-to-Cradle (C2C), and (3) the Principles of Green Chemistry and Green Engineering.

24.2 Design for Environment

With growing environmental consciousness around the world in the 1970s, manufacturers began to incorporate environmental considerations into their designs. Energy efficiency became a desirable trait and designers began to screen out chemicals with noted toxic or ecotoxic effects. Eventually, these design criteria became formalized under a generic framework called 'Design for Environment' (DfE). Each product and sector has different requirements and environmental considerations, many of which have been further standardized into eco-labels and certifications, but there are some overall DfE guidelines that are widely applicable.

Product designers, engineers, and architects must balance multiple attributes of their designs. In addition to potential environmental considerations, a product must of course technically perform well to fulfill its desired function, and the designer must also think about aesthetics, ergonomics, durability, repairability, manufacturability/constructability, cost, regulatory compliance, supply chain logistics, compatibility with existing systems, and many other constraints. This is clearly a lot to keep track of, and so designers have developed a large suite of tools and methods to aid in the design process such as computer aided design (CAD). The past decade has seen tremendous progress in integrating environmental considerations into such design tools so that DfE is a core goal.

One of the most common manifestations of DfE in manufacturing is material selection guides. Often, multiple materials might be satisfactory for a given performance specification, and the designer can then choose which one has the lowest cost, most attractive look, or in our case, the best environmental profile. One popular material selection guide is the Ashby Plot, named for materials scientist Michael Ashby at Cambridge University (example in Figure 24.1). Once the technical strength requirements for a material have been calculated by the design engineer, say at 10^2 MPa on the y-axis, then the product design team can consider options starting with the lowest embodied energy on the x-axis, first technical ceramics, then composite materials, then metals. Ashby plots can be generated for a large number of physical, chemical, and environmental attributes in order to consider other parameters.

Another important DfE application is in guiding choices among chemical products and evaluating alternatives, with a focus on safety. For example, the U.S.

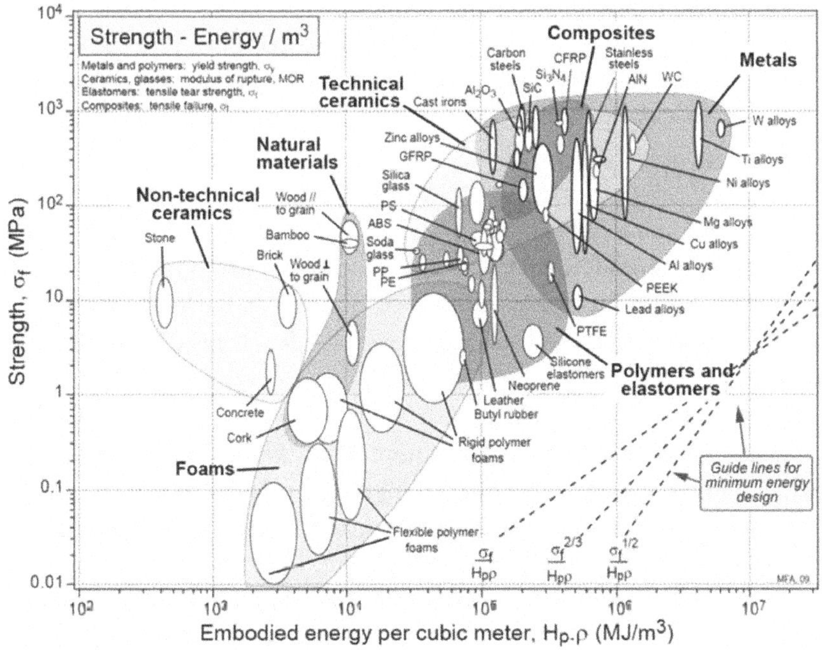

Figure 24.1. Ashby Plot for material strength versus volumetric embodied energy (CES EduPack, Granta Design) (Ashby, 2012).

Environmental Protection Agency DfE program began in the early 1990s and has evolved into the Safer Choice program. Products that carry the Safer Choice eco-label must have constituent chemicals reviewed for carcinogenicity, reproductive/developmental toxicity, ecotoxicity, and persistence in the environment. In addition, the product must meet performance, pH, low volatile organic carbon content, and sustainable packaging standards. Many similar eco-labels and safer chemicals screening tools exist around the world in order to guide chemical formulators.

The most widespread and important legislative approach to track hazardous chemicals in commerce is the EU's Registration, Evaluation, and Authorisation of Chemicals (REACH) program. This action, which came into force on June 1, 2007, requires that chemicals manufactured in quantities greater than one metric ton must be registered, those manufactured in quantities greater than 100 metric tons be evaluated for toxicity, and those of high concern (such as carcinogens or mutagens) must be specially authorized for use. Downstream users of the chemicals are involved as well as those who make the chemicals. REACH has made it much more difficult and expensive to utilize unauthorized materials and provides a substantial incentive for attempts to meet product technical specifications without utilizing problematic new materials or new material combinations.

Various authors have devised DfE principles for specific sectors; we will later provide a general list that has wide applicability. As in industrial ecology, DfE recognizes the importance of a life cycle perspective and the desire to avoid unintended consequences. Accordingly, general DfE principles can be organized according to life cycle stage.

Materials and Manufacturing:

- Choose readily available materials, e.g., those that avoid issues of supply risk.
- Minimize or avoid hazardous substances both in the product itself and in the manufacturing process, and utilize closed loops for necessary but hazardous constituents.
- Minimize materials diversity.
- Minimize energy and resource consumption in the production phase.
- Design the manufacturing process to minimize material waste.
- Use structural features and high-quality materials to minimize product weight and improve material efficiency.
- Employ materials that are renewable, reusable, and recyclable and/or recycled, whenever possible.

  Packaging is another important area of environmental design. IE research has shown that the environmental impacts of some products is dominated by packaging choice, and that single-use packaging makes up about 40% of the entire market for plastics.

Packaging:

- Minimize packaging materials by reducing redundant layers, size, and thickness.
- Use recycled content packaging materials.
- Label packaging materials so that they can be identified for recycling or composting.
- Design packaging in standard sizes or geometries that can be efficiently packed for transport.
- Avoid the use of packaging materials such as polyvinyl chloride (PVC) that may have harmful emissions during incineration.

Use:

- Design for energy and water use efficiency.
- Minimize the use of consumables, emissions, and material dissipation during use.
- Choose materials, surface treatments, and structural arrangements to inhibit dirt, wear, and corrosion, thereby enhancing product life.

- Design products for durability that matches their technological lifetime.
- Design products that can be maintained, repaired, upgraded, reconfigured, or refilled through accessibility, labeling, modules, breakage points, and manuals.

End-of-Life:
- Design for disassembly by using as few joining elements as possible and emphasizing screws, snap fits, and geometric locking.
- Avoid material combinations that force time-consuming and otherwise unnecessary separation at the recycling stage.
- Avoid designs that make it impossible to separate materials or lead to cross-contamination, thus inhibiting recyclability.
- Minimize painting and fillers.

In most situations, designers following these DfE principles will produce much "greener" products than would otherwise be the case and will contribute to the sustainability of their firm and their planet.

As mentioned earlier, numerous product-specific DfE lists have been developed across nearly every sector. Many of these DfE lists have been formalized into certifications, standards, or eco-labels. DfE lists of generally agreed-upon environmental criteria provide manufacturers with an easy-to-follow map to designing greener products. Products that become DfE certified have a market advantage, as industries and consumers use DfE eco-labels to make environmentally conscious choices. DfE programs also bolster the reputation of companies and their environmental, social, and corporate governance (ESG), making them more attractive to investors and business partners. As such, DfE standards and labels are extremely popular, with hundreds of programs currently running around the world.

24.3 Cradle-to-Cradle

In the LCA section presented in Chapter 11, we learned about different scopes and life cycle phases, including end-of-life. While reuse and recycling are certainly included in a cradle-to-grave LCA, the terminology implies that eventually materials and products come to the end of their life cycle. In the early 2000s, German chemist Michael Braungart and American architect William McDonough came together to espouse an approach they called 'Cradle-to-Cradle' (C2C), which aims to mimic nature by eliminating the very notion of waste and end-of-life (Cradle to Cradle Products Innovation Institute (n.d.)). The goal of C2C is to design products and systems that can be infinitely cycled, either as biological nutrients that can be returned to nature, or as technical nutrients that be recycled repeatedly without being 'downcycled' to products of lesser quality. By closing nutrient loops, all materials eventually return to their cradle to be remade into new products.

The C2C design framework shares much of its vision with industrial ecology, especially the notions of 'waste = food' and closing material loops as much as possible. But where IE largely studies the interplay of production, consumption, and technology as it exists today, C2C is about designing the products and systems of the future. Where LCA might be used to choose the least bad option (i.e., with lowest environmental impact) or what is sometimes called 'eco-efficiency', C2C envisions products that are instead 'eco-effective', with designs focused on creating environmental benefits. C2C designs are renewable and rejuvenating, rather than depleting; for example, green buildings treat water, purify air, and provide attractive spaces for people to come together to work, rather than simply being more energy efficient. The C2C framework reminds us that it is important to focus on the positive power of technology and design and not just on negative consequences. Accordingly, a new approach in LCA put forward by Greg Norris is called product 'handprinting' (as opposed to footprinting), which quantifies the net environmental benefits that a product has over its life cycle.

The C2C design framework is stricter than many of the DfE principles listed in the previous section, and there is some criticism that it cannot realistically be applied to many categories of products and is more inspirational than practical. Nevertheless, a certification scheme has been created based on the C2C framework and numerous products have gone through the certification process, which sets required performance tiers across the five areas of material health, product circularity, clean air and climate protection, water and soil stewardship, and social fairness (Cradle to Cradle Products Innovation Institute (n.d.)).

24.4 Principles of Green Chemistry and Green Engineering

A major theme of green design frameworks is chemical toxicity and product safety. As with the C2C framework, we must not only avoid substances of concern, but also innovate to create new substances that have the functional properties we want but without negative side effects. The most influential paradigm in this area is the 12 Principles of Green Chemistry (GC) (Table 24.1), created by chemists Paul Anastas and John Warner in the late-1990s. Their work launched the field of green chemistry, now a worldwide community of academic and industrial chemists with hundreds of academic programs.

The Principles of Green Chemistry are just that: principles that can guide chemists in designing new molecules and syntheses. However, most by themselves

Table 24.1. Twelve Principles of Green Chemistry (Anastas and Warner, 1998).

1. It is better to prevent waste than to treat or clean up waste after it has been created.
2. Synthetic methods should be designed to maximize incorporation of all materials used in the process into the final product.
3. Wherever practicable, synthetic methods should be designed to use and generate substances that possess little or no toxicity to human health and the environment.
4. Chemical products should be designed to preserve the efficacy of function while reducing toxicity.
5. The use of auxiliary substances (e.g., solvents and separation agents) should be made unnecessary wherever possible and, innocuous when used.
6. Energy requirements should be recognized for their environmental and economic impacts and should be minimized. Synthetic methods should be conducted at ambient temperature and pressure.
7. A raw material or feedstock should be renewable rather than depleting whenever technically and economically practicable.
8. Unnecessary derivatization (use of blocking groups, protection/deprotection, temporary modification of physical/chemical processes) should be minimized or avoided if possible because such steps require additional reagents and can generate waste.
9. Catalytic reagents (as selective as possible) are superior to stoichiometric reagents.
10. Chemical products should be designed so that at the end of their function they break down into innocuous degradation products and do not persist in the environment.
11. Analytical methodologies need to be further developed to allow for real-time, in-process monitoring and control prior to the formation of hazardous substances.
12. Substances and the form of a substance used in a chemical process should be chosen to minimize the potential for chemical accidents, including releases, explosions, and fires.

do not convey information about how well a certain chemical or process performs. To operationalize the Principles, various metrics have been proposed that can track green chemistry performance quantitatively. For example, one way to track Principle 2 is the metric of *Atom Economy*, defined as the percentage of reagents that make it into the final product. Increasing Atom Economy means that fewer reagents are required per unit of target chemical and that less chemical waste is produced in the process. However, chemical reactions involve more than just reagents; solvents, catalysts, buffers, and other auxiliary inputs are almost always involved. So, a broader green chemistry metric was developed called the *Process Mass Intensity (PMI)*, defined as the percentage of all chemicals involved in a process or step that make it into the final product. An ideal process that green chemists should strive for is when PMI is equal to 100%.

From the perspective of LCA, a mass-based metric like PMI is unsatisfying because it does not differentiate process chemicals in terms of their environmental implications; essentially, all process chemicals are equally important in the metric. But it is important to remember that green design frameworks strive to provide simple heuristics to guide practice. Other GC tools such as green solvent selection guides can be used in concert to incorporate other life cycle considerations such as embodied emissions or toxicity.

Following the success of the GC Principles, a parallel framework was developed for engineering by Paul Anastas and Julie Zimmerman, called the 12 Principles of Green Engineering (Table 24.2).

The Principles of Green Engineering are broadly applicable to product engineering and design, and some are closely related to the DfE criteria described earlier. Together, the Principles of Green Chemistry and Green Engineering have been taken up and promulgated by various groups, including the American Chemical Society. Unlike other design frameworks that have been developed and

Table 24.2. Twelve Principles of Green Engineering (Anastas and Zimmerman, 2003).

1. Designers need to strive to ensure that all materials and energy inputs and outputs are as inherently non-hazardous as possible.
2. It is better to prevent waste than to treat or clean up waste after it is formed.
3. Separation and purification operations should be designed to minimize energy consumption and materials use.
4. Products, processes, and systems should be designed to maximize mass, energy, space, and time efficiency.
5. Products, processes, and systems should be "output pulled" rather than "input pushed" through the use of energy and materials.
6. Embedded entropy and complexity must be viewed as an investment when making design choices on recycle, reuse, or beneficial disposition.
7. Targeted durability, not immortality, should be a design goal.
8. Design for unnecessary capacity or capability (e.g., "one size fits all") solutions should be considered a design flaw.
9. Material diversity in multicomponent products should be minimized to promote disassembly and value retention.
10. Design of products, processes, and systems must include integration and interconnectivity with available energy and materials flows.
11. Products, processes, and systems should be designed for performance in a commercial "afterlife."
12. Material and energy inputs should be renewable rather than depleting.

commercialized into certifications or eco-labels, the Principles maintain their status as a set of guidelines, where chemists and engineers at companies strive to follow those Principles and associated metrics that best map onto their situations.

24.5 A Cautionary Note on Design Guidelines, Heuristics, and Principles

Design frameworks and the heuristics and principles that comprise them are so useful because they provide general guidance, saving time and effort. But it is important not to be complacent and over-generalize. There are so many different types and uses of products in the world that there are bound to be exceptions to each framework. For example, bio-based plastics may not be environmentally preferable to petrochemical plastics, depending on the particular feedstock, where it was cultivated, and the processing route that is used. IE tools, especially LCA, should still be used in concert with sustainable design frameworks to ensure that new products and materials do in fact have more sustainable outcomes.

We live in a rapidly changing world. What was environmentally preferable five years ago may not be five years from now, and vice-versa. Sustainable design frameworks, eco-labels, and sustainability certifications must be updated regularly, informed by a life cycle and systems perspective, so that they continue to point in the correct direction.

Further Reading

Anastas, P. T., and Warner, J. C. (1998). *Green Chemistry: Theory and Practice*. New York: Oxford University Press.

Anastas, P. T., and Zimmerman, J. B. (2003). Design through the 12 principles of green engineering. *Environmental Science & Technology*, *37* (5), 94A–101A.

Ashby, M. F. (2012). *Materials and the Environment: Eco-Informed Material Choice*. London: Elsevier.

Cradle to Cradle Products Innovation Institute (n.d.). *Introducing the Cradle to Cradle Certified Product Standard Version 4.0*. Retrieved from: https://www.c2ccertified.org/get-certified/cradle-to-cradle-certified-version-4.

McDonough, W., Braungart, M., Anastas, P. T., and Zimmerman, J. B. (2003). Applying the principles of green engineering to cradle-to-cradle design. *Environmental Science & Technology*, *37* (23), 434A–441A.

Norris, G. A., Burek, J., Moore, E. A., Kirchain, R. E., and Gregory, J. (2021). Sustainability Health Initiative for NetPositive Enterprise handprint methodological framework. *The International Journal of Life Cycle Assessment*, *26* (3), 528–542.

U.S. Environmental Protection Agency (n.d.). *Safer Choice*. Retrieved from: https://www.epa.gov/saferchoice.

Zimmerman, J. B., Anastas, P. T., Erythropel, H. C., and Leitner, W. (2020). Designing for a green chemistry future. *Science*, *367* (6476), 397–400.

Chapter 25

Remanufacturing

Chapter Concepts

- The remanufacturing of products preserves the productive life of materials and products, saves energy, and minimizes waste
- Product remanufacturing is aided by close links between the original manufacturer and the remanufacturer, links that have often not been well established
- The business of remanufacturing, despite its materials and environmental benefits, is quite challenging and is poorly supported by government policy in general

25.1 The Remanufacturing Concept

The typical sequence for the manufacture and use of a technological product is pictured in Figure 25.1. The process begins with the extraction of raw materials, transformation into usable forms of the materials, fabrication of components, and product assembly. After distribution and use, the owner discards the product, thus meeting the traditional sequence of "make it, use it, and discard it when it is no longer useful". Some of the discarded materials might be recovered and reused. A more extensive approach, termed *remanufacture*, is to disassemble the product, clean it, replace any components as appropriate, upgrade the product to at least the original level of performance, refinish or repaint as appropriate, and return the product to service. This process is the subject of the present chapter.

Why might one wish to go through the remanufacturing process as opposed to acquiring a replacement for a product that is no longer satisfactory for one reason

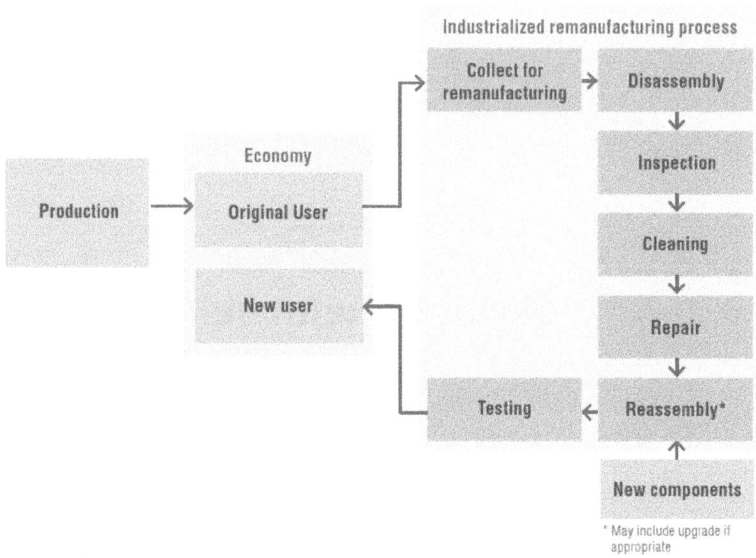

Figure 25.1. The detailed remanufacturing flow chart (Nasr *et al.*, 2018).

or another? One reason that has resonance for industrial ecologists is that remanufacturing retains 80–90% of the materials in the product for another use, thus avoiding the not uncommon practice of extracting materials, processing them, using them once, and then disposing of them. A second reason is that the remanufacturing process typically retains 60–70% of the embedded energy in the original product. While these reasons seem enough to justify routine remanufacture, it turns out nonetheless that technical, operational, and economic factors can render remanufacture problematic.

25.2 The Remanufacturing Process

Remanufacturing begins with a recovered product, termed a *core*, which is "a previously sold, worn, or non-functional product or part, intended for the remanufacturing process." During reverse logistics, a core is protected, handled, and identified for remanufacturing to avoid damage and to preserve its value. A core is not waste or scrap, but it is not intended to be reused until it has been remanufactured (European Remanufacturing Network, 2020). There are two unstated implications in this definition. The first is that the core must be perceived by its owner to have value even when obsolescent. The second is that the owner must know how to access a reverse logistics network and transfer the core to that network. In practice, these two aspects require almost all remanufacturing to involve the large-scale industrial organizations who own the core or know who does and thus have substantial knowledge of the core design, manufacture, and current locale. This is the case, in fact, with the most commonly remanufactured products: automobile components, heavy-duty equipment, aerospace products, medical devices, machinery, photocopiers, and the like.

Despite the perceived virtues of remanufacture, some product types are not suitable targets for this process for one or more of the following reasons:

- The product technology is not stable from one product cycle to the next (e.g., new medical devices are so different from previous devices as to make remanufacturing impractical).
- The process technology is not stable from one product cycle to the next.
- The product typically fails by dissolution or dissipation rather than functionally (e.g., brake linings).
- Critical components typically fail before the product itself fails.
- The cost to remanufacture is too high to justify the process.

Some of these challenges can be surmounted by more thoughtful design, as described later; others cannot be readily overcome.

Remanufacturing, as with initial manufacturing, is a multi-step process, as shown in Figure 25.1. The first stage, disassembly, immediately presents the case for "design for remanufacturing", because remanufacturing is unlikely to occur unless it is relatively straightforward to disassemble the core with tools routinely available and to do so while doing no damage. The core, and perhaps its subparts, are then inspected and cleaned to the level typical of new components. If repair is needed, it is taken to the new component level. The core is then reassembled and tested to make sure that the remanufactured product is the equivalent of the original in appearance and performance.

An alternative that can be useful for some cores, such as office furniture suitable for reuse, may be to remanufacture the product into a form useful to consumers seeking a lower but still useful level of performance. This creates the "cascade loop", which retains the product and its materials for at least one additional product cycle before remanufacturing becomes impractical or too expensive to justify.

25.3 Product Evolution during Remanufacturing

The rapid progress of modern technology enables manufacturers to incorporate new performance features into their products at an increasingly rapid rate. As a consequence, many products become obsolete quickly. Unless a used product can be upgraded during remanufacturing, there is little opportunity for future product employment rather than scrappage at the end of life.

It is interesting from the "suitability for remanufacturing" perspective to return to the ecological perspective that lies at the basis of much of industrial ecology, and to ask why some natural organisms evolve more rapidly than others, and what properties (phenotypes) of those organisms make such evolution possible. This issue has been termed *evolvability*, a characteristic that favors three phenotypic characteristics that influence variation (see Calcott, 2014):

1. Minimize disadvantages resulting from changes.
2. Increase the variety of results enabled by the changes.
3. Reduce the number of changes needed to evolve into a more useful form.

Calcott (2014) suggests that the application of these concepts in engineering product design and product manufacture, as with natural selection, has the potential to generate more highly evolvable and useful technological artifacts. What might be some of the characteristics of the products that could be produced by such a procedure? The answer to this question can perhaps be best illustrated by the example of the Xerox photocopy machine, probably the first product for which the design team seriously considered remanufacture. The team first determined the aspects of the design that were likely to remain relatively stable over future generations of the product, determining that the size of the frame (i.e., the overall product dimensions) and the electrical interface, among other attributes, fell into that category. They next asked what aspects of the product might well evolve, that is, what might be the transformative features of the photocopier of the future. The list included the following:

- improved optics and electronics for better image scanning;
- transformed image generation (better inks or electronic paper);
- improved paper handling (higher speed, fewer paper jams, etc.).

Given that foundational thinking, the copiers were then designed to be capable of evolving during remanufacturing, especially as regards those features expected to significantly change over time and to do so as new features were introduced to subsequent designs. The anticipated transformable parts were made readily accessible, readily removable, and readily replaceable. Following this design philosophy over decades has enabled Xerox to maintain photocopy systems for as many as seven remanufacturing cycles.

The message for remanufacturing is obvious: anticipate product evolution and then design today's systems to facilitate upgrading them during future remanufacturing.

25.4 The Business of Remanufacturing

Despite the obvious benefits of remanufacturing — preserving the productive life and embedded energy of materials and products, minimizing waste, capturing critical materials that would otherwise be lost, etc. — the remanufacturing business is quite challenging. A list of the characteristics of the business includes the following:

- Increasing the first lifetimes of many new products renders core designs and manufacturing approaches often a distant memory, the need for replacement parts is often smaller than it had been historically, and the time to initial remanufacture is longer.
- Parts needed for remanufacturing are often not available if design evolution is very rapid.
- Electrical components are ever more numerous and are increasingly difficult to locate, identify, evaluate, and replace.
- The increasing complexity of products makes them challenging to remanufacture.
- Cores that have not been handled carefully after they leave service may require major initial processing because of contamination, grease, rust, or other impediments.
- A remanufacturing business that is not highly specialized (e.g., one that only remanufactures aircraft engines) must deal with a very wide range of products, including appliances, electronic and network equipment, heavy-duty components (earth movers and mining equipment), machine tools, medical equipment, military defense platforms (armored vehicles, etc.), and more. Accordingly, a large and diverse stock of parts must be kept on hand.
- The financial picture for a remanufacturer can be very challenging because of such issues as legal constraints, the modest size of the market for remanufactured goods, warranties on some products, and the supply chain for recovered products suitable for remanufacturing.

Notwithstanding these challenges, remanufacturers can provide real service not only to society but also to original product manufacturers. Remanufacturers often know more about how equipment performs over time — how equipment tends to be used, what parts tend to fail and when, how amenable particular equip ment is to multiple cycle of reuse, and the like. As with recycling in general, increasingly close links between the original product designer and the remanufacturer are likely to benefit all parties and to better preserve products and their constituents for multiple uses over time.

25.5 Linking Remanufacturing and Recycling

It is worth noting that Figure 25.1 includes recycling as an option at the end of the product life. It is significant that if recycling occurs, it implies that the next step will occur at the material transformation stage rather than at the components and fabrication stage, that is, that it will be elements that will be recycled, not products or product modules. Consequently, the embedded energy in the assembled components will be lost even if recycling is perfect. And, as noted in Chapter 9, end-of-life recycling rates for many materials are very small, so the recycling and reuse of end-of-life materials is problematic. Therefore, from both a materials and energy standpoint, remanufacturing to the degree possible, is likely to be far preferable to recycling at the end of a product's first (or even second or third) life.

Nonetheless, for situations in which remanufacturing is not the ideal approach, such as major design upgrades that make it unlikely that a remanufactured product would be competitive in the marketplace, design choices focused on recycling may be desirable. As an example, Ford Motor Company chose for its F-150 trucks a particular set of aluminum alloys designed to optimize recycling and avoid problems with tramp elements that made recycling difficult or impossible (Figure 25.2).

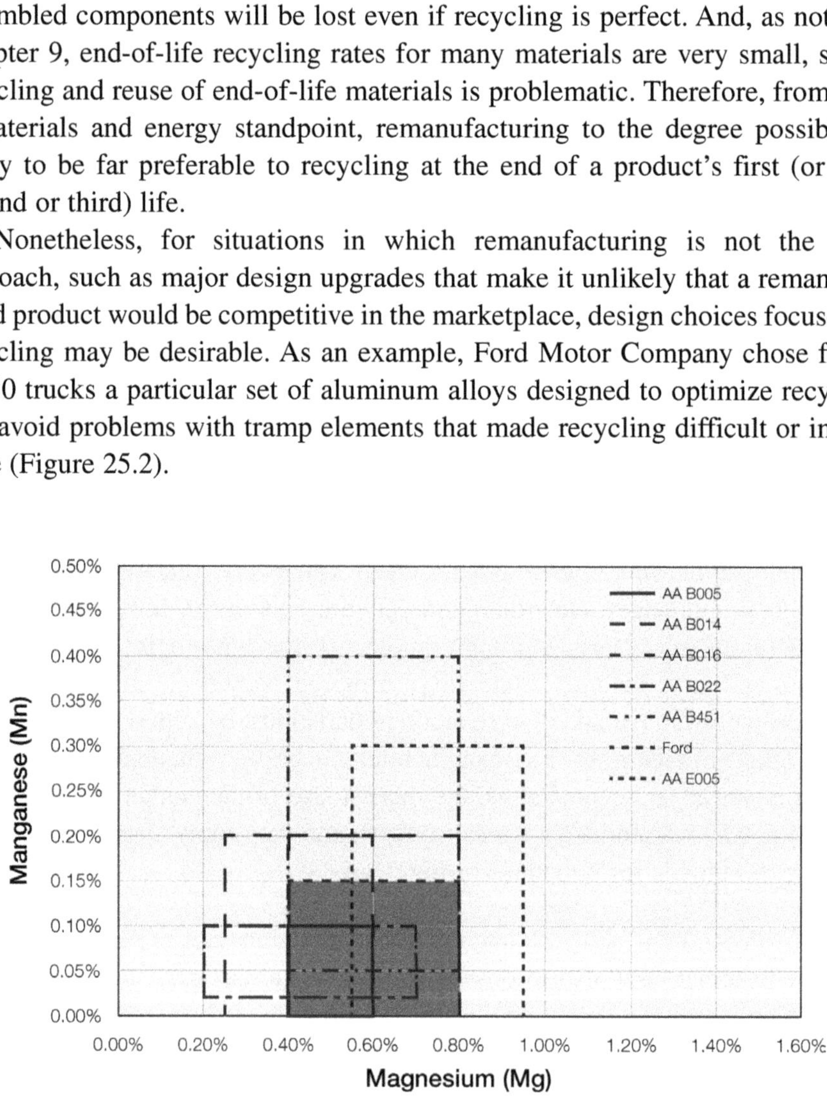

Figure 25.2. Aluminum alloy selection for Ford F-150 trucks (adapted from Chappuis, 2015).

Remanufacturing has, in general, not been well aided by government policies and regulations. An important future job for governments and manufacturers is to support the remanufacturing industry so as to fully realize its benefits for society and the environment.

Further Reading

Allwood, J. M. (2014). Squaring the circular economy. In E. Worrell and M. A. Reuter, (Eds.) *Handbook of Recycling*, 445–477. Amsterdam, The Netherlands: Elsevier.

JBey4you. (2018, April 20). Apple introduces Daisy, a new robot that disassembles iPhone to recover valuable materials [Video file]. Retrieved from: https://www.youtube.com/watch?v=2Bu-gl7v-P8.

Calcott, B. (2014). Engineering and evolvability. *Biology and Philosophy*, *29*, 293–313.

Chappuis, L. B. (2015). *Material Specifications & Recycling for the 2015 Ford F-150*. Retrieved from: https://www.lbcg.com/media/downloads/events/411/day-2-laurent-b-chappuis-technical-expert-lightweight-stampings-vehicle-program-engineering-sbu-ford.6735.pdf.

Cooper, D. R., and Gutowski, T. G. (2015). The environmental impact of reuse. *Journal of Industrial Ecology*, *21* (1), 38–56.

European Remanufacturing Network. (2020). Retrieved from: https://www.remanufacturing.eu.

Klausner, M., Grimm, W. M., and Hendrickson, C. (2015). Reuse of electric motors in consumer products. *Journal of Industrial Ecology*, *21* (1), 38–56.

Matsumoto, M., Yang, S., Martinson, K., and Kainuma, Y. (2016). Trends and research challenges in remanufacturing. *International Journal of Precision Engineering and Manufacturing-Green Technology*, *3* (1), 129–142.

Nasr, N., *et al.* (2018). *Re-defining Value — The Manufacturing Revolution. Reuse, Refurbishment, Repair, and Direct Reuse in the Circular Economy*. Paris, France: International Resource Panel.

Rahito, D. A. W., and Azman, A. H. (2019). Additive manufacturing for repair and restoration in remanufacturing: An overview from object design and systems perspectives. *Processes*, *7* (11), 802.

Tolio, T., *et al.* (2017). Design, management and control of demanufacturing and remanufacturing systems. *CIRP Annals — Manufacturing Technology*, *66*, 585–609.

Chapter 26

Corporate Sustainability

Chapter Concepts

- The analytical tools of industrial ecology (MFA, LCA, EEIO, Criticality, etc.) offer an approach complementary to those of traditional business tools
- Improving resource efficiency in industry often leads to economic savings
- Businesses are increasingly adopting environmental, social, and corporate governance (ESG) policies and strategies, often including reporting on sustainability topics

26.1 Industrial Ecology in the Business World

A major goal of this book is to enable scientists, engineers, and managers in the private sector to gain perspective, develop targets, and learn to use industrial ecology tools that can help businesses to achieve their environmental and sustainability goals. The 25 chapters that precede this one all, in some way or another, relate to the inherent relationships that link the manufacturing sector to issues of the environment and sustainability. These issues can be encapsulated in five themes:

- *Evolving perspectives*: Industrial ecology thinking adds new vistas to the thinking of the firm: the interactions of technology with society, the concept of factories and service facilities as industrial organisms, and the relevance and importance of sustainability.

- *Evolving tools*: Industrial ecology tools are becoming increasingly diverse and robust: metabolic analysis, sustainable engineering, design for environment and sustainability (DfES), life cycle assessment (LCA), and material flow analysis (MFA).
- *Resource limitations*: For various resources and in various locales, limitations are beginning to be felt and must be addressed by corporate product design and overall management.
- *Changing markets*: The traditional markets for industrial products, long directed to industrial customers or wealthy suburbanites, are evolving toward the less-developed countries and especially the large urban regions within those countries.
- *Links to larger systems*: Industry increasingly realizes that it acts within larger socio-technical systems, both through its own supply chains, the communities in which it operates and into which it sells, and other cultural, political, and policy drivers.

These themes form the basis of a corporate manager's to-do list of issues related to the environment and sustainability.

The environmental program of a firm is seldom equivalent to what we would term a true industrial ecology program. In the past, environmental issues for the firm arose from practices involving emissions or waste streams, and the traditional governmental response — remediation and compliance — was driven by regulation. Even now, many corporate sustainability programs still fall within environmental health and safety (EHS) programs. Moreover, the focus of environmental activity within and outside the firm was primarily on the manufacturing process and generally characterized by end-of-pipe controls on emissions. Regulations were concerned with single media, individual substances, specific sites, or particular process emissions. There was little recognition (or incentive to recognize) that impacts arising from the firm's activities were fundamentally linked with regional and global natural, technological, and economic systems. Industrial ecology, however, requires that recognition and encourages a fundamental integration of these issues at the level of the firm.

Although regulation is undoubtedly important, market demand tends to be far more influential in the cultural model of businesspeople. Sophisticated customers drive much commercial procurement, and many (including government purchasing agencies) increasingly demand environmentally preferable products and services. Business customers and suppliers are also increasingly forging relationships based on the environmental performance of potential partners. A major shift has taken place over the past 30 years in how many businesses view environmental concerns, from treating them as compliance-related overhead unrelated to core decision-making, to viewing them as a source of competitive advantage and societal responsibility.

Corporations are in business to make money for their stockholders or owners. This goal conflicts with proactively spending money on equipment to minimize pollution, the traditional interaction of managers with environmental issues. As a result, many managers have historically regarded anything related to the environment and sustainability as detrimental to financial performance. However, the opposite is often true, and this anti-environment bias is changing rapidly. Ab Stevels of Philips Electronics has described five ways (slightly revised by us) in which considering the environment and sustainability can be financially beneficial:

1. *Making money by Design for Environment and Sustainability (DfES)*: A primary goal of DfES is to use less of everything — materials, energy, and water — and to emit less — discards, byproducts, etc. All of these actions have the potential to be financially beneficial because they reduce supply costs.
2. *Making money by using the supply chain*: Suppliers provide what they are asked to provide, but enlisting them in green design efforts brings benefits to both supplier and purchaser, such as lower-weight components, elimination of problematic chemicals, and environmentally superior packaging.
3. *Making money by green marketing*: It is a truism that few customers will pay extra for green products, but green attributes such as energy minimization during product use or added durability can tip the scales in the direction of a purchase. Note the transition from simply "green products", which include only environmental concerns, to "sustainable products". The latter not only addresses environmental considerations but, as in the case of energy efficiency, also enhances social benefit (e.g., by providing the customer with an energy-efficient, and therefore preferable, product).
4. *Making money by increasing quality*: A design review focused on environment and sustainability can often minimize complexity and thereby improve product yield, with fewer rejects at either the supplier or manufacturer level.
5. *Making money by paradigm shifts*: Paradigm shifts are changes in mindset — realizing that a design or an operation doesn't need to be done the way it always has been done in the past. DfES activities sometimes initiate these shifts, as in using folded cardboard instead of Styrofoam in packaging (more compact, easier to recycle) or the marketing of services instead of products (enhances takeback, retains customer contact).

The bottom line for corporate actions related to the environment and sustainability is that they need to make economic and/or strategic sense. When these actions are seen to be economically beneficial, they are much more easily integrated into corporate functions, top to bottom.

In this chapter, we discuss how businesses are using industrial ecology concepts, tools, and results to advance corporate sustainability. We have earlier discussed the design of products and the role of businesses in managing their use and end-of-life. Here, we focus on the raw material extraction and manufacturing stages of the life cycle, focusing on (1) management of internal manufacturing operations, (2) management of external operations with suppliers, and (3) management of the firm's role in global system of production and consumption and the role of standards and reporting.

26.2 Greening the Industrial Facility

Each industrial organization that is involved in the processing, combining, and/or assembly of materials and components has a unique pattern of operations dictated by its starting materials, processing sequences, and product spectrum. As a consequence, the starting point for evaluating and optimizing the overall operations begins with the construction of a flow chart that illustrates the overall sequence of operations. Just as with the process flow diagrams used in LCA (Chapter 12), internal environmental managers start with generic diagrams for each process and then link them together to form a full process flow diagram, as shown in Figures 12.1(a) and (b). The next step is to quantify each of the flows based on measurements and purchasing/waste management records, thereby quantifying requirements and generating information relevant to discard and disposal flows, and calling attention to possible environmental and safety hazards.

For example, a facility designed to separate the individual rare earth elements from monazite ore might have a process flow diagram like that shown in Figure 26.1. The sequence begins with the separation of non-rare earths from the ores, as well as the separation and disposal of the radioactive thorium that is generally present. The resulting chemical solution is then forwarded to the sequential extraction of ten or more of the rare earth elements.

The process flow diagrams of a particular industrial facility may be more or less complex than that of Figure 12.1(b), but constructing and quantifying such flow diagrams is essential and has been common in industry for many years. They provide vital information for process and supply chain management and for a variety of facility oversight functions such as life cycle assessment (LCA) (Chapters 11 and 12), material flow analysis (MFA) (Chapter 8), material

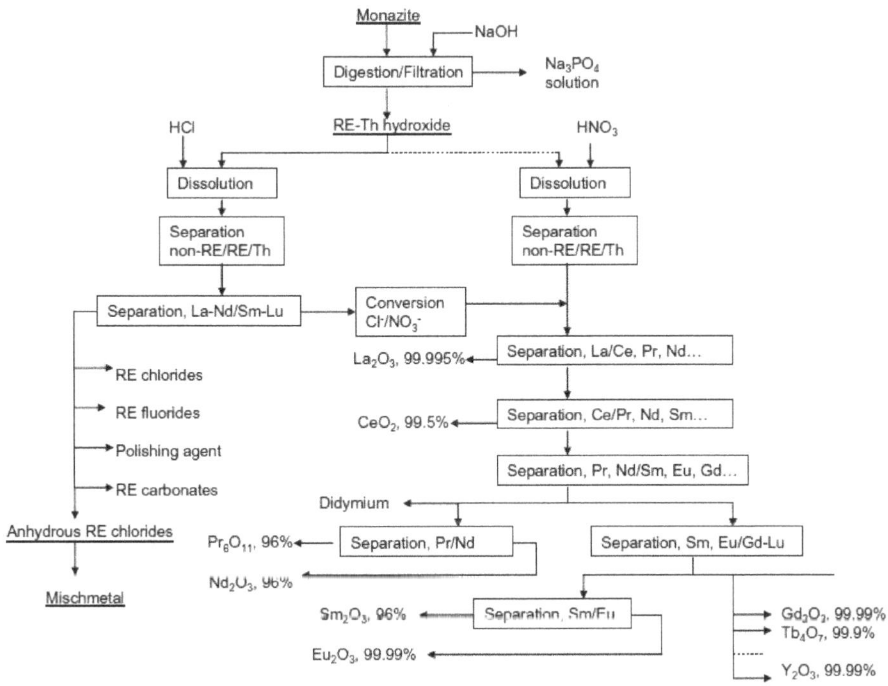

Figure 26.1. A process flow diagram for the separation of rare earth elements (Xie *et al.*, 2014).

recycling (Chapter 9), techno-economic analysis (Chapter 13), and green chemistry/green engineering (Chapter 24). The periodic updating of these evaluations is crucial to the maintenance and improvement of high-quality corporate and environmental performance.

For example, facility managers might see a large flow of electricity required for a specific process and decide to invest in new, energy-efficient equipment. Or they might notice a large waste flow and decide to re-design their process in order to recirculate that flow back into the process. Investments such as these require capital, of course, but often those actions that improve material and energy efficiency also lead to economic savings from avoided feedstock and waste management costs and the output of more salable products.

26.3 Greening the Organization

The kinds of organizational restructuring required to implement the principles of industrial ecology and sustainability in existing firms involve changes in both the internal context — within the firm's existing divisions and organizational entities — and externally, as the firm seeks to achieve new relationships with customers and suppliers. While it is true that developing more socially, environmentally, and economically efficient technologies is an important support for this evolution, it is also true in many firms that the most difficult barriers are cultural, not technological.

Among the obvious challenges is determining where responsibility for various components of an industrial ecology program rests. In some cases, such as traditional environmental compliance, ownership is relatively clear; in other cases, such as the implementation of global worker policies, from pay to safety issues, ownership may be fragmented. Implementing industrial ecology and sustainability in the firm involves the types of activities and programs usually associated with any culture change in complex organizations:

1. Individual champions, who ideally come from several appropriate organizations within the firm and are willing to take the associated risks, must be identified and supported. For example, green accounting issues would be the province of a financial organization champion, while Design for Environment and Sustainability (DfES) programs might be championed by someone from the product research and development group.
2. Barriers to change, especially those which arise from corporate culture and informal patterns of behavior, must be identified and reduced. Support from the highest levels of the corporation, combined with extensive training programs, is essential in this regard.
3. The least threatening method of introducing new techniques, tools, and systems, such as design teams that include specialists from many disciplines, should be identified and used. The best changes are often those that are never recognized by those that implement them.
4. Managers and firms are more likely to be comfortable with practices and procedures that are widely accepted, even if there are some economic penalties involved. For example, an important component of an industrial ecology implementation program at many firms is the reliance on a formal environmental management system, or EMS. EMSs range from the relatively simple to the quite complex, as illustrated by ISO 14000, a set of environmental systems requirements developed by the International Organization for Standardization (see References). In many cases, internally developed EMSs are more effective, more efficient, less bureaucratic, and less expensive than external systems, especially where certification requires extensive audits by outside consultants. Still, because the ISO is a recognized process, it may be easier to implement a 14000 process than to develop an internal EMS.

5. External partnerships and reporting can be an important motivator, but the ultimate goal is making change, not just the reporting itself. Spurred by customers and competitors, and pushed by a cultural shift toward corporate sustainability, many organizations are announcing new environmental, social, and corporate governance (ESG) strategies and reporting on their environmental performance to groups like the Carbon Disclosure Project and the Global Reporting Initiative, using standards such as the Greenhouse Gas Protocol (see References). These actions are increasingly expected among investors and business partners, and are a welcomed shift, enabled in part by industrial ecology tools. Under the maxim that "you cannot manage if you cannot measure it", organizations are undertaking great efforts to account for their energy and water use, waste generation, and especially greenhouse gas emissions. But it is critical that these measurement efforts actually get used for management. A typical approach is to set targets for performance, such as energy intensity for dollar of revenue, or absolute metrics such as total corporate emissions. Scenarios (Chapter 34) can help determine what these targets should be in order to effect meaningful environmental change; a notable approach for setting decarbonization targets is the Science-Based Targets initiative (SBTi) which ensures that pledges adhere to the goal of limiting global warming to less than 2°C, and ideally less than 1.5°C.

Especially for those beginning their corporate career, it is important to remember that most firms are open to environment and sustainability initiatives but that those initiatives must align with the economic interests and internal culture of the firm if they are to be successful. Policies that encourage such alignment, and reward such initiatives on the part of firms, reinforce internal sustainability champions and reward appropriate institutional innovation and are thus important components of any successful corporate governance portfolio.

Philips Electronics' Ab Stevels has discussed three alternative approaches that corporations can take toward issues of environment and sustainability. One is the defensive approach classically adopted by most industry over the years. Driven by legislation, it aims solely for compliance, is invariably costly, and is not industrial ecology. The second approach is cost-oriented. It aims toward evolutionary change, saving money by minimizing resource use and involving suppliers. We might term this "industrial ecology light". The third approach is transformational and proactive, aimed at fully integrating industrial ecology and sustainability throughout the firm. In this era of drastic environmental change, it is obvious that this third approach is what we wish and need for every firm.

Further Reading

Esty, D. C., and Winston, A. (2009). *Green to Gold: How Smart Companies Use Environmental Strategy to Innovate, Create Value, and Build Competitive Advantage*. New York: John Wiley & Sons.

Carbon Disclosure Project (n.d.). Retrieved from: https://www.cdp.net/en.

Greenhouse Gas Protocol (2015). *A Corporate Accounting and Reporting Standard*. Retrieved from: https://ghgprotocol.org/corporate-standard, and https://ghgprotocol.org/sites/default/files/standards/ghg-protocol-revised.pdf.

Global Reporting Initiative (n.d.). 25 years of empowering sustainable decisions. Retrieved from: https://www.globalreporting.org/.

Graedel, T. E., and Howard-Grenville, J. (2005). *Greening the Industrial Facility*. New York: Springer.

International Organization for Standardization (ISO), ISO 14001 — Environmental Management. Retrieved from: https://www.iso.org/iso-14001-environmental-management.html.

Science Based Targets (n.d.). Ambitious Corporate Climate Action. Retrieved from: https://sciencebasedtargets.org/.

Stevels, A. L. N. (2007). *Adventures in Ecodesign of Electronic Products*. The Netherlands: Delft University of Technology.

Xie, F., Zhang, T. A., Dreisinger, D., and Doyle, F. (2014). A critical review on solvent extraction of rare earths from aqueous solution. *Minerals Engineering*, *56*, 10–28.

Industrial Ecology of Socio-Technical Systems

Chapter 27

Telecoupling in Industrial Ecology

Chapter Concepts

- Telecoupling is the study of human and natural interactions over distances near and far
- Telecoupling enables the interchange of natural and industrial products that have the potential to improve people's lives but also results in the generation and translation of environmental impacts
- From an industrial ecology perspective, telecoupling presents challenges in acquiring and interpreting data, but the process provides insights not otherwise available

27.1 Introduction

Organisms, biological and industrial, survive and prosper by utilizing resources close at hand — a tasty leaf for a rabbit, rich loam for a farmer. For humans, this localized picture took on a different aspect with the invention of the watercraft, for then both people and goods could be transported over distances previously difficult or impossible to imagine. The means of transport have, of course, evolved over time, and today routinely include trucks, trains, aircraft, and perhaps the biggest game-changer of them all, the container ship. No longer is a valued resource in a distant country beyond one's grasp — it can readily be transported to those who desire it and can afford it, no matter where the source or the destination may be located.

On a local scale, human and natural systems are coupled by their employment of resources and the consequential implications of those actions. The study of human–natural interactions over a distance is termed *telecoupling* ("tele" reflecting distance, "coupling" reflecting interaction, from the original use of the term to refer to voice connectivity enabled by the telephone). The implications of telecoupling are not merely significant; they have become transformative, allowing large populations to live where local resources are not sufficient to sustain them, manufacturing products in one part of the world and selling them in other parts, and generating environmental damage in distant regions for the purposes of satisfying demand that exists elsewhere.

Because the modern world is extensively telecoupled, and because much of that telecoupling involves materials of interest to industrial ecologists, the study of the implications of the telecoupling of resources has become a part of industrial ecology. In this chapter, we discuss the general aspects of telecoupling, the implications thereof, and the analytical approaches that have demonstrated the benefits and encumbrances of the telecoupling process.

27.2 Telecoupling Methodology

The process of telecoupling can be illustrated in a simplistic way in Figure 27.1. A product is produced in one locale, creating some local and perhaps regional environmental impacts. The product is shipped by any of a variety of means to a second location, where it is put into use and the purchase price is sent to the supplier. After an appropriate period of time, the product exits use. At that stage and perhaps also during use, a variety of environmental impacts may occur at the receiving location.

As with much of industrial ecology analysis, the devil is in the details. How were the data acquired? Governmental or intergovernmental agencies are possibilities, as are industrial associations, individual corporations, databases assembled by researchers, and other sources. How reliable are the data? Reliability can

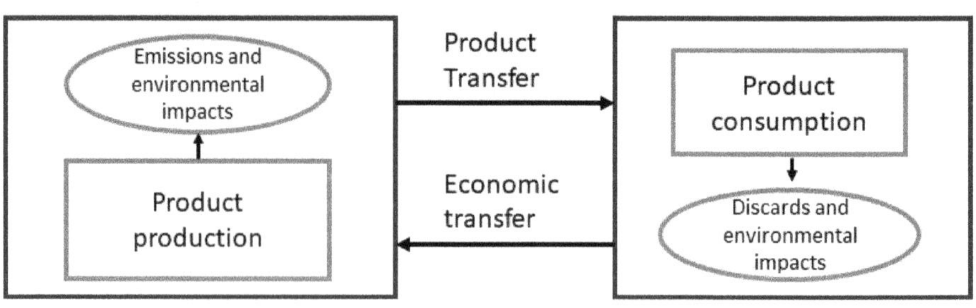

Figure 27.1. Defining telecoupling by specifying flows of materials and/or environmental impacts from one place to another. Note that immediate environmental impact occurs at the point of origin while the delayed environmental impacts occur at the point of receipt.

vary widely. In a telecoupling study, suitable data must come from both the originating and receiving locations. This requires, of course, that a suitable mechanism has been set up to monitor flows at the appropriate entry and exit points. Such arrangements almost universally exist at national entry and exit points, and the United Nations (UN) Comtrade database is thought to be relatively accurate, but the categories may not be exactly what the analyst wishes.

At scales smaller than that of nations, data must be acquired by more innovative means. Estimates have involved interviews with local businesses and residents, motor vehicle registration information, and analyses of satellite images. In the upcoming chapter on urban material flows in Singapore, in fact, the research team attempting to quantify changes in in-use stocks of building-related materials surveyed those material flows throughout the city by conducting driver surveys for trucks that pick up and deliver construction materials.

It is often the case that we wish to focus on products that contain materials rather than the materials themselves. This can be done (again using UN Comtrade) for specific products or for groups of products.

The methodology then can be expressed in a series of steps:

1. Specify the resource(s) of interest.
2. Define the boundaries — What spatial entity (country, region, etc.) produces the resource? What spatial entity receives the resource?
3. Is the resource embodied in one or more products? If so, what is the resource quantity per unit of product?
4. What data are available to quantify the telecoupling?
5. Are there "spillover" effects, i.e., influences occurring in other than the sending and receiving entities as a consequence of the telecoupling?

A final data need in a telecoupling analysis is information on the environmental impacts that result from the manufacture of products for export. There are two challenges here. The first is sufficient understanding of crop production, manufacturing, or other activities that generate export flows (often quite challenging, especially for researchers in other locales), and the second is still more challenging: to quantify the related emissions. In some countries and regions, data on emissions by product sector may be available. If not, one must estimate the required information as best one can.

27.3 Embodied Flows of Materials

The following sections discuss telecoupling analyses for a number of different targets of study as examples, not only of the diversity of telecoupling but also of some of the approaches that have been used to quantify, express, and display telecoupling information.

27.3.1 *The telecoupling of food*

The importance of quantitatively understanding the implications of actions at a distance can be illustrated by food production, transport, and consumption. Some parts of the Earth are particularly suited to the growing of crops, the raising of animals, or the harvesting of fish or shellfish. Human populations, however, may thrive elsewhere. The transport of food and related technological products enables human populations to live where they choose, in regions of our planet that would otherwise be sparsely settled.

The quantitative production and trade information has been recorded over many years by the Food and Agriculture Organization (FAO) of the UN. Using FAO data, Porfirio *et al.* (2018) studied the transport of four principal food grains (wheat, rice, coarse grains, and oilseeds — four key commodities that provide some two-thirds of global caloric intake) from and to 14 global regions. The results are presented in a cloropleth diagram in Figure 27.2. (This is a diagram that requires careful study but contains much useful information.) Several features of the results are readily apparent: (1) the European Union (EU) largely feeds itself by intercountry transfer, (2) the U.S. is almost exclusively a food exporter (to East Asia, China, Latin America, and non-U.S. North America), not an importer, (3) Latin America indeed imports grains from the U.S., but exports grains (perhaps

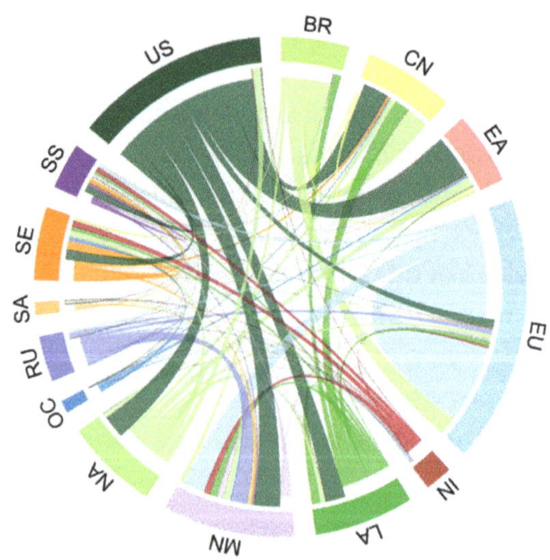

Figure 27.2. The global trade of food grains in 2015 (Porfirio *et al.*, 2018).

different ones) to China and elsewhere. Other exchanges of grain throughout the world can also be qualitatively derived from the diagram and assessed quantitatively through the datasets cited in the publication.

27.3.2 *The telecoupling of water*

For food to be grown, of course, water for irrigation is generally required. The water can be accessed from precipitation ("green water") and ground and surface water ("blue water"), and polluted water ("gray water") is generally produced. Around 20% to 25% of all water used worldwide goes to the growing of food. The demand for food to feed the planet's growing population is such that the extraction of surface and (especially) groundwater occurs at rates that threaten long-term supply.

When food raised in one place is exported to another, a physical product undergoes relocation. Food cultivation is enabled by the use of water for irrigation, but the water is not directly transported from the originating region to the receiving region. Rather, that water (known as virtual water) is what enabled the production of the food that is consumed elsewhere. The largest virtual water flows from sending to receiving countries are shown in the diagram in Figure 27.3 (Hoekstra and Mekonnen, 2012). The large flows can readily be seen: U.S. to Mexico, U.S. to Japan, Argentina to Brazil, Canada to the U.S., and so forth. In a second diagram in this research, these and other flows are broken down by product category: sugar crops, fibers, fruits, and so forth. Some crops can be deduced to have a greater water footprint than others — important information in a time of growing global population and higher incomes producing enhanced demand for water-intensive food of all types.

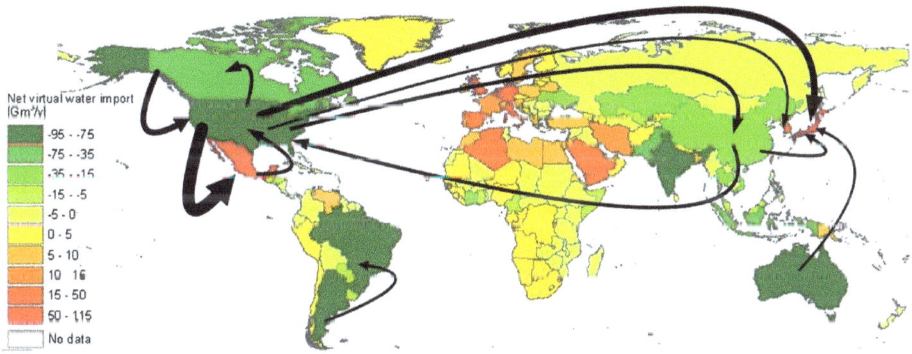

Figure 27.3. The global trade of embodied ("virtual") water related to trade in agricultural and industrial products, summed from 1996–2005 with only the largest flows shown (Hoekstra and Mekonnen, 2012).

27.3.3 *The telecoupling of materials*

As databases have expanded in recent years, it has become possible to study the telecoupling of a variety of trade flows other than those related to food. The diagram in Figure 27.4(a) is similar to that for food trade (Figure 27.2) but depicts value-added exports of metal products. The data are from multiregional

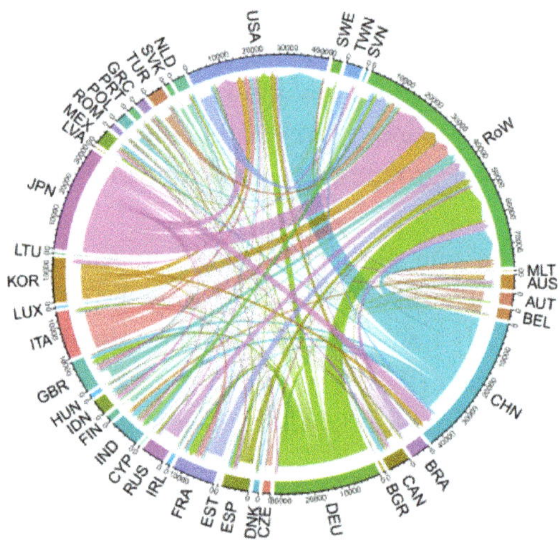

(a)

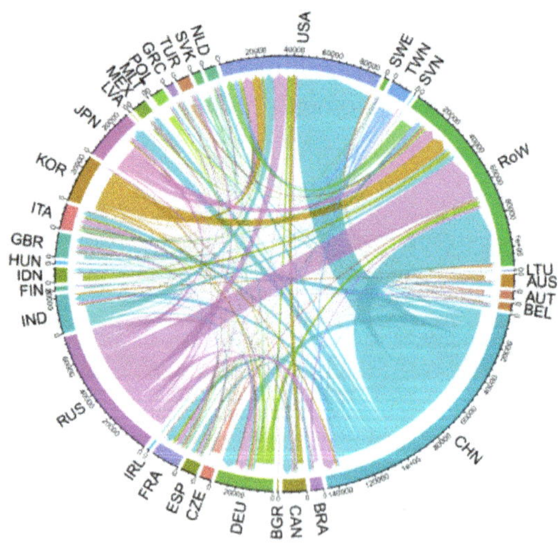

(b)

Figure 27.4. (a) Metal products in world trade, 2009. (b) Embedded greenhouse gas (GHG) emissions related to metal products in world trade, 2009 (Xiong *et al.*, 2018).

input–output tables, specifically that of the World Input–Output Database (WIOD). Any of the trade flows included there could have been studied, but this analysis applies strictly to "basic metals and fabricated metal products". The results are of particular interest because many could probably not have been predicted in detail. Japan is seen to be the world's largest net exporter of metal goods, followed by Germany and China. The U.S. is almost entirely a recipient. Italy and South Korea are also major participants, largely as exporters.

In addition to being of interest to industrial ecologists, such diagrams have high importance to businesses and governments because they reveal potential opportunities for industrial growth or governmental agreements related to the material flows that have been analyzed.

27.3.4 *The telecoupling of environmental impacts*

As mentioned earlier and in Chapter 10, the fairly recent creation of multiregional input–output tables has helped greatly in analyzing telecoupling flows. More recently, those tables have been extended with environmental impacts in mind to create the Environmentally-Extended Multiregional Input–Output (EE MRIO) tables. The results require assumptions and estimations, of course, but are regarded as reasonably accurate and quite useful for environmental impact assessments related to telecoupled flows.

An example of a telecoupled evaluation is that related to the environmental impacts of the metal flows study of the international transport of metal products. Using EE MRIO tables, the authors generated the flow data illustrated in Figure 27.4(b). The arrows on the diagram refer to the importing countries responsible for the manufacturing and related impacts and express in this case the greenhouse gas (GHG) emissions that occurred in the exporting countries. The results are quite dramatic. China is the source of about half of the GHG emissions embedded in products. Russia is also a major source, with a number of other countries being significant as well.

The manufacturing resulting from out-of-country product demand may not only influence the manufacturing country. Figure 27.5 illustrates regional air pollution impacts in the northern hemisphere that can be credited to Chinese manufacturing of goods destined for export to the U.S. (The analysis combined manufacturing data with atmospheric chemical transport modeling.) In the cases of sulfate (upper left) and ozone (upper right), the major impacts occurred within China. For carbon monoxide (lower right), however, and especially for black carbon (lower left), there were significant environmental implications for the eastern Pacific Ocean and the western edge of the U.S. Country-to-country impacts in

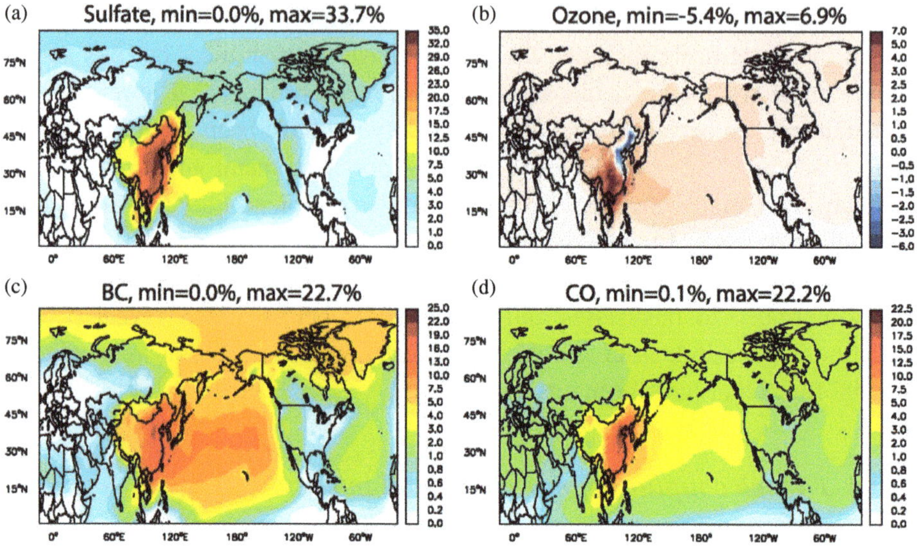

Figure 27.5. Fractional surface concentrations of four pollutants resulting from Chinese production of goods for foreign export. The results are for 2006 and combine economic and emissions data with results from atmospheric chemical transport modeling. The dominant air flow direction is from Asia to western North America (Lin *et al.*, 2014).

more geographically constrained regions (Europe, South Asia, etc.) would be expected to be even more severe.

27.4 Spillover Effects

Spillover effects are environmentally-related impacts in a third locale that result from a telecoupled flow between two other locales. Not all spillover effects are amenable to quantification, but some are. In one well-characterized example (Figure 27.6), exports of soybeans from Brazil to China, Spain, Thailand, and the Netherlands were possible only because of exports of potassium fertilizer from Canada, Belarus, Russia, and Germany to Brazil, with associated environmental impact in the fertilizer-supplying countries and from global transport.

Transport emissions are an important source of spillover effects in general because they occur in intermediate locations. Researchers have mapped out direct emissions of global trade from road, rail, pipeline, barge, and ship. For example, shipping contributes about 3% of global GHG emissions (Figure 27.7) and is an important source of harmful air pollution, particularly affecting residents who live nearby busy ports.

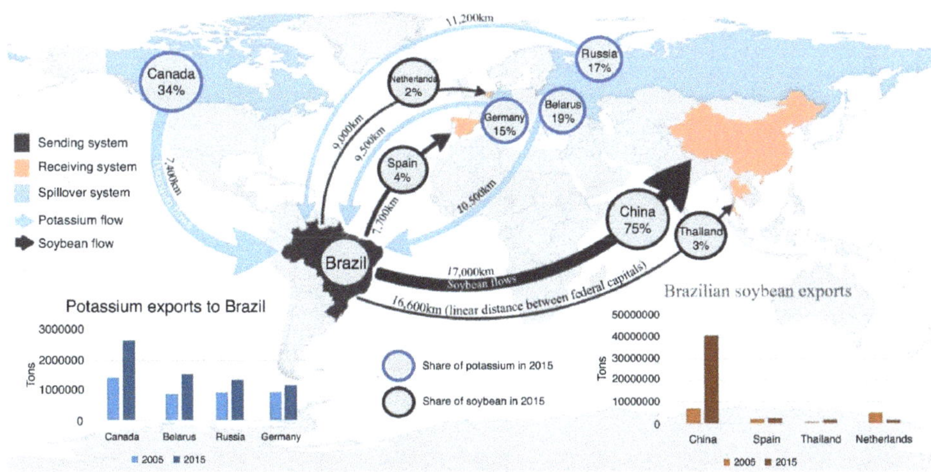

Figure 27.6. Flows of soybeans from Brazil to Spain, China, and Thailand, and the spillover systems providing potassium to Brazil (Liu *et al.*, 2018).

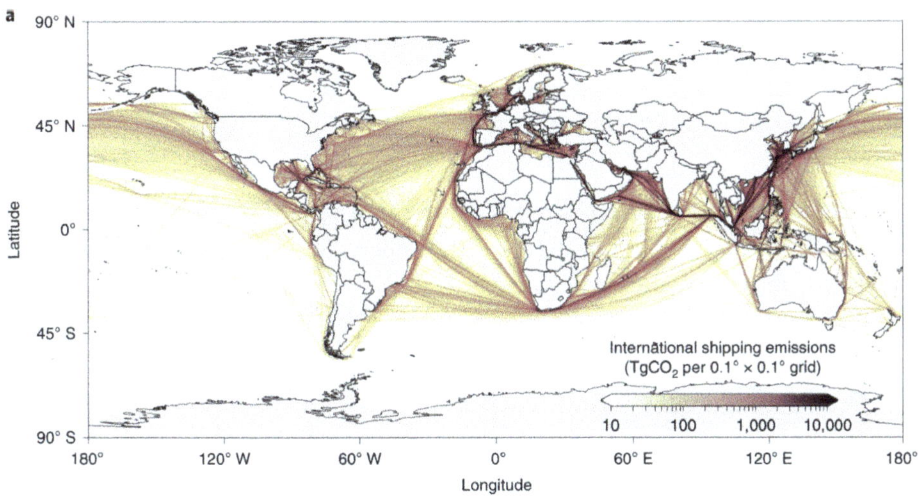

Figure 27.7. Carbon dioxide emissions from ocean shipping where the color intensity indicates the level of emissions (Wang *et al.*, 2021).

27.5 Dynamic Analyses of Trade-Related Environmental Impacts

With the increased standardization of data in national statistical offices over time, it has become possible to examine telecoupling implications over time. Wood and colleagues did so for the period 1995–2011, taking advantage of the multiregional

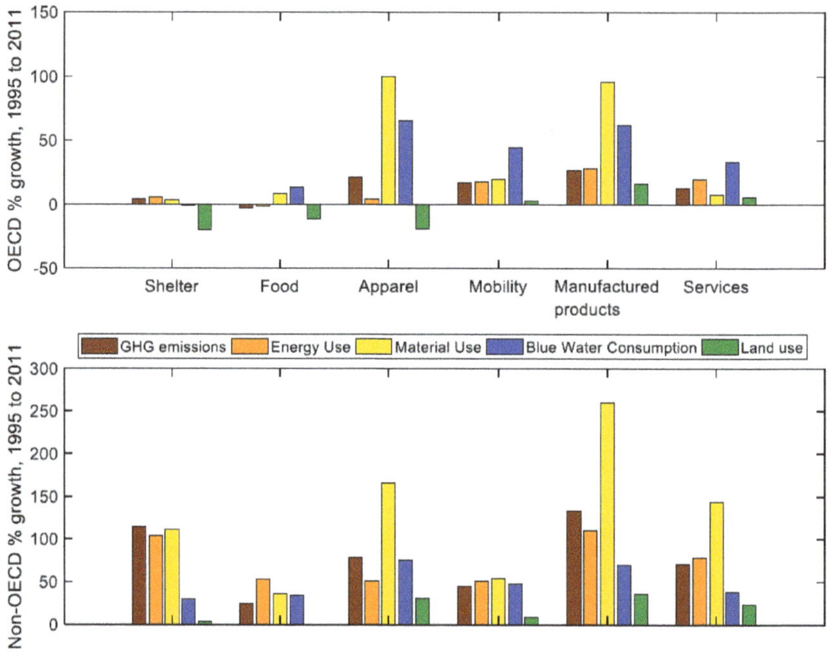

Figure 27.8. Growth in environmental impacts from 1995 to 2011. OECD is the Organization for Economic Cooperation and Development (Wood *et al.*, 2018).

input–output database EXIOBASE 3 (a database discussed by Stadler *et al.* (2018). Their results from the major effort are encapsulated in Figure 27.8. where percentage growth in a variety of environmental impact categories related to trade are presented for the OECD's 37-country group and for the remainder of the world's countries (generally the poorer and less politically stable). Several features of the data are of particular interest:

- Percentage growth in virtually all environmental footprint categories is obvious, with material use, GHG emissions, and energy use being generally the highest.
- The 2011 footprints in all categories and for all purposes are higher in non-OECD countries than in the OECD countries. In OECD countries, the footprints for shelter and food are actually little changed or a bit lower than was the case in 1995.
- The most rapid increases are seen for apparel and manufactured products, perhaps reflecting a shift into higher volumes of cheaper goods.

Results such as these demonstrate that environmental impacts related to international trade and its expansion over time can be well quantified and will increasingly be part of global discussions related to the environmental impacts of global development.

27.6 A Telecoupled World

Connections at a distance, small or large, directly related to industrial activity or not, are increasingly recognized in data sets derived for many purposes. The observing systems that are employed, various data repositories that have been developed, and impact quantification that has emerged and/or improved over the last several decades increasingly reveal a telecoupled world. Some of the connections are problematic. In others, the teleconnections may be beneficial, as when efficient large-scale manufacturing can be shown to be preferable to multiple small-scale, inefficient operations. In any case, where telecoupled relationships arise from industrial activity, as many do, the location of sources and the quantification of implications are within the scope and analytical capabilities of industrial ecology. Whether our telecoupled world is beneficial or problematic, it is with us to stay.

Further Reading

Carlson, A. K., Taylor, W. W., and Liu, J. (2019). Using the telecoupling framework to improve Great Lakes fisheries sustainability. *Aquatic Ecosystem Health and Management, 22* (3), 342–354.

Hoekstra, A. Y., and Mekonnen, M. M. (2012). The water footprint of humanity. *Proceedings of the National Academy of Sciences, 109* (9), 3232–3237.

Hovis, C. L., Dou, Y., Herzberger, A., and Liu, J. (2021). Through the lens of telecoupling and metacoupling: New perspectives for global sustainability. *Sustainability, 13*, 2953.

Lin, J., *et al.* (2014). China's international trade and air pollution in the United States. *Proceedings of the National Academy of Sciences, 111* (5), 1736–1741.

Liu, J., *et al.* (2018). Spillover systems in a telecoupled Anthropocene: Typology, methods, and governance for global sustainability. *Current Opinion in Environmental Sustainability, 33*, 58–69.

Porfirio, L. L., Newth, D., Finnigan, J. J., and Cai, Y. (2018). Economic shifts in agricultural production and trade due to climate change. *Palgrave Communications, 4*, 111.

Stadler, K., *et al.* (2018). EXIOBASE 3: Developing a time series of detailed environmentally extended multi-regional input-output tables. *Journal of Industrial Ecology*, *22* (3), 502–515.

Wang, X. T., *et al.* (2021). Trade-linked shipping CO_2 emissions. *Nature Climate Change*, *11* (11), 945–951.

Wiedmann, T., and Lenzen, M. (2018). Environmental and social footprints of international trade. *Nature Geoscience*, *11*, 314–321.

Wood, R., *et al.* (2018). Growth in environmental footprints and environmental impacts imbedded in trade. *Journal of Industrial Ecology*, *22* (3), 553–564.

Xiong, H., Millington, J. D. A., and Xu, W. (2018). Trade in the telecoupling framework: Evidence from the metals industry. *Ecology and Society*, *23* (1), 11.

Chapter 28

Islands: Geographically Constrained Systems

Chapter Concepts

- Because island ecosystems are largely isolated from external influences, they are useful in studies of species interactions and resource use and loss
- The capabilities of monitoring material stocks and flows in systems largely decoupled from other systems make islands also a good resource for industrial ecology studies
- Simple industrial systems mimic the advantages of islands as objects of study given their restricted input and output flows

28.1 Islands as Systems of Study

Islands have been recognized as units ideally suitable for study since the time of Charles Darwin in the mid-1800s. Darwin took advantage of what Peter Vitousek (2002) terms a *model system*: a system that displays a general process or property of interest in a way that makes it easier to comprehend than more complex versions of itself. Industrial ecologists are interested in islands because they offer tractable approaches to the study of such topics as material inputs, material types and uses, energy generation and use, the generation and fate of waste, and many more.

Most cities and countries of interest to industrial ecologists for one reason or another are large, multifaceted, and challenging to analyze. London has scores of entry points through which resources flow, for example. Quantitative data may exist in the records of trucking companies, railyards, airports, and the like, but the data are largely proprietary and thus immensely challenging and time-consuming to acquire and validate. Similar challenges exist with respect to islands, of course, but ports of entry are few, the materials spectrum is often less diverse, the incoming and outgoing shipments easier to follow, and the necessary personal contacts tend to be easier to arrange. In many respects, islands are as much model systems to the industrial ecologist as they are to the ecosystem ecologist.

28.2 Material Flow Data for Islands

In Chapter 8, material flow analyses (MFA) were shown to depend upon the availability of numeric data on material stocks and flows within and external to a well-defined entity such as a country. However, little was said in the chapter about the challenges involved in acquiring and evaluating the data necessary for the analysis. As one might imagine, some data are easier to acquire than others, and some are easier to evaluate than others. In these respects, islands appear to have the potential to be useful locations for industrial ecology analyses.

Until perhaps the beginning of the 20th century, it was far from routine to travel to most islands not located near land masses. As a consequence, such islands tend to be lightly populated, with inhabitants who tended to live largely off their surroundings — forests for trees, fish from the sea, and so forth. In such a setting, there was minimal materials diversity and minimal material exchanges with the outside world. In a few places such conditions have existed until quite recently. A well-characterized example is that of Trinket, one of the Nicobar Islands in the Bay of Bengal. In 2000–2001, graduate student Simron Singh spent considerable time on Trinket, an island of some 3,600 hectares and with about 400 inhabitants. He measured the material stocks on the island and worked with the small number of importers and exporters to quantify the material flows to and from Trinket. Singh's results are shown in Figure 28.1; they reveal that little material crossed the island boundary, and almost all that did consisted of minimal amounts of fossil fuels, and minerals (imports) and copra (export).

Singh was able to accomplish his material flow analysis because of his dedication to the task but especially because the island was small, the population was light, and imports and exports were minimal and easily categorized. Other islands and island systems are more complex, as we shall see, but still have many of the same characteristics of the place that enabled Singh's work.

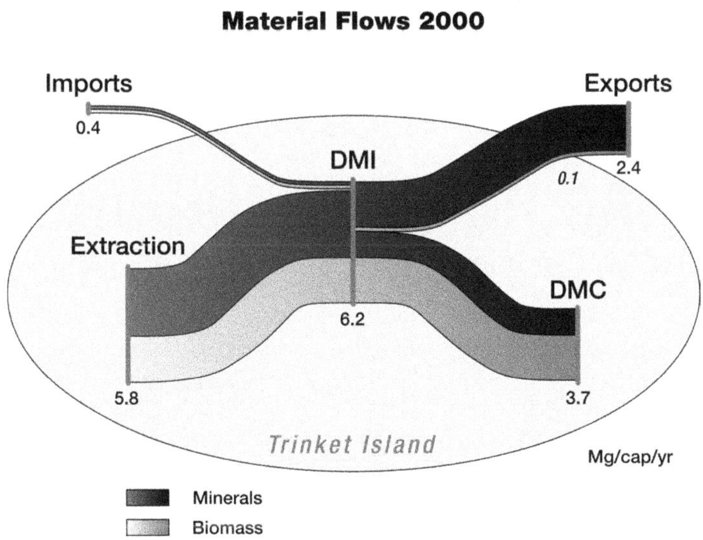

Figure 28.1. Flows of materials and energy on Trinket Island, Indian Ocean, 2000 (adapted from Singh and Haas, 2016).

28.3 Islands in Transition

One of the first transitions that tends to occur on islands that for one reason or other are moving from isolation to increased interaction with the rest of the world is that they acquire electricity. It is generally challenging to bring this about, however. Most oceanic islands are distant from resources such as fossil fuels, which have traditionally been the enablers of an electricity grid, no matter how small the grid may be. In the Galapagos Islands, well to the west of Ecuador, the initial electrical sector included only San Cristobal (the islands' capital) and the Charles Darwin Research Station on Santa Cruz Island. By 2007, the four central islands had been electrified, all by diesel fuel systems. A decade later, diesel fuel was still providing 85% of the electricity in the islands. The long-term plan is to transition to a system of 100% sustainable energy (wind power, solar power, and perhaps wave energy). The pace of the transition will depend largely on the level of monetary support provided by the Ecuadorean government (Llerena-Pizarro *et al.*, 2019).

A different challenge to transitional development is presented by islands that are, from time to time, in the paths of extreme weather events. The island of Grenada in the Caribbean region provides a case study. Travel and tourism are

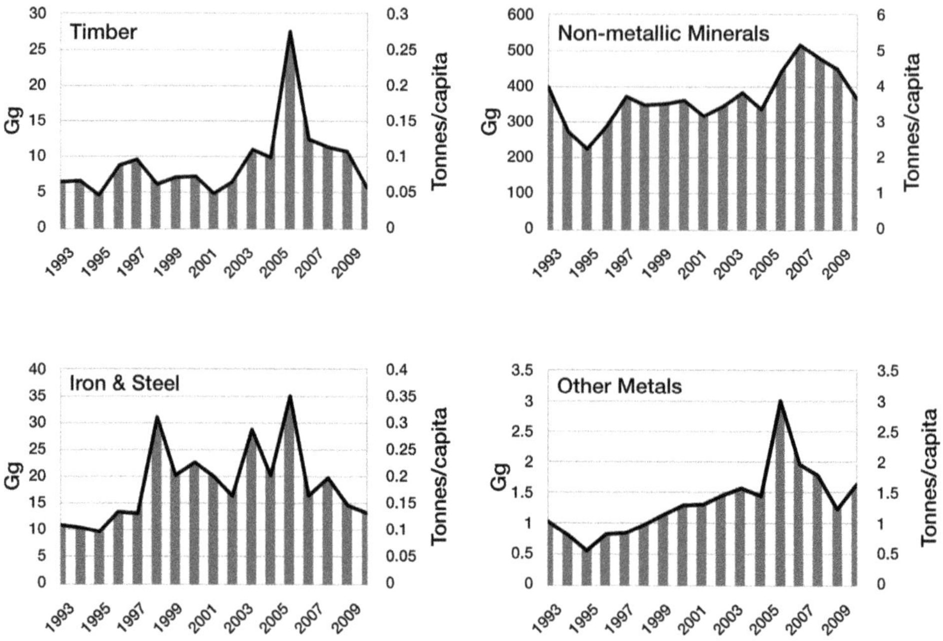

Figure 28.2. Gross additions to in-use stocks from 1993 to 2009 in Granada in the Caribbean. Hurricane Ivan in 2004 stimulated enhanced stock buildup in 2005. Absolute values are given in the left axes and per capita values in the right axes; gross addition to stock (GAS) (adapted from Symmes *et al.*, 2020).

important to the Grenadian economy, and thus tourist accommodation and associated infrastructure are important as well. A material flow analysis for Grenada has demonstrated the irregular change in the stocks of construction materials (Figure 28.2). In 2004, Hurricane Ivan caused severe damage, and construction material imports and stock additions sharply increased in response. The material flow analyses that have been conducted over the years provide useful quantification should such a response again be required in the future (as is likely).

Island studies of ecological systems often benefit from the stability of the ecosystem, whereas industrial ecology studies of islands tend to focus on systems where rapid change is occurring. Island systems in flux are not only dealing with materials and energy but also with the resident human populations, their traditional lifestyles, and their aspirations. Studies of that type within industrial ecology fall within the sociometabolic framework discussed in Chapter 35. All islands and all island inhabitants are involved in sociometabolic processes, the pressures of which are generally increasing over time. In some cases, change is resisted by island inhabitants even if their traditional lives are difficult. A case in point is that of the Greek island of Samothraki. As described by Fischer-Kowalski *et al.* (2020), traditional agriculture provides a poor living for the Samothraki, especially as livestock numbers in recent years have increased past the possibility of the land to support the herds (recall the Tragedy of the Commons story in Chapter 1). Nonetheless, a major transition to a focus on tourism was not embraced with enthusiasm by the islanders.

A diametrically opposed transition is that of the Hawaiian Islands, where a major transition from agriculture to tourism has provided many jobs and other benefits but has largely neglected the impacts of the changes on the native Hawaiians and on the islands themselves. Statistics suggest the magnitudes of the transition. With a mostly stable population of about one million people, Hawaii's million people hosted over 10 million visitors in 2019, with major impacts on land use, traditional land holdings, employment, and the environment. Not all of the resources needed to support Hawaii's economy can be produced on the islands, and so imports make up the shortfall. Figure 28.3 shows results from an MFA for the Hawaiian island of Oahu in 2005. The largest import category by far is fossil fuel needed to generate electricity, followed by finished goods, the large variety of which the island's economy does not have the manufacturing base to produce.

28.4 Dealing with Island Outflows

Before modern sources of transportation made islands easily available, domestic agricultural practices generated food and reused the nutrients retained in agricultural and biological wastes. In the 21st century, however, most islands are not agriculturally self-sufficient, so waste material must be dealt with in other ways. In addition, the spectrum of waste on islands has expanded to include metals, plastics, and a variety of synthetic materials that require other than traditional waste treatment. The question then becomes how islands in transition can and should approach waste management.

Island geography turns out to be a significant factor in waste management decisions. Islands generally (though not always) are one of two topographical types, either mountainous (which restricts most of the population, industry, and agriculture to the lower ground near the sea) or flat, with a limestone-based geology characterized by shallow water tables (again providing constraints to the land area suitable for human use). A significant aspect of these situations is that the competition for suitable land makes it very difficult to site new landfills. In addition, the shortage of land means that suitable disposal or incinerator sites are likely to be in the proximity of housing or tourist activities, often making disposal decisions problematic.

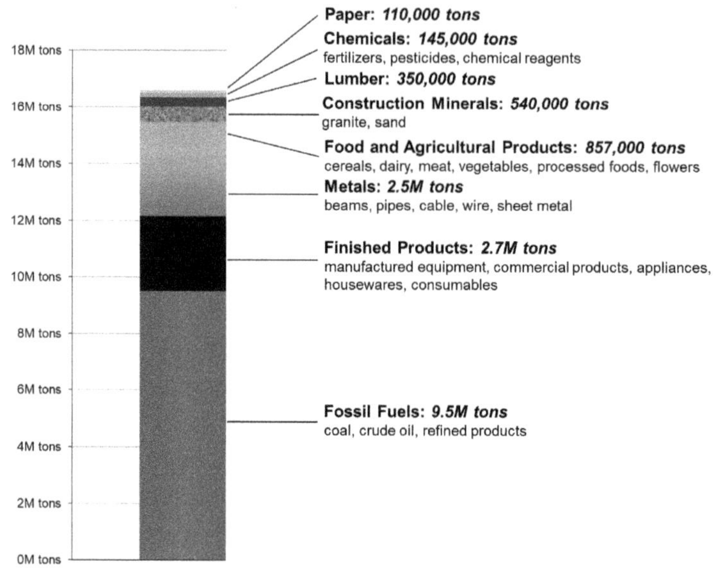

(a)

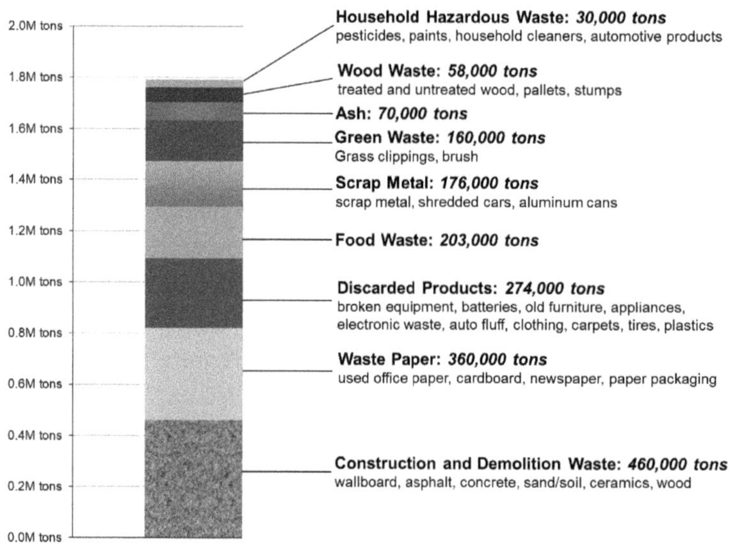

(b)

Figure 28.3. (a) Island-wide net import material flows for Oahu, Hawaii, 2005, adjusted for re-export to other islands; (b) island-wide waste flows for Oahu, Hawaii, 2005 (Eckelman and Chertow, 2009).

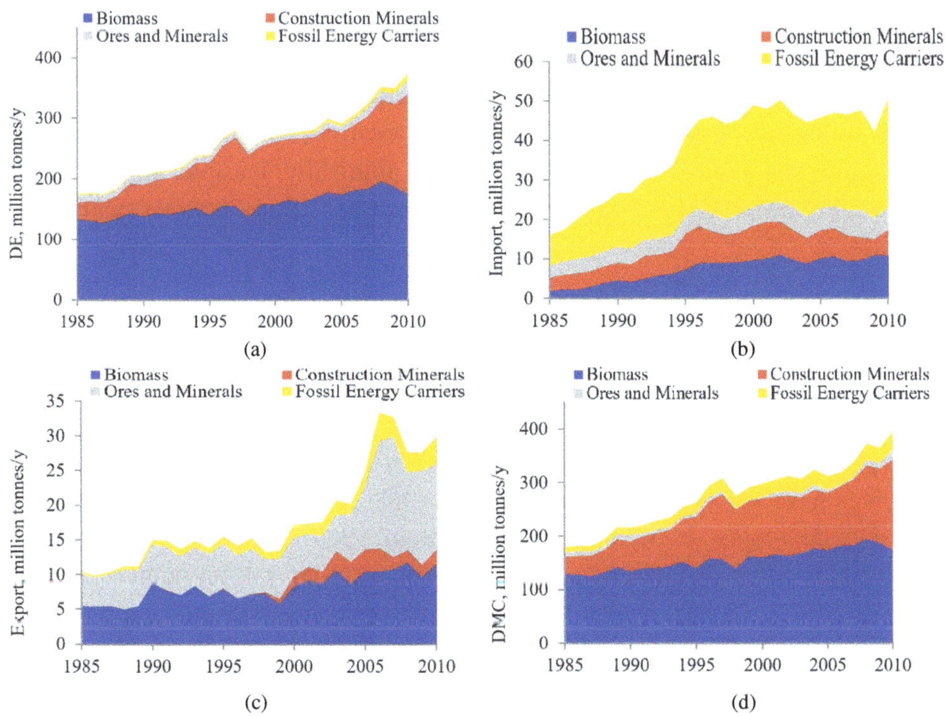

Figure 28.4. Material flows in the Philippines, 1985–2010: (a) domestic extraction; (b) imports; (c) exports; (d) domestic material consumption. Note the different scales on the four plots (Martinico-Perez *et al.*, 2017).

A perspective on island waste management is provided by a study of waste flows on the Hawaiian island of Oahu (Eckelman and Chertow, 2009). Data for 2005 are shown in Figure 28.4. The largest flow is for construction and demolition waste, a reflection on the rapid expansion of the tourist industry. Nearly as large, however, is the flow of waste paper — office paper, cardboard, and other packaging. Discarded product flow was also significant — an indirect signature of the transition from an agrarian to an industrial economy.

A more general look at waste management on islands was performed by Eckelman *et al.* (2014). They point out the complexities not only of limited land resources and social and political dynamics, but also of high energy costs and large seasonal fluctuations in waste volumes. Waste flow results for 41 islands indicated that per capita waste generation was weak for land areas, population density, and present agricultural land, but was strong for gross domestic product

(GDP) per capita in purchasing power parity (PPP). Thus, affluence is a useful determination of waste generation rates, but little other useful results emerged.

All in all, it is difficult to see ready solutions for waste management challenges on islands, especially those in remote locations. It is clear, however, that addressing this issue in detail and forging politically feasible approaches is increasingly necessary as island populations and per capita incomes continue to increase.

28.5 Island Nations and Continents

28.5.1 *The island country of Singapore*

Singapore is a modest-sized island connected by bridge to Malaysia. Its principal resources have always been a natural harbor and a central location along major global trade routes. It became independent in 1965 and has since developed rapidly, expanding its GDP by some 300 times from its starting point in 1965 to the year 2015, largely by reliance on import-export trade and export-driven industrialization.

Schulz (2007) makes the case that for Singapore and probably for other cities undergoing rapid transitions that the level of material use can be directly linked to that of urban development. If this is a general case, as seems likely, the development of cities and urban populations throughout the world (see Chapter 33) will require major flows of material resources in the next several decades.

28.5.2 *The multi-island country of the Philippines*

Just as the Hawaiian Islands are comprised of several islands with both inter-island and external-island flows, the Philippines is a composite of islands but on a wholly different scale: more than 7,000 islands, some 2,000 of them inhabited. As a country interested in its material flows, the Philippines maintains an extensive set of statistical yearbooks of resource extraction and materials trade, on which an industrial ecology analysis can be based. These data have permitted an extensive material flow analysis for the period 1985–2010 for the major material groups (Martinico-Perez *et al.*, 2017). The results for domestic extraction, imports, exports, and domestic material consumption are shown in Figure 28.4; they demonstrate a doubling of construction mineral extraction over the period. Imports rose during the end years of the 20th century but were relatively stable thereafter. Exports rose sharply from 2005 onward due to the expansion of nickel ore shipments. The combination of these factors resulted in a continued increase in domestic material consumption.

28.5.3 *The island continent of Antarctica*

The only continent for which a material flow analysis can be conducted is Antarctica, and that opportunity exists because permitted activities on the continent are defined by the Antarctic Treaty, which was signed in 1959 by all nations active on the continent. The treaty stipulates that Antarctica can be used only for peaceful purposes such as scientific research, and mandates the keeping of records of activities on the continent. Drawing upon that information, Klee (2001) conducted a 1998 material flow analysis for Scott Base, New Zealand's Antarctic research station. He found that there was no recycling of materials taken from New Zealand to Scott Base, although some of the material was transshipped, perhaps for later recycling.

In another industrial ecology study in Antarctica, Crossin *et al.* (2020) conducted a life cycle analysis of the environmental impacts of operating Australia's Casey Station in 2015. They then studied the impact reduction potentials of six potential improvement scenarios. As Table 28.1 shows, the scenarios predicted reductions of between 2% and 19%, depending on the level of deployment.

These results demonstrate that both material flow analysis (MFA) and life cycle assessment (LCA) can be applied to islands and their entities, provided the needed data can be acquired.

28.6 Industrial Facilities as Islands

An island is a good industrial ecology laboratory if it has few entry and exit locales and the flows through those locales can be monitored and quantified. But a system need not be an island to share these characteristics. For example, remote cities and settlements can also be analyzed by drawing a geographic boundary around them and examining the limited points where there are inputs and outputs. Industrial ecologists have applied this kind of approach to studying even research stations on Antarctica! Industrial facilities, too, can be examined in this way, where input flows, material transformations, and output flows are routinely monitored for logistical, financial, and regulatory reasons (see Chapter 26).

Thinking of a business as an island that receives, transforms, ships, and discards materials provides a useful mental image of a limited environment within which material transformations and implications can be better understood and improvements devised. The approach need not be restricted to isolated industrial facilities, however. Indeed, a multi-facility business might be regarded as resembling the Hawaiian Islands in concept, with inter-island flows offering the opportunity to optimize connections within the business itself. Indeed, such an analysis was carried out by Toyota in 2007. Material flow analyses of islands certainly provides the opportunity to better understand and perhaps optimize individual and

Table 28.1. Potential Reductions in Environmental Impacts from Australia's Casey Station in Antarctica (Crossin *et al.*, 2020). The Baseline Data are for Year 2015.

Impact category	Unit	Baseline	Temperature setpoint reduction	Building benchmarking	New icebreaker	Solar at Casey Station	Solar at Wilkins Aerodrome	Load leveling battery	Combined scenarios
Climate change potential	tonne CO_2 eq	4,784	4,766	4,709	4,046	4,766	4,772	4,772	3,909
	Saving		0.4%	1.6%	15.4%	0.4%	0.3%	0.3%	18.6%
Cumulative energy demand	GJ LHV	59,711	61,313	59,489	58,829	50,057	59,497	59,560	59,563
	Saving		0.4%	1.5%	16.2%	0.4%	0.3%	0.2%	19.1%
Marine eutrophication	kg N eq	1,257	1,257	1,138	1,257	1,228	1.249	1,237	1,079
	Saving		<0.1%	9.5%	—	2.3%	0.7%	1.6%	15.8%
Marine ecotoxicity–ReCiPe	kg 1,4-DB eq	17,332	17,301	17,047	17,322	17,255	17,290	17,275	16,882
	Saving		0.1%	1.6%	—	0.4%	0.2%	0.3%	2.8%
Marine ecotoxicity–CML 2001	kg 1,4-DB eq	3,607	3,606	3,214	3,608	3,512	3,581	3,541	3,023
	Saving		<0.1%	10.9%	—	2.6%	0.7%	1.8%	18%
Marine ecotoxicity–AUPLCI	day	2.48×10^{-7}	2.47E-07	2.44E-07	2.47E-07	2.47E-07	2.47E-07	2.47E-07	2.43E-07
	Saving		0.1%	1.1%	—	0.3%	0.1%	0.2%	2.0%

governmental decisions regarding resources, but the approach can just as well serve to improve business operations if applied with a similar analytical approach.

Further Reading

Crossin, E., Verghese, K., Lockrey, S., Ha, H., and Young, G. (2020). The environmental impacts of operating an Antarctic research station. *Journal of Industrial Ecology*, *24* (4), 791–803.

Eckelman, M. J., and Chertow, M. R. (2009). Using material flow analysis to illuminate long-term waste management solutions in Oahu, Hawaii. *Journal of Industrial Ecology*, *13* (5), 758–774.

Eckelman, M. J., Ashton, W., Arakaki, Y., Hanaki, K., Nagashima, S., and Malone-Lee, L. C. (2014). Island waste management systems: Statistics, challenges, and opportunities for applied industrial ecology. *Journal of Industrial Ecology*, *18* (2), 306–317.

Fischer-Kowalski, M., Low, M., Noll, D., Petridis, P., and Skoulikidis, N. (2020). Samothraki in transition: A report on a real-world lab to promote the sustainability of a Greek island. *Sustainability*, *12*, 1932.

Klee, R. J. (2001). Industrial metabolism on ice: A case study of industrial material flows and environmental management alternatives for Scott Base, New Zealand's Antarctic research station. *Journal of the Royal Society of New Zealand*, *11* (2), 393–409.

Llerena-Pizarro, O. R., Micena, R. P., Tuna, C. E., and Silveira, J. L. (2019). Electricity sector on the Galapagos Islands: Current status, renewable sources, and hybrid power generation proposal. *Renewable and Sustainable Energy Reviews*, *108*, 65–75.

Martinico-Perez, M. F. G., Fishman, T., Okuoka, K., and Tanikawa, H. (2017). Material flow accounts and driving factors for economic growth in the Philippines. *Journal of Industrial Ecology*, *21* (5), 1226–1236.

Matsuoka, J., and Kelly, T. (1988). The environmental, economic, and social impacts of resort development and tourism on native Hawaiians. *Journal of Sociology and Social Welfare*, *15*, 29–44.

Schulz, N. B. (2007). The direct material inputs into Singapore's development. *Journal of Industrial Ecology*, *11* (2), 117–131.

Singh, S. J., and Haas, W. (2016). Complex Disasters on the Nicobar Islands. In Habert, H., Fischer-Kowalski, M., Krausmann, F., and Winiwarter, V. (Eds.). *Social Ecology*, pp. 523–542. Cham: Springer.

Symmes, R., *et al.* (2020). The weight of islands: Leveraging Grenada's material stocks to adapt to climate change. *Journal of Industrial Ecology*, *24*, 369–382.

Vitousek, P. M. (2002). Oceanic islands as model systems for ecological studies. *Journal of Biogeography*, *29*, 573–582.

Chapter 29

Holarchic Industrial Ecology

Chapter Concepts

- Holons are self-organizing interconnected systems of organisms (natural, technological, or societal)
- A holarchy is a system of systems in which information but not control is exchanged between holons
- Holarchic analysis is increasingly used in such varied areas as health care and manufacturing. In industrial ecology, holarchic analysis is used to assess materials networks, interactions among organizations, and the energy implications of development

29.1 Introduction to Holarchies

A system is a group of interacting and/or interdependent elements that form a collective entity. Such a system in ecosystem ecology as well as industrial ecology is characterized by exchanges of resources such as materials or energy. It is common in ecosystem ecology, industrial ecology, and many other fields as well for a system to be connected to other systems to form a "system of systems" (SoS). Understanding an SoS can be a complex and lengthy task, but it is a central feature of much research activity.

In complex systems theory, an individual entity in a system of systems is termed a *holon*, and the network of holons is a *holarchy*. Because the behavior of some or all of the holons may not be predictive, a holarchy cannot in general be the product of a comprehensive, anticipatory design. Rather, a holarchy tends to be a self-organizing, hierarchical, open entity (SOHO) — self-organizing because

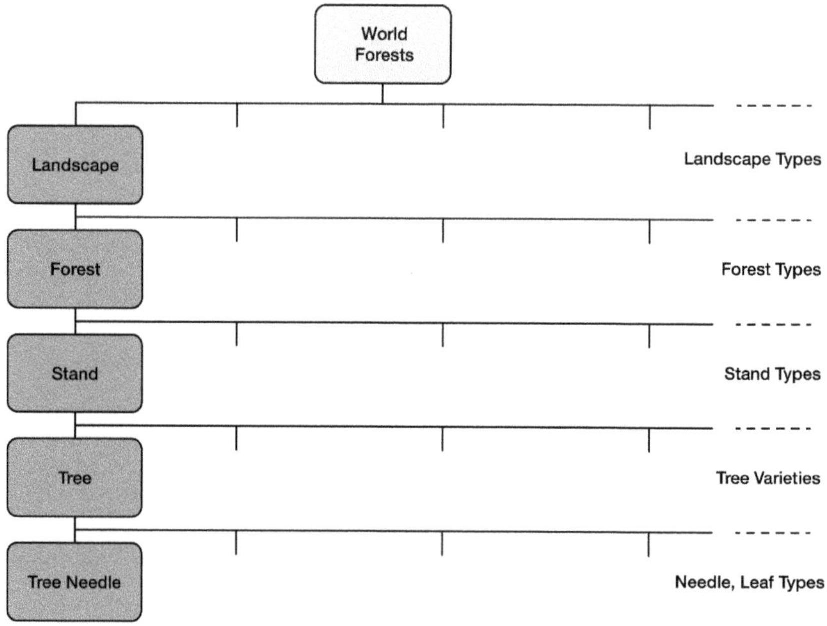

Figure 29.1. From needles and leaves to global forests — a classical holarchy of nature.

it can organize itself independently through its own actions; hierarchical in the sense of being a conceptually or causally linked system of grouped objects or processes along an analytical scale; and open in the sense of requiring external inputs of materials and energy.

Forests and their components provide an example, as shown in Figure 29.1. Here, the low-level holons are needles or leaves; they share resources through material exchanges. This holarchy has within it an additional (though hidden) property: that of an increase in typical lifetime as one moves up the holarchic sequence, and the forest holarchy can be analyzed to quantify the flows of nutrients from one holonic level to the next. This diagram suggests an important property of holarchies: that size tends to increase as one moves from lower holons to higher, and speed tends to decrease. In a forest, it is easy to recognize an individual tree and to note that each tree's needles have cycles much smaller and faster than those of a tree itself. The evolution of a stand or forest is clearly much slower than that of a single tree. In the cardiac case, cells are small and their physical and chemical processes are rapid; the opposite is true of the organ itself. The result is that the organ level sets constraints and boundaries for the cell coupling level. The cell level itself provides a mechanistic foundation for the cell coupling level.

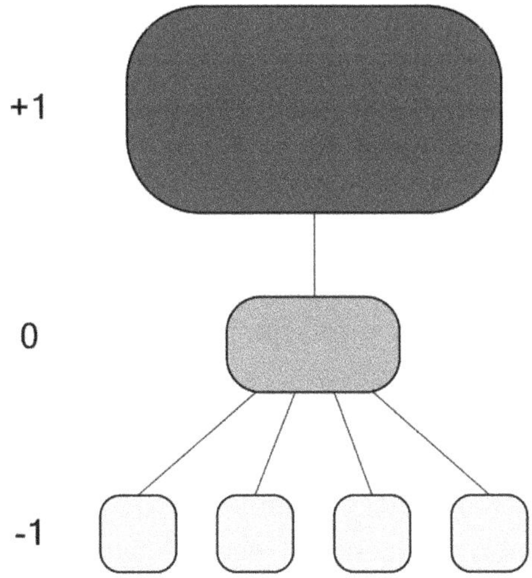

Figure 29.2. Relationships between holons in a holarchy. The level of interest (Level '0') is in the center of the diagram. The dynamics of Level '0' are constrained and bounded by the larger and slower-moving Level '+1'. Interactions at the small, fast-moving Level '–1' generate the signatures felt in Level '0' (adapted from O'Neill, 1988). This is a network diagram.

In any holarchy in a metastable state, the details of the '–1' holon's processes are too small and fast to be anything but background noise to Level '0'. When the system is disrupted, however, the rapid dynamics of the '–1' holon may overcome the constraints imposed by the '+1' holon, and the system adopts a new metastable state (Figure 29.2).

All holarchies have a number of properties in addition to their hierarchical structure that are worth noting. Perhaps the most obvious is that they are nonlinear and thus are not strictly predictive. A second is that they are dynamic systems that exist in quasi-steady states rather than at equilibrium. They often exhibit feedback loops involving materials or energy cycling. Finally, there is the potential for chaotic and catastrophic behavior if they find themselves removed from their quasi-steady state.

A feature of holarchies that frequently confounds analysts is that of emergent behavior, in which even a detailed knowledge of one holon is insufficient to predict behavior at a different holonic level. An obvious example from the realm of nature is the beating heart. The heart at its lowest level consists of cells, of course, which can be described extensively from physical and chemical perspectives. Nothing at the cell level suggests electrical activity leading to rhythmicity, however. Rather, rhythmicity arises as a consequence of the electrical properties of numerous intracellular gap junctions, as modified by the three-dimensional architecture and structure of the organ itself. However, a major disturbance at one holonic level has the potential to disrupt the overall holonic system: in the cardiac example, damage to an artery of the wall of the heart can lead to an irregular heartbeat, or perhaps to no heartbeat at all.

29.2 Ecological Holarchies

Ecosystem ecologists generally relate the holarchical concept to that of "trophic levels", which are the levels in a food chain or food web in a natural ecosystem, discussed in Chapter 6. There are typically five levels in such holarchies, as illustrated in Figure 29.3: (1) Primary Producers, (2) Primary Consumers, (3) Secondary Consumers, (4) Tertiary Consumers, and (5) Top Predators. Organisms at each level tend to eat from the next lower level and be eaten by the next higher level, but some are Omnivores, which can feed at several trophic levels. Discarded materials (defecation, dead species, etc.) are utilized by Decomposers to return material resources to the system, which is driven ultimately by solar energy.

Each holon of an ecological ecosystem is a SOHO: much of the activity is self-ordained, members of each species generate societal structure, hunt together, reproduce, and so forth. Sometimes, they will interact with a higher or lower holon,

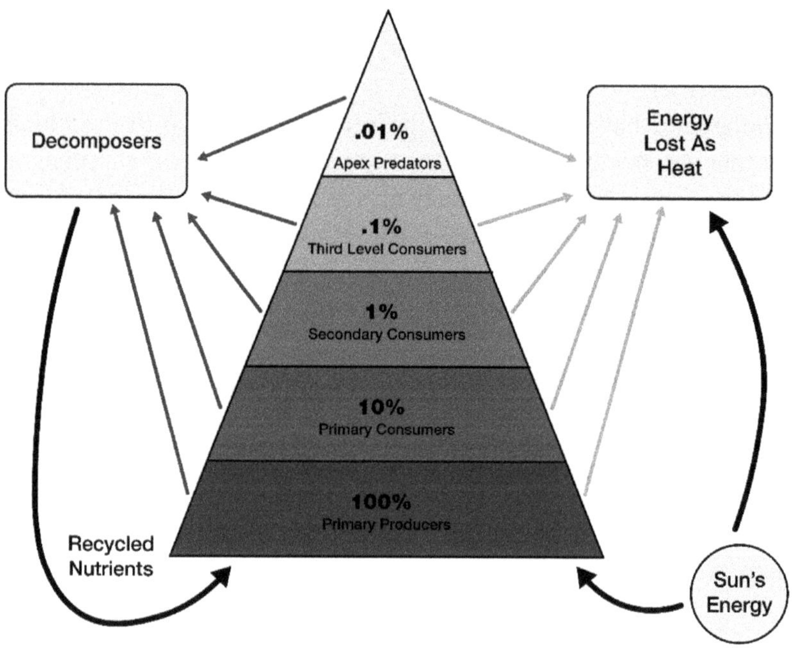

Figure 29.3. Trophic levels in an ecological ecosystem (adapted from "Trophic Level", 2019).

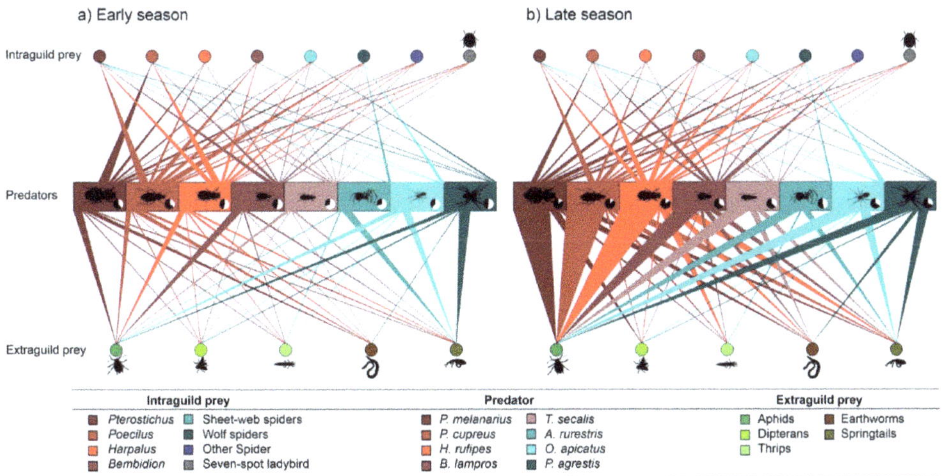

Figure 29.4. Early season (a) and late season (b) food webs for an agricultural ecosystem. These line widths indicate fractions of total predator consumption of individual prey species (Roubinet *et al.*, 2018).

but not in an encompassing way. They help sustain higher level holons, for example, but their material resources do not sum to those of the next higher level holon, nor are their material resources the sum of those of the next lower level holon.

How do ecosystem ecologists study the interholonic interactions in a food web? A common approach is to determine the predator–prey interactions of the species by the capture and analysis of the gut contents of the predators. The results can then be shown as resource transfers from one holonic level to other levels. Figure 29.4 provides an example for the invertebrate predator species in agricultural fields. The line thicknesses here relate to the number of prey species eaten by the predators. Other analyses express these interactions in terms of flows of material (generally carbon). The holonic levels are thus linked by specific resource exchanges, and those exchanges can then be understood as functions of ecological properties of interest such as abundance of species, weather, season of the year, and so forth.

The concepts of holons, holarchies, and nutrient exchanges are of great utility in ecosystem ecology and in the next section to show how those concepts guide some of the most useful and widespread of the analytical approaches of industrial ecology and industrial/societal ecosystems.

29.3 Industrial Holarchies

Industrial ecology has technology at its core but also recognizes technology's tight links to human actions (the social system) and to nature (the environmental system). In other words, industrial ecology deals with systems of systems, as does nature. Industrial products as a feature of the comprehensive built environment emphasize stocks of materials in use. Additionally, the industrial process as a component of societal action emphasizes flows of materials and energy. All these examples are systems of systems.

The study of holarchies within biological ecology thinking has a fairly long history, but its incorporation into technologically related fields is more recent and less developed. Unlike the nature-centered perspective, however, holarchies related to technology can be regarded as purposeful. This has enabled Mark Maier of the Aerospace Corporation to list five characteristics of technological holarchies:

- *Operational independence* of the holons, each holon being useful in its own right.
- *Managerial independence* of the holons; they exchange information but not control.
- *Evolutionary development*, a result of the incorporation of new resources, technologies, and ideas.
- *Emergent behavior* rather than planned behavior (as discussed in the following in more detail).
- *Geographic distribution*, rendering close linkages of material and energy flows, and perhaps information flows, problematic.

Maier goes on to define and exemplify three types of technological holarchies, as follows:

- The *Directed* Technological Holarchy. An example is the integrated air defenses of modern military forces. In such a system, the holons (the networks of surveillance radars, missile launch batteries, fighter aircraft) operate independently, but their normal operational mode is subordinated to the centrally managed purpose as required.
- The *Collaborative* Technological Holarchy. An example is the urban transit systems. In these systems, the holons (private vehicles, taxis, subways, railroads, aircraft) voluntarily collaborate to fulfill the agreed-upon central purpose.
- The *Virtual* Technological Holarchy. An example is the World Wide Web, with its many users and connections. Virtual holarchies lack any central authority, other than the basic enabling protocols.

Technological holarchies are obviously very difficult entities to study, but undeniably meet the industrial ecology criterion of involving human transformations of mass and energy, and thus can be addressed from a systems perspective. Hence, any SOHO entity in which a technological product or a societal structure utilizes materials and/or energy can be properly termed an industrial ecology holarchy.

Emergent behavior is also a feature of industrial ecology holarchies. Consider the example of cellular telephony. This complex technology was developed in the 1980s and 1990s. The fixed-location base stations that were originally needed were few, and the telephones expensive and briefcase-sized. The cell phone holon and the infrastructure holon that supported it were largely predictable, and users were anticipated to be a modest number of physicians, traveling salespeople, and others not having convenient access to a landline phone.

Around the year 2000, improved technology made cell phones much smaller and more affordable. Parents began to buy cell phones for their children, and firms for their employees. Suddenly, it was possible to call anyone from anywhere. Demand skyrocketed, especially in developing countries where the technology made it possible to avoid installing landline phones almost completely. A decade later, "smart" phones merged voice telegraphy with broad information access. An entirely new pattern of social behavior had emerged, unpredicted and certainly unplanned.

The smartphone story is relevant to industrial ecology because it involves humans, resources, energy, and the environment. The hundreds of millions of smartphones require an incredible diversity of and large quantities of materials to enable optimal functionality. At one point in mobile phone evolution, tantalum came into short supply, and the mineral coltan was mined in Africa by crude technological approaches in order to fill this supply gap, doing significant environmental damage in the process. This resource challenge continues, and the smartphone network worldwide is now trying to implement a new emergent behavior: the recovery of precious metals from discarded cell phones by primitive "backyard" technologies and more recently by advanced technological approaches. This social/technological resource recovery activity did not exist when smartphones were few; as they became abundant, the recycling networks flipped into a new and unanticipated state.

In recent years, the emergence of ever more complex systems that cross traditional boundaries has made for explosive advances in the design and analysis of SoS structures that are firmly technological but also firmly human. The planning and execution of healthcare networks is a common example. As Figure 29.5

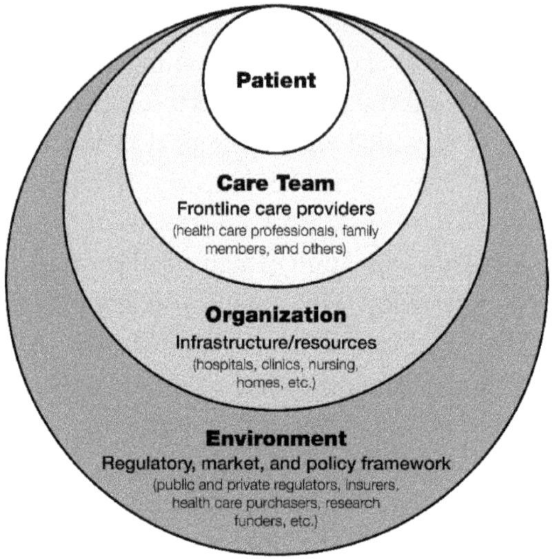

Figure 29.5. Holonic structure in the human health system of systems (adapted from Hata *et al.*, 2009).

indicates, there are at least four holonic levels: (1) the patient, (2) the "care team" (immediate service providers), (3) the organization (ambulances, hospitals, etc.), and (4) the societal environment (regulatory, public policy, etc.). Each holon obviously must be capable of semi-independent operation, but is also linked to other holons. An additional aspect not shown in Figure 29.5 is that aspects of Holons 2 and 3 have a very strong industrial ecology component — materials and products that must be available, the potential for recycling instruments and diagnostic machinery no longer wanted, etc. In a sense, this is a cross-holonic linkage of two holarchical systems.

29.4 The Hawaiian Islands: An Industrial Ecology Holarchy

An aspect of holarchies not yet discussed, but clearly integral to many holarchies, is the spatial aspect. An archetypical example is the Hawaiian Islands (Figure 29.6(a)), which comprise an archipelago of six fairly large islands and several small islands in the north-central region of the Pacific Ocean. We presented some IE results for Hawaii in Chapter 28. Each holon can be studied as an individual entity or as part of the larger holarchic system, as shown in Figure 29.6(b) and in the sections as follows.

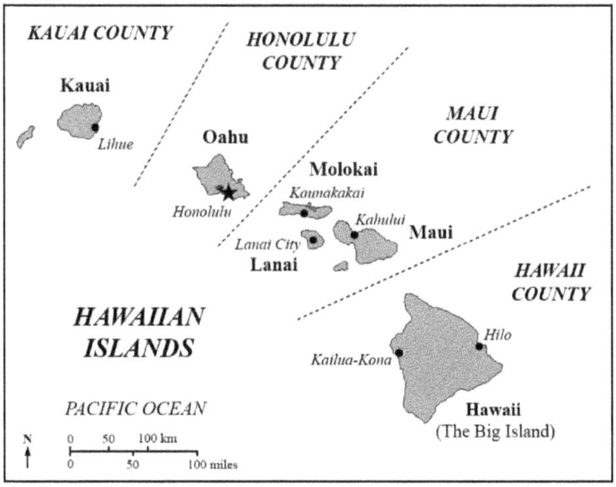

(a)

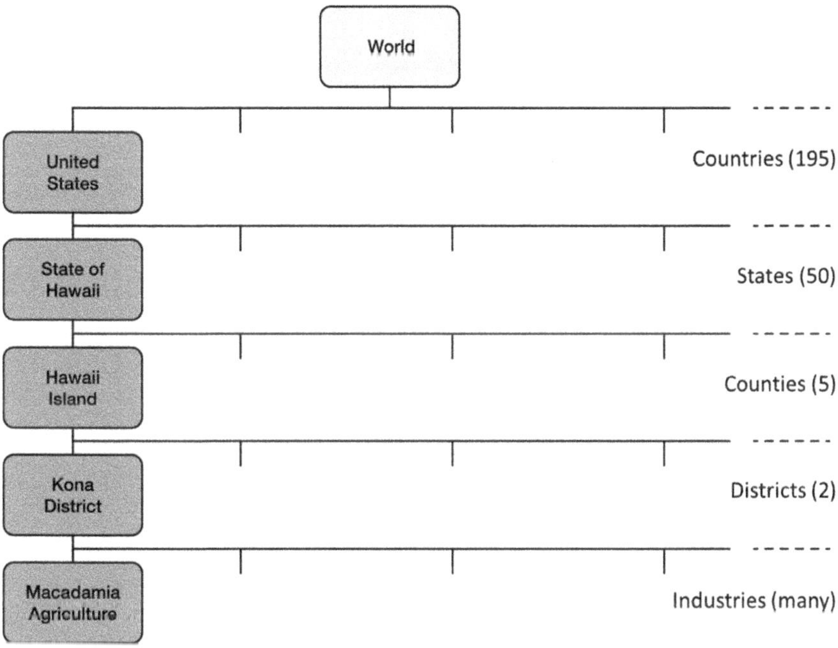

(b)

Figure 29.6. (a) Map of the Hawaiian Islands Archipelago (courtesy of Koichi Kanaoka). (b) An industrial holarchy for the Hawaiian Islands. (c) Material and energy flows in the macadamia nut industry on Hawaii Island (2006), in Gg per year. (d) Material flows into the Hilo region (2006), in Gg per year,). (e) Material flows for Hawaii Island (2006), in Gg per year (adapted from Houseknecht, 2006). (f) on the Hawaiian Archipelago (2007), in Tg per year (adapted from Chertow and Seto, 2011). (g) Material flows for the U.S. (1996), in Pg per year (adapted from Matthews *et al.*, 2000). (h) Material flows for the world (2005), in Pg per year or Gt per year (adapted from Haas *et al.*, 2015).

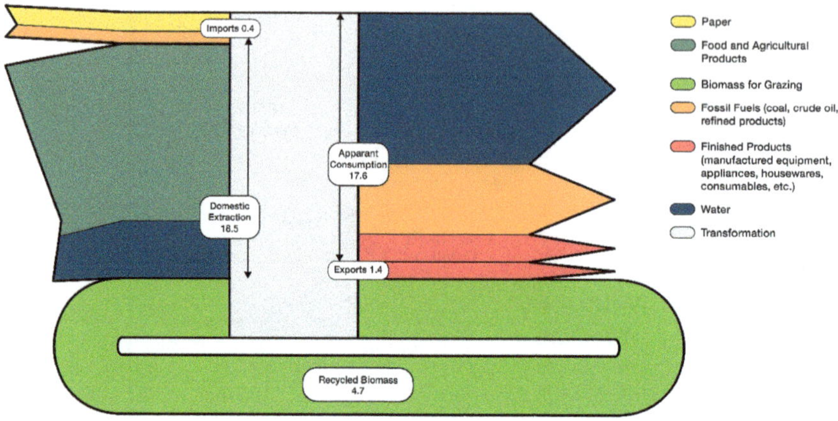

(c)

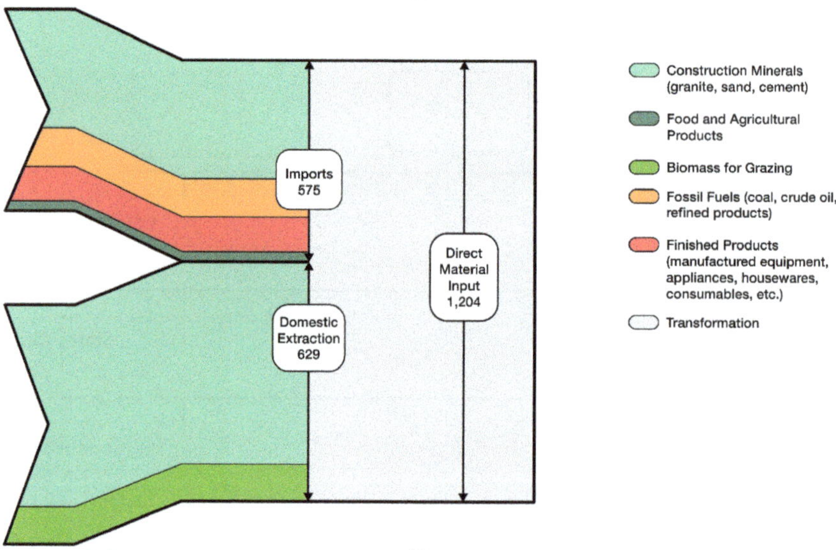

(d)

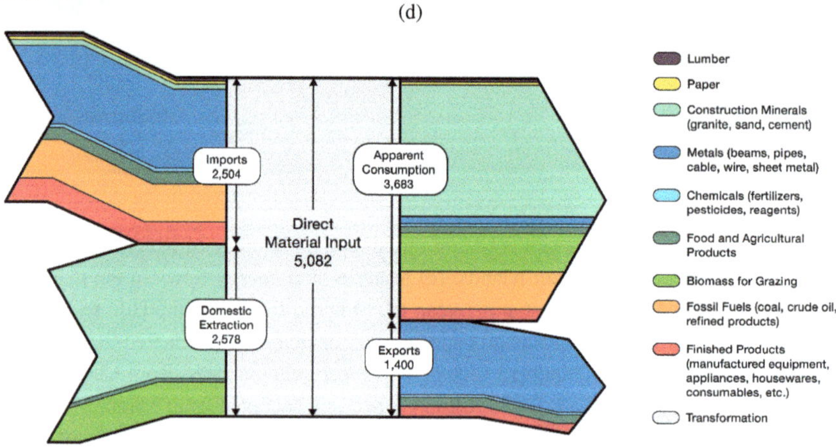

(e)

Figure 29.6. (*Continued*)

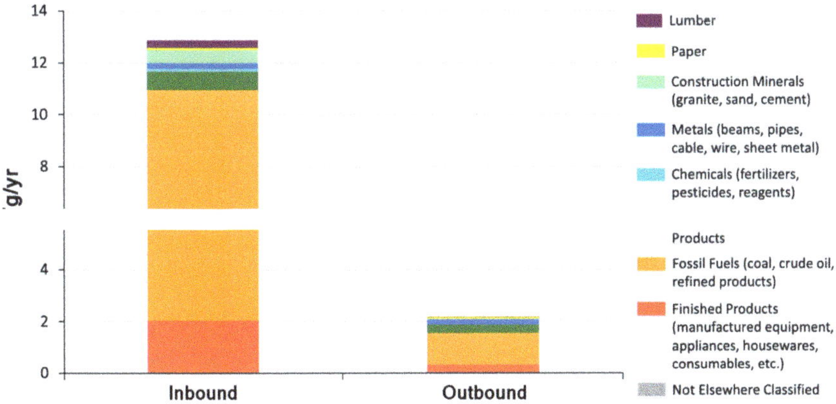

(f)

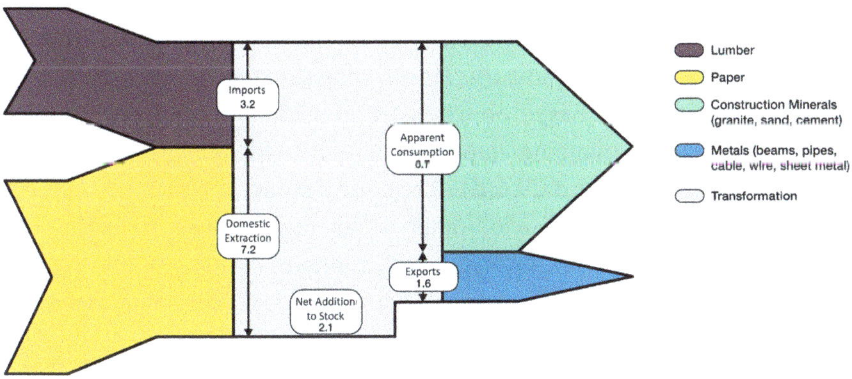

(g)

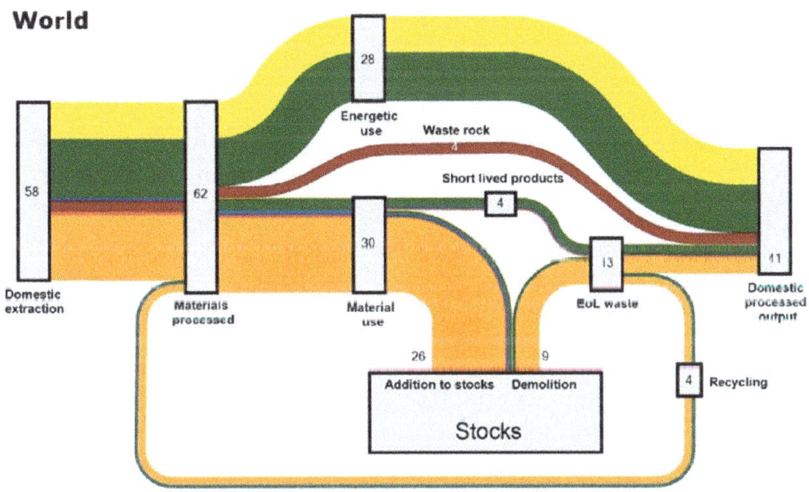

(h)

Figure 29.6. (*Continued*)

29.4.1 *Holon Level 1: An industry on Hawaii Island*

The macadamia nut industry is centered on Hawaii Island (sometimes termed the "Big Island" to minimize confusion). Its material and energy flows in 2007 are shown in Figure 29.6(c). Most of the resources originate on the island, with only small flows of packaging and fuel coming from imports. The nuts are sold both in Hawaii and by export. Air pollutants as products of fuel oil combustion for energy are quantized in the figure. Water and biomass are extensively recycled, and the overall industry cycle is quite efficient.

29.4.2 *Holon Level 2: The Hilo District: A municipal region on Hawaii Island*

The example holon for Level 2 is the Hilo region municipality, within which exists most of the macadamia nut industry. Hilo's spatial extent is defined as the Hilo Census Designated Place, based on extensive consultation with geographers and the County of Hawaii planning departments. The Hilo District material flows in 2009 are shown in Figure 29.6(d). They incorporate, of course, the flows of Figure 29.6(c) as well as those of other urban regions and industrial entities. Domestic extraction provides biomass and minerals, while imports from other islands provide a variety of food products, energy sources, and manufactured goods.

29.4.3 *Holon Level 3: Hawaii Island*

Hilo is the capital city of Hawaii Island. However, its material and energy flows constitute only a portion of those of the entire island (Figure 29.6(e)) because the flows of Kona, other communities and industries, and rural settlements need to be added in. Both imports and exports are more diverse than was the case for Hilo, especially as it relates to a variety of products for construction and agriculture.

29.4.4 *Holon Level 4: The Hawaiian Archipelago (State of Hawaii)*

The example holon for Level 4 consists of the six largest islands of the Hawaiian Archipelago. The incoming material flows are most easily tracked by determining the inflows to Honolulu by ship (these are the vast majority of import/export materials by weight, and all enter the archipelago through a single port). In Honolulu, the materials for Oahu Island are off-loaded and the remainder is then shipped to the other islands, including the Big Island. Figure 29.6(f) illustrates the overall flows of materials for the archipelago.

In 2005, the State of Hawaii imported a total of 15.6 Tg of materials and exported 8.5 Tg. Fossil fuels were the most imported materials, accounting for over half of the state's total imports. Other large import flows were finished products (2.4 Tg), metals (2.2 Tg), and food and agricultural products (1.3 Tg), which together comprised another 38% of all imports to the archipelago. On the exports side, fossil fuels were the second largest export from Hawaii by mass, amounting to roughly 20% of imported fossil fuels. Much of the exported fossil fuels consisted of naphtha — a solvent and feedstock for plastics — for which no significant market exists in Hawaii.

29.4.5 *Holon Level 5: The United States of America*

The material flows and stocks of the 50 states are shown in Figure 29.6(g). The diagram is a simple one, as more detailed and consistently reported information appears not to have been quantified and assembled. Nonetheless, features similar to those of Hawaii are evident: dominant domestic extraction, significant addition to in-use stocks, imports exceeding exports, and a large level of "apparent consumption" (waste generation).

29.4.6 *Holon Level 6: The World*

The material and energy flows for the world are shown in Figure 29.6(h). The analysis showed that only four gigatonnes (Gt) per year were recycled, compared with the 62 Gt per year of processed materials and 41 Gt per year of outputs. This discrepancy largely results from the fact that 27 Gt per year of processed materials were used to provide energy and could not therefore be recycled, and net additions to stock totaled another 17 Gt per year.

Note in Figure 29.6 that each holon could be shown to be (or at least should be) the sum of all components of the next lower holon of the holarchy (e.g., Level 4 is the sum of all flows in the six islands in holon Level 3). In practice, this may not be the case because of incomplete evaluations or different characterizations of material stocks and flows. In principle, however, if all but one flow analysis exists at any level, that missing analysis can be mathematically derived from the existing flows at that level and the holonic level above.

29.5. Relating Holoarchies to Networks

The previous discussions in this chapter strongly suggest that there are significant similarities between holarchies such as the Hawaiian Islands and the ecosystems and industrial networks described in Chapter 14. The three frameworks share common approaches of graphical display, mathematical tools, and structural analysis, and increasingly methods are crossing disciplinary boundaries. Despite their structural similarities, they can be distinguished as follows:

- *Ecological Ecosystems* — The majority of interactions between nodes of these systems are largely opportunistic, taking advantage of such factors as proximity or fortuity. Although hierarchies exist in these systems, there is no overall control or guidance.
- *Managed Technological Systems* (such as supply chains) — In a managed technological system, the decisions are made at the top holonic level and resulting activities occur at lower levels. In an electric vehicle, for example, the overall design and performance specifications stimulate the establishment of component supply chain, module manufacturing organizations, and overall assembly teams.
- *Societal Holarchic Systems* — In these systems, a number of semi-independent entities share resources and information and may use those assets in ways anticipated to be beneficial — building new roads, establishing social structures, etc. As a result, decision-making is dispersed among the different holonic levels. Such systems embody relatively high levels of managerial independence at all levels, with human preferences and relationships retaining much of the control at each level. These systems are mergers of societal and technological interconnections.

29.6 Adaptive Management of Technological Holarchies

Humans desire to control their surroundings for the benefit of themselves and others, of course. As technology becomes more and more deeply ingrained and integrated into human life, the challenges faced by control-minded humans increase. This is particularly true when the challenges are recognized as holarchical. As Table 29.1 demonstrates, the distinctions between the management of simple systems and of holarchies are great.

Because of the human element, industrial ecology holarchies display dynamics that are characteristic of social as well as engineered systems, such as:

Table 29.1. Distinctions Between Unitary Systems and Holarchies.*

Area	Unitary system	Holarchy
Focus	Single complex system	Multiple complex integrated holons
Objective	Achieve a purpose	Optimize holarchic functions
Expectation	Solution	Appropriate response to situation
Boundaries	Fixed	Fluid
Problem	Defined	Emergent
Goals	Unitary	Pluralistic
Approach	Intellectually rigorous	Intellectually adaptive

Note: *Adapted from Keating *et al.*, 2003.

- Problems are unlikely to be definitively described or defined. Any boundaries drawn around such systems necessarily oversimplify other aspects of the system, making even basic steps such as system definition somewhat arbitrary.
- There is nothing like an indisputable public benefit. In other words, not only is there no definitive prioritization of values that can guide design and analysis, but every "benefit" identified by the systems analyst necessarily carries with it a "disbenefit". For example, policies that encourage foreign direct investment in underdeveloped countries necessarily also change the culture and institutions of that country, which some people regard as "cultural imperialism".
- There are no objective definitions of equity. Even in the case of sustainability, the assumption of egalitarian values necessarily means that libertarian values are de-emphasized (that is, equality of outcome is elevated over equality of opportunity).
- Policies for social problems cannot be meaningfully correct or incorrect. They may or may not achieve their stated goals, but even if they do, they will frequently create perturbations elsewhere in the complex fabric of technological, environmental, economic, and social networks.
- There is no optimality. One can certainly design better or worse solutions, but a single perfect state is neither definable nor achievable.

The holarchy of the automotive system exemplifies many of these points, wherein holons might be designated as auto parts, automobiles, street and road infrastructure, and societal entities related to the automobile, such as parking lots, auto repair shops, and auto dealerships. Even a cursory evaluation of the current system indicates that attention may be focused on the wrong holon, and illustrates the fundamental truth that a strictly technological solution is unlikely to fully mitigate a culturally influenced problem. The engineering improvements of the vehicle, its energy use, its emissions, its recyclability, and so forth, on which much attention has been lavished, are truly spectacular. Nonetheless, and contrary to the usual understanding, the greatest attention so far as the system is concerned should probably be directed to the highest levels — the infrastructure technologies and the social structure of mobility.

Consider the energy and environmental impacts that result from just two of the major system components required by the use of automobiles. The construction and maintenance of the built infrastructure — the roads and highways, the bridges and tunnels, the garages and parking lots — involve huge environmental impacts (discussed in detail in Chapter 32). The energy required to build and maintain that infrastructure, the natural areas that are perturbed or destroyed in the process, the amount of materials demanded, from aggregate to fill to asphalt are all required by the automobile culture and attributable to it. In addition, the primary customer for the petroleum sector and its refining, blending, and distribution components — and, therefore, causative agent for much of its environmental impacts — is the automobile. Efforts are being made by a few leading infrastructure and energy production firms to reduce their environmental impacts, but these technological and management advances, desirable as they are, cannot in themselves begin to compensate for the increased demand generated by the cultural patterns of automobile use.

The final and most fundamental effect of the automobile may be in the geographical patterns of population distribution for which it has been a primary impetus. Particularly in lightly populated and highly developed countries such as Canada and Australia, the automobile has resulted in a diffuse pattern of residential and business development which is otherwise unsustainable. The lack of sufficient population density along potential mass transit corridors makes public transportation uneconomic within many such areas, even where absolute population density would seem to augur otherwise (e.g., in densely populated suburban New Jersey in the U.S.) This transportation infrastructure pattern, once established, is highly resistant to change in the short term, if for no other reason than that residences and commercial buildings last for decades.

Ludwig *et al.* (1993) suggest five principles for adaptive management of human societal systems; they are worth listing here:

- Include human motivation and responses as part of the system to be studied and managed.
- Rely on scientists and engineers to recognize problems but not to remedy them.
- Act before scientific consensus is achieved.

- Distrust claims of sustainability.
- Confront uncertainty.

In the industrial ecology holarchic sense, flexible adaptive management, properly informed and conducted with a few clearly stated goals in mind, has the potential to much improve our approaches to these complex, intertwined entities that are so important to sustainability.

Further Reading

Chertow, M. R., Graedel, T. E., Kanaoka, K. S., and Park, J. (2020). The Hawaiian Islands: Conceptualizing an industrial ecology holarchic system. *Sustainability*, *12*, 3104.

Chertow, M. R. and Seto, K. (2011). Materials flows and land use change in two urban areas of Hawai'i Island. In Ecological Society of America (Ed.), *Proceedings of the Ecological Society of America Annual Convention*. Austin, TX: Ecological Society of America.

Eckelman, M. J. and Chertow, M. R. (2009). Using material flow analysis to illuminate long-term waste management solutions in Oahu, Hawaii. *Journal of Industrial Ecology*, *13*, 758–774.

Geels, F. W. (2005). Processes and patterns in transitions and system innovations: Refining the co-evolutionary multi-level perspective. *Technological Forecasting & Social Change*, *72*, 681–696.

Haas, W., Krausmann, F., Wiedenhofer, D., and Heinz, M. (2015). How circular is the global economy? *Journal of Industrial Ecology*, *19*, 765–777.

Hata, Y., Kobashi, S., and Nakajima, H. (2009). Human health care system of systems. *IEEE Systems Journal*, *3*(2), 231–238.

Houseknecht, M. (2006). *Waste Management on the Big Island: Mapping a Vision for an Economically Sustainable Hawaii*. Hilo, HI: The Kohala Center.

Keating, C., *et al.* (2003). System of systems engineering. *Engineering Management Journal*, *20*(3), 36–45.

Ludwig, D., Hilborn, R., and Walters, C. (1993). Uncertainty, resource exploitation, and conservation: Lessons from history. *Science*, *260*, 17, 36.

Maier, M. W. (1998). Architecting principles for systems-of-systems. *Systems Engineering*, *1*, 267–284.

Maier, M. W. (2019). Architecting a portfolio of systems. *Systems Engineering*, *22*(4), 335–346.

Matthews, E., *et al.* (2000). *The Weight of Nations: Material Outflows from Industrial Economies*. Washington, DC: World Resources Institute.

O'Neill, R. V. (1988). Hierarchy theory and global change. In Rosswall, T., Woodmansee, R. G., and Risser, P. G. (Eds.), *Scales and Global Change*, New York: John Wiley.

Chapter 30

Industrial Symbiosis

Chapter Concepts

- A symbiosis is an interaction between organisms, biological or industrial, usually with benefits to one or both of the entities involved
- Industrial ecosystems emerge, often by accident, when industries located near each other discover that resource exchanges are mutually beneficial
- Industrial symbioses, especially if planned by governments, often merge resource flows between public and private entities

30.1 Biological Symbiosis

A symbiosis is an evolved interaction or close living relationship between organisms from different species, usually with benefits to one or both of the individuals involved. Symbioses may be *obligate*, in which case, the relationship between the two species is so interdependent that each of the organisms is unable to survive without the other, or *facultative*, in which the two species engage in a symbiotic partnership through choice but can survive individually. Obligate symbioses are often evolved over a long period of time, while facultative symbioses may be more recent behavioral adaptions. Given time, facultative symbioses may evolve into obligate symbioses.

Mutualisms are a form of symbiosis in which both symbiotic partners benefit from the interaction, often resulting in a significant fitness gain for either one or both parties. Mutualisms can take the form of resource–resource relationships, service–resource relationships, or service–service relationships.

Resource–resource mutualisms happen through the exchange of one resource for another between the two organisms involved. Such mutualisms most often occur between an autotroph (a photosynthesizing organism) and a heterotroph (an organism that must absorb or ingest food to gain energy). Most plants have a trophic mutualism that is a symbiosis between the roots of the plants and a fungus. The fungus colonizes the plant's roots and is provided with carbohydrates, sucrose, and glucose. In exchange, the plant benefits from the fungi's higher water and mineral *absorption* capabilities.

Service–resource mutualisms occur when the symbiotic partner provides a service in exchange for a resource reward. One of the best-known examples of this is the exchange between plants and their pollinators. While visiting the plants to gain a supply of energy-rich nectar, the pollinator (insects, birds, moths, bats, etc.) provides the plant with the service benefit of being pollinated, thus ensuring their own pollen is distributed when the pollinator visits more plants of the same species.

Commensalism is a symbiosis in which one organism benefits from, and is often completely dependent on, the other for food, shelter, or locomotion, although there is no obvious effect on the host. The relationship between whales and barnacles is an example of commensalism. The barnacles attach themselves to the tough skin of whales, and benefit from the widespread movement and exposure to currents, from which they feed, while the whale is seemingly unaffected by their presence.

30.2 Industrial Symbiosis

Industrial symbiosis (IS) refers to behaviors by groups of industrial actors that mimic many of the characteristics of biological symbiosis. In these *eco-industrial parks* (EIPs), firms in proximity exchange by-products such as materials, water, and energy, with the result that one company's waste can become another company's feedstock. Variations on this concept are the sharing of electric utilities or of infrastructure such as wastewater treatment plants, or the joint provisioning of services such as transportation or worker training. Industrial symbiosis would thus appear to offer the promise of developing environmentally superior industrial systems compared with systems not developed along this paradigm.

IS activities can be divided into three stages of development:

1. *Emerging* — In this stage, advantageous material exchanges are identified and exchanges begin, generally by a small number of innovative participants.

2. *Developing* — In this stage, additional participants and exchanges are identified, the geographic scope of the EIP may increase, and interactions grow in scale and diversity.
3. *Maturing* — In this stage, most participants and exchanges are stable, but the EIP continues to evolve as new participants join and existing participants leave — the natural consequence of changing business conditions, corporate restructuring, etc.

Many EIPs benefit from a variety of support mechanisms. For example, independently established networks can provide introductions to potential new entrants and discuss the potential benefits of joining. Governments can provide targeted regulatory environments and support networks, and regulated material flows. Both governments and self-organized networks often provide training to corporate employees, establish regular discussion groups, and in other ways, facilitate the goals of the EIP.

An exemplary example of collaborative private sector–government collaboration is provided by the United Kingdom's National Industrial Symbiosis Network. This activity that begun in 2001 was partially supported by the British government for more than a decade and is now self-organized. The goal was to aid in establishing partnerships between industrial facilities so that wastes or by-products from one become raw materials for another, thereby reducing waste and costs, and helping initiate new sources of revenue. Sponsored meetings bring together participants from industries willing to explore interactions, and many collaborative projects result. Peter Leybourn, who began the British program, extended it internationally in 2007; some 30 countries including Austria, China, France, and South Africa now have similar programs in varying stages of development.

30.3 Designing and Developing Symbiotic Industrial Ecosystems

How might one design, or at least recognize, an efficient industrial ecosystem? How can one decide whether several actors who exchange resources constitute an industrial ecosystem? One approach is the "3-2 heuristic" shown in Figure 30.1 as the minimum criterion: at least three different entities must be involved in exchanging at least two different resources. None of the three entities may be primarily engaged in a recycling-oriented business. Thus, emerging-stage activities can be designated as "kernels" (having bilateral or multilateral exchanges) or "precursors" (having resource exchanges with a public goods component) that

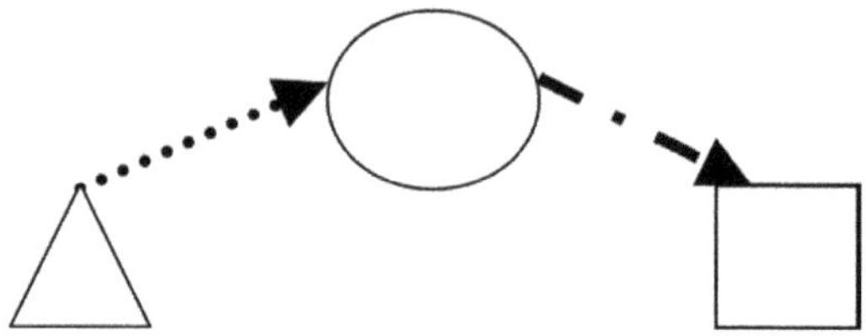

Figure 30.1. The 3-2 symbiosis defining a minimal industrial ecosystem (adapted from Chertow, 2007).

may not (yet) meet the minimum industrial ecosystem requirement but have the potential to evolve into that status.

It seems clear that a successful EIP system needs to involve a broad sectoral and spatial distribution of participants and, at the same time, be flexible and innovative. A key to an effective broad scale IS is a high degree of synergy between input and output flows of resources. Industrial ecosystems are in earlier evolutionary states than biological ecosystems (Chapter 6), in that BE organisms all exchange organic nutrients of similar chemical forms. IE organisms, however, have much broader feeding habits: some desire petrochemicals, some metals, some forest products. In such a system, mismatches between output flows and input requirements can make it difficult to find suitable nearby partners.

Analyses of a large number of industrial food webs may reveal characteristics not demonstrated by other approaches — a missing sector or type of industrial activity that could increase connectivity, for example. Conversely, such studies might help address a question that remains as difficult for BE as for IE — what will be the response of a particular ecosystem to a particular type of perturbation? These topics provide a rich area for detailed investigation and thoughtful planning.

As can be seen from this discussion, industrial ecosystems are evolutionary, not static. Emergence systems can become Development systems or even Maturation systems over time, for example. An industrial symbiosis can, in principle, be consciously developed, often around a key tenant such as a power plant that can readily begin the resource exchange process with a variety of potential industrial partners.

As with biological exchanges of resources, symbiotic exchanges within industrial ecosystems possess characteristic properties. One of these properties refers to the types of resource exchanges and is consistent with the biological analog: it is usually *facultative*, but could also be *mutualist* in the form of service–resource relationships. The second property is the degree of maturity of an industrial symbiotic system: *emerging* (where only two or three exchanges have been achieved), *developing* (where a functioning eco-industrial park exists but significant room for enhancement remains), and *mature* (where an eco-industrial park has achieved a solid and extensive resource exchange system while allowing for ongoing evolution in participants and exchanges). In the examples discussed in the following sections, these properties will be discussed and described for each.

30.4 Examples of Industrial Symbiosis

30.4.1 *The Kymenlaakso Industrial Park, Finland*

Finland is a country of forests, and the forest industry is vital to the country. As a result, forest-related industries and industry clusters are common. One such cluster, in the Kymenlaakso region in southern Finland, is a modest EIP centered around a pulp and paper plant (the "keystone species").

As is not uncommon in Finland, the Kymenlaakso forest product-centered EIP connects with a local municipal power plant and waste treatment facility. The pulp and paper plant is, however, the keystone species in the EIP. A few other wood-related industries are included, such as a hydrogen peroxide plant. Analysis of the energy and CO_2 emissions has demonstrated the financial and environmental benefits of the resource exchanges among the participants.

The Kymenlaakso EIP is clearly an emerging EIP. As pictured in Figure 30.2, it is relatively small and not industrially very diverse. Nonetheless, the benefits of the facultative resource exchanges are clear to see and provide a framework for further development.

30.4.2 *The Campbell Industrial Park, Oahu, Hawaii*

The Campbell Industrial Park, created in 1958, is the only heavy industry site in Hawaii, located on the island of Oahu. It is comprised of some 250 companies. A core group of these companies, anchored by the only coal-fired plant on Oahu, self-organized to form an industrial symbiosis. Other members include the oil-fired Kalaeloa cogeneration plant, two large oil refineries, a cement company, a private construction and demolition (C&D) waste landfill, the municipal water authority, and a local golf course. Figure 30.3 depicts the symbiosis among 11 companies as of the late-2000s; the figure highlights some of the major materials being exchanged: various grades of water and steam, unconventional fuels such as granular activated carbon (GAC), waste oil, and shredded tires. Also shown in the figure are industries outside the physical boundaries of the industrial park that participate in the flows of resources to and from in-park industries.

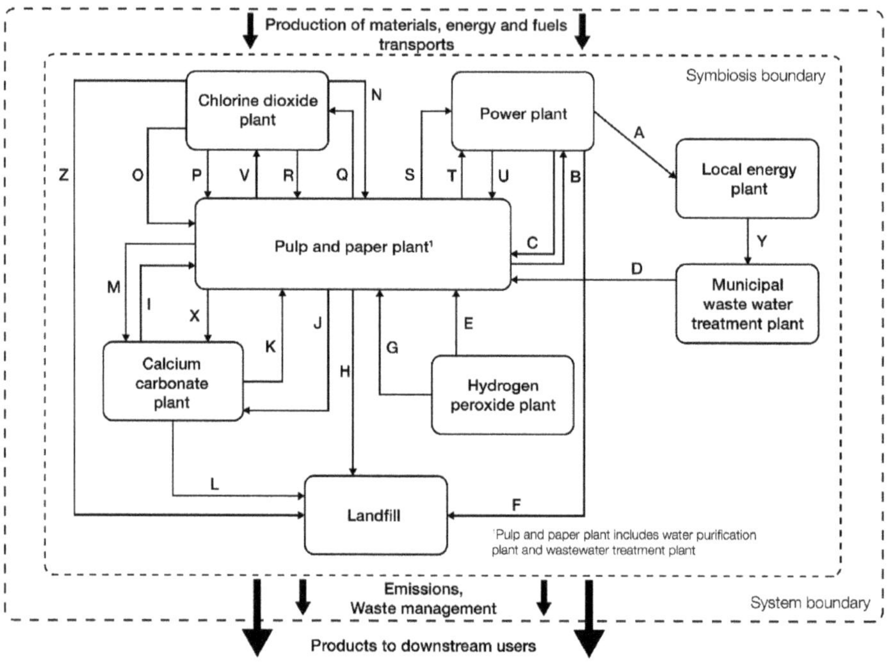

A District heat and electricity
B Sewage sludge
C Wastewater
D Sewage sludge
E Wastewater
F Ash
G Hydrogen peroxide (H_2O_2)
H Miscellaneous inert waste
I Wastewater
J Water
K Calcium carbonate ($CaCO_3$)
L Miscellaneous waste
M Carbon dioxide (CO_2)
N Chlorine (Cl_2)
O Wastewater
P Chlorine dioxide (ClO_2)
Q Water
R Sodium hydroxide (NaOH)
S Sodium hydroxide (NaOH)
T Biomaterials used as fuel
U Steam, electricity, heat
V Steam
X Steam and electricity
Y Electricity
Z Waste

Figure 30.2. Material flow map of the forestry-based industrial ecosystem in Kymenlaakso, Finland (adapted from Sokka *et al.*, 2011).

30.4.3 *The Norrköping industrial system, Sweden*

On an island off Norrköpping in the Baltic Sea is an eco-industrial system that was established in the early 21st century in a cooperative venture between industry and government. Farming communities in the vicinity have become strong partners as well. The EIP in its current form is illustrated in Figure 30.4. Many material and energy exchanges, often involving the combined heat and power plant anchor tenant, link the industrial and municipal facilities and agencies, and expansion of the EIP to include CO_2 capture and storage has received funding from the European Commission.

This EIP is an example of a Developing EIP rapidly becoming a Mature EIP. Most of its exchanges are facultative but some partnerships, especially those involving government agencies, can be service–resource relationships.

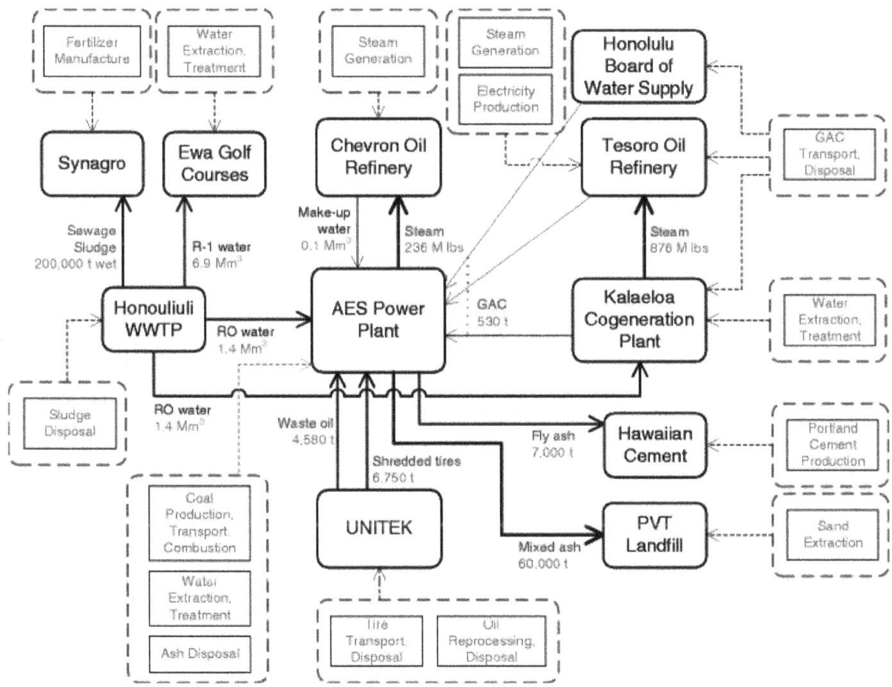

Figure 30.3. The resource interactions within the Campbell Industrial Park in Oahu, Hawaii, and firms outside the park that exchange resources with firms inside the park (Eckelman and Chertow, 2013).

30.4.4 *The Kalundborg industrial complex, Kalundborg, Denmark*

The Kalundborg complex is an industrial area containing more than a dozen independent companies that in 1972 began sharing material and energy flows among themselves (Figure 30.5). It is the archetypical example of a mature industrial ecosystem, widely recognized and widely written about.

As in many EIPs, Kalundborg's anchor tenants are the power plant and the oil refinery, which supply heat and power to most of the industries and to the city nearby. Among other Kalundborg's facultative material exchanges are wastewater, gypsum, yeast, sulfur compounds, and many others, all of which are residues to some and resources to others. All exchanges are agreed upon in contracts among those involved — there is overall encouragement but no overall control. Because the EIP has been in existence for half a century, there are excellent records that demonstrate that the exchanges are highly valuable economically, and that the emissions reductions and other natural ecosystem interactions have provided substantial environmental benefits.

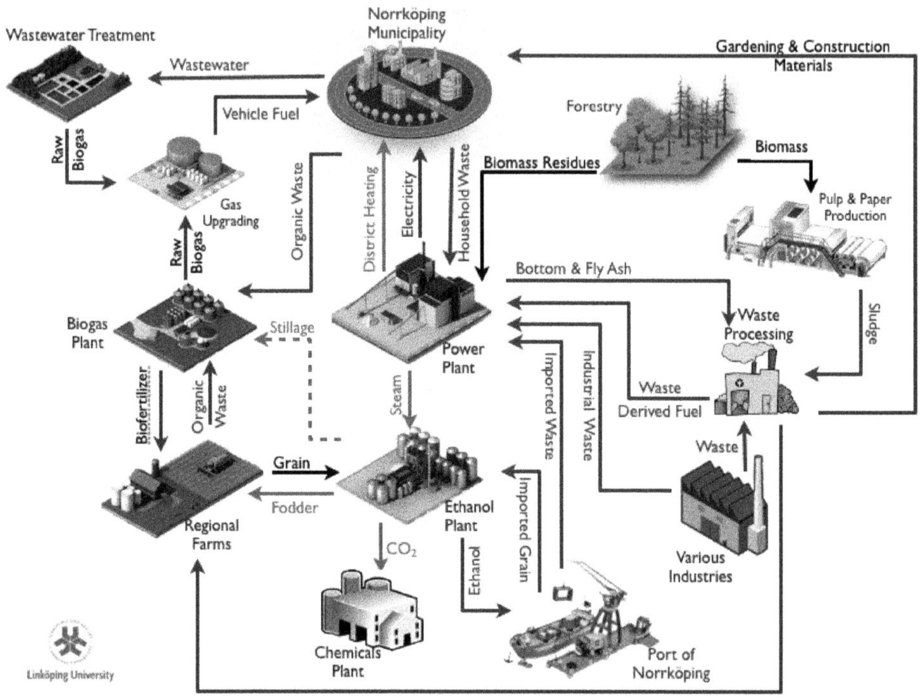

Figure 30.4. The industrial symbiosis at Norrköping, Sweden (Hatefipour *et al.*, 2011).

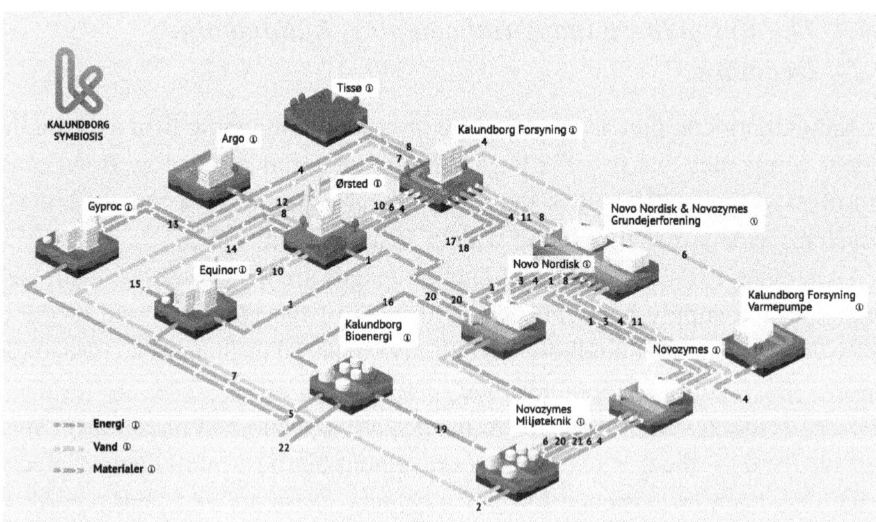

Figure 30.5. Flows of resources in the eco-industrial system at Kalundborg, Denmark (Reproduced with permission from Kalundborg Symbiosis, Kalundborg, Denmark.)

Perhaps because the Kalundborg EIP was established by agreements among independent companies, the interactions over the years have demonstrated that mutual management understanding and cooperative commitment is essential. Extensive networking and regular meetings and classes among the participants have played a major role in Kalundborg's success.

30.5 Discussion

The flourishing or demise of EIPs around the world has revealed a number of drivers that tend to stimulate them, as well as a number of barriers that tend to inhibit them. Some of the more important are summarized in Table 30.1. What is evident from the table is that industrial ecosystems are not merely the creations of technology, that is, symbioses do not arise just because of the technical potential to share resources; rather, they result from the complex interplay of technology, economics, government, and society. It is also clear, however, that triggering events may sometimes initiate ecosystem behavior — a regional water shortage, the technical obsolescence of a facility, or a new environmental regulation, for example.

Notwithstanding all of the above, industrial ecosystems are potentially fragile creatures. It is common that some industries embrace the plan while others, for a

Table 30.1. Drivers for and Barriers to Industrial Ecosystems (adapted from Gibbs *et al.*, 2005).

Drivers	Barriers
Financial opportunities: Most IS exchanges make good business sense in terms of lower input costs, lower operating costs, and/or increased revenues.	*Informational*: A lack of understanding of the process inputs and outputs of potential symbionts often inhibits exchange possibilities.
Resource scarcity: Water is the most common example.	*Economic*: Exchanges or joint approaches to acquisition or disposal may carry with them excessive economic costs.
Reduced liability: Potentially problematic discards or by-products that are exchanged become the responsibility of others.	*Technical*: The set of potential symbionts may not "fit together" so far as inputs and outputs are concerned.
Sustainability focus: Industrial symbiosis is a natural complement of increasing corporate attention to sustainability.	*Regulatory*: In some cases, regulations may prevent or inhibit the exchange of potentially hazardous resources.
Staff mobility: Technically trained people moving from one facility to another, especially in a different industry, often see symbiotic opportunities not visible to others.	*Motivational*: Firms, regulators, and others must be willing to commit to symbiotic relationships.

variety of reasons, are less engaged or do not participate. Perhaps one of the symbionts may go out of business, or a resource flow valued by one symbiont may no longer be produced by another, or a symbiosis champion may retire without a suitable successor on hand. When industrial symbiosis connections succeed, however, the participants find the interactions to be financially beneficial, environmentally sound, and socially welcome. These attributes make it worthwhile to discover, stimulate, and propagate industrial ecosystems of all kinds, in all kinds of places.

Further Reading

Chertow, M. R. (2007). "Uncovering" industrial symbiosis. *Journal of Industrial Ecology, 11*, 11–30.

Eckelman, M. J., and Chertow, M. R. (2013). Life cycle energy and environmental benefits of a US industrial symbiosis. *International Journal of Life Cycle Assessment, 18*, 1524–1532.

Gibbs, D., Deutz, P., and Proctor, A. (2005). Industrial ecology and eco-industrial development. *Regional Studies, 39*(2), 171–183.

Hatefipour, S., Baas, L., and Eklund, M. (2011). The Händelö area in Norrköping, Sweden: Does it fit for industrial symbiosis development? In World Renewable Energy Congress (Ed.), *World Renewable Energy Congress 2011 Sweden*. Linköping, Sweden: World Renewable Energy Network, DOI: 10.3384/ecp110572468. Retrieved from: https://ep.liu.se/ecp/057/vol12/060/ecp57vol12_060.pdf.

Lombardi, D. R., and Leybourn, P. (2012). Redefining industrial symbiosis: Crossing academic-practitioner boundaries. *Journal of Industrial Ecology, 16*, 28–36.

Magnusson, A., Andersson, H., and Ottesson, M. (2019). Industrial ecology and the boundaries of the manufacturing firm. *Journal of Industrial Ecology, 23*(5), 1211–1225.

Sokka, L., Pakarinen, S., and Melanen, M. (2011). Industrial symbiosis contributing to more sustainable energy use — An example from the forestry industry in Kymenlaakso, Finland. *Journal of Cleaner Production, 19*, 285–293.

Veleva, V., Todorova, S., Lowitt, P., Angus, N., and Neely, D. (2015). Understanding and addressing business needs and sustainability challenges: Lessons from Devens eco-industrial park. *Journal of Cleaner Production, 87*, 375–384.

Chapter 31

Industrial Ecology and Healthcare

Chapter Concepts

- Healthcare is energy- and material-intensive and highly connected to other sectors of the economy
- Environmentally extended input–output modeling of healthcare consumption has shown that it is a major contributor to resource use and environmental impact, comprising about 5% of greenhouse gas emissions
- Many healthcare organizations and clinicians want to make greener choices and are using life cycle assessment to compare different medical products and procedures
- Major disruptions to healthcare supply chains have led to shortages of vital medicines and materials, and alternate investments are informed by criticality assessment

31.1 Healthcare and Sustainability

The healthcare sector plays a critical role in society and in local and national economies. Healthcare systems may be public or private, but they share similar features in physical terms that make them an important topic for industrial ecology. The delivery of healthcare is commonly considered a service, but it is both energy- and material-intensive and relies on a complex supply chain to source medical devices, supplies, drugs, and equipment. Large hospitals are essentially small cities, with their own energy infrastructure, food management, transportation systems, and often solid waste and wastewater treatment. Hospitals are also one of the most energy-intensive building types, in part because of elevated requirements for lighting and ventilation and the presence of high-powered medical equipment that is rarely turned off. Healthcare is not just resource-intensive physically but also economically. The U.S. spent a mind-boggling $3.8 trillion on healthcare in 2019, representing 18% of its gross domestic product (GDP). Other high-income countries spend less, but still sizable proportions on their own healthcare systems.

Despite its resource intensity, the healthcare system has lagged well behind many of the other sectors we have considered in terms of environmental sustainability efforts and industrial ecology research. Why is this? The causes of healthcare unsustainability are structural and complex, and mostly well outside the scope of this textbook, but one reason might be the misguided perception that environmental issues are irrelevant to the purported core mission of healthcare, which is mistakenly seen as *protect the patient, regardless of cost*. The primacy of patient health over all other considerations has historically led to skepticism in some parts of the community about using environmental issues as another set of performance measures.

Such concerns echo those aired across manufacturing firms over the past several decades from people who complained that environmental issues were tangential to their core businesses, but who are now leading the charge on environmental technology innovations and decarbonization. An important lesson that manufacturers learned and that is starting to pervade healthcare is that there are many 'win–win' opportunities to improve both environmental and operational performance, without compromising quality (or patient care). "Green to gold" became a mantra of companies who found economic savings in improving resource efficiency in manufacturing and accessed new markets of customers, business partners, and investors who valued their environmental performance.

In the case of hospitals and healthcare organizations, reducing waste and pollution has another important benefit, namely improving public health. The Lancet Commission on Pollution and Health reports that air pollution is responsible for nine million premature deaths, and is the largest environmental cause of disease and premature death in the world. Global warming is affecting and will affect the health of millions more. Yet, some of these emissions that cause these health damages are due to healthcare itself.

A better representation of the core mission of healthcare derives from the Hippocratic Oath that all physicians take: *First, do no harm*. This considers harm to the patient, of course, but can be broadened to consider harm to the planet that indirectly affects public health. Healthcare professionals and administrators have a responsibility to reduce their own contribution to that toll. This prime directive, in place for more than two millennia, is strikingly similar to current guidelines from environmental scientists on living within planetary boundaries and absolute sustainability measures.

Even though, as a whole, healthcare has lagged behind other sectors in addressing sustainability, there have been dedicated campaigns on certain topics such as eliminating the use of mercury and other toxic chemicals in medical products, and much progress has been made on waste management and energy efficiency. Additionally, in recent years, there has been an explosion in interest and action on healthcare sustainability, so perhaps healthcare sustainability is becoming mainstream. In 2020, England's National Health Service (NHS), the largest public health system in the world (and the largest employer in Europe), declared that it was going to become the first net-zero health system and be carbon neutral by 2045, including its entire supply chain. Health systems and organizations around the world are following suit in setting environmental targets, particularly for carbon emissions.

Industrial ecology tools are increasingly being brought to bear on the healthcare sector, to benchmark performance, to compare products, to map supply chains, and to create scenarios for future improvements. This chapter discusses some of the most visible efforts, with the hopes of inspiring the next generation of industrial ecologists to work in this critical area.

31.2 EEIO for Environmental Footprinting of Healthcare

A basic question for those working in healthcare sustainability is as follows: 'How large are the impacts of healthcare?' Environmentally extended input–output (EEIO) models are ideal for answering this question when posed at a national or global level. Recall that EEIO models are constructed from monetary input–output tables representing the structure of the economy and the monetary flows among sectors and satellite tables with information on resource inputs, emissions, and other considerations such as labor. Some EEIO models consider multiple countries and regions and the trade among them; these are multi-region input–output (MRIO) models. Once the tables are constructed, EEIO calculations are typically performed by specifying final demand (consumer expenditures) and observing the resulting economic activity and emissions in the supply chain. Final demand for healthcare services can be derived from national health expenditure accounts or from international health statistics collected by the World Health Organization (WHO).

Using this approach, numerous national-level studies have been carried out to evaluate GHG emissions and other environmental impacts associated with healthcare. Starting with a paper by the U.S. health researchers Chung and Meltzer in

2009, there have been estimates to date for Austria, Australia, Canada, China, England, Japan, and several updates for the U.S., with more work coming out regularly. National-level results can reveal which healthcare activities are driving emissions as a necessary first step in creating decarbonization plans, such as the NHS in England's net-zero plans. In some cases, such as for Austria and England, researchers have been able to incorporate bottom-up data on transportation distances, building energy, or the use of specific medicines in order to create a hybrid model with higher accuracy.

Figure 31.1 shows national-level results for the most recent 2020 U.S. study by Eckelman *et al.* (2020). Emissions shift over time due to factors such as the decarbonization of background systems (e.g., electricity generation), overall spending on healthcare, and relative shifts in how that spending is allocated among goods and services. For the U.S., increases in overall health spending are the dominant trend, leading to modest growth in both absolute and per-capita emissions over the past decade. In 2018, the U.S. healthcare system represented approximately 8.5% of national emissions, making it one of the most important targets for future decarbonization efforts.

Though healthcare systems vary considerably in structure and management, supply chain emissions (represented as Scope 3 in Figure 31.1, following the GHG Protocol scopes framework) have been dominant across the board. Figure 31.2 shows similar results for Austria from Weisz *et al.* (2020) divided now into

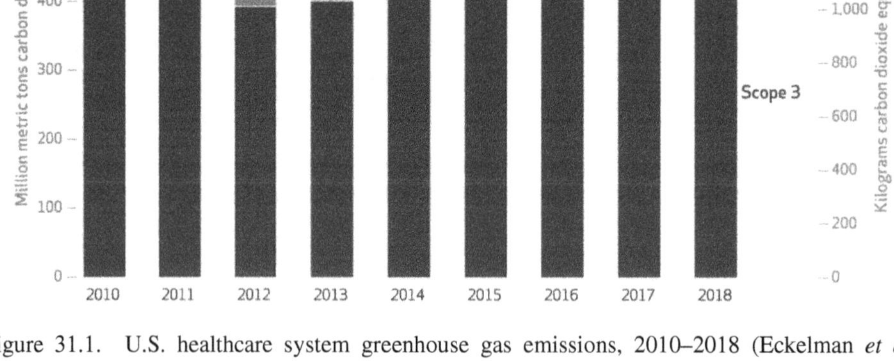

Figure 31.1. U.S. healthcare system greenhouse gas emissions, 2010–2018 (Eckelman *et al.*, 2020).

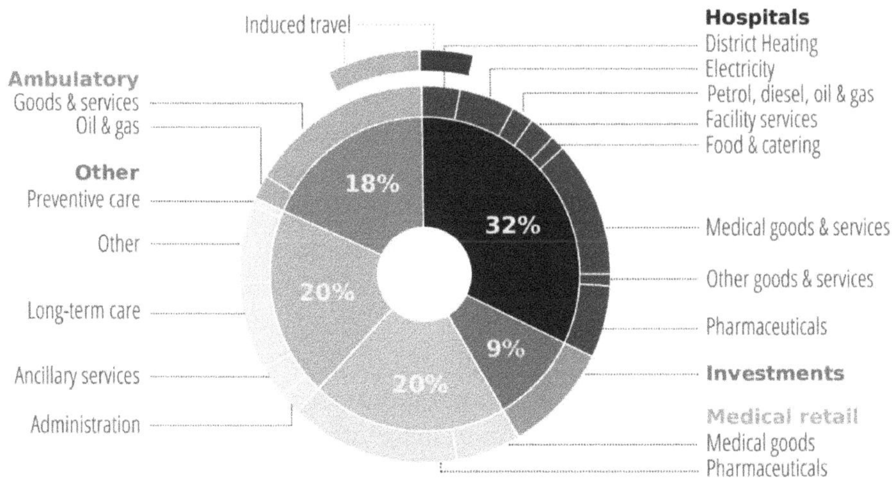

Figure 31.2. Austria healthcare system greenhouse gas emissions, 2010 (Weisz *et al.*, 2020).

different branches of healthcare and showing the large contributions of purchased items such as medical goods and services and pharmaceuticals.

The more detailed the expenditure data, the more insights can be gained for implementing change. National-level results have been disaggregated by state/province (the U.S.), age and disease type (Japan), and healthcare organization (England). The trend is now for individual hospitals and even clinicians to be able to link purchasing and clinical decisions to EEIO tools to be able to understand the carbon implications.

Climate change was declared by the WHO to be "single biggest health threat facing humanity," causing health damages through wildfires, flooding, extreme temperatures, vector-borne diseases, and malnutrition, among other causes. National EEIO studies have shown that the healthcare sector itself contributes to these health damages through its direct and indirect GHG emissions. But GHG emissions are just one type of pollution; EEIO models can also track emissions to air, water, and soil that have deleterious health effects, and use endpoint life cycle impact assessment (LCIA) methods to express impacts in terms of health damages like premature mortality or disability-adjusted life-years (DALYs). Studies for Canada and the U.S. have taken this approach in order to communicate results in the language of public health. For the U.S., healthcare-related emissions in 2018 lead to an estimated 244,000–531,000 DALYs in health damages, dominated by direct damages from $PM_{2.5}$ exposure and future damages from climate change. For perspective, it is interesting to note that these results are on par with the damages associated with the preventable medical errors that are a major focus of healthcare improvement efforts worldwide.

It is difficult to compare national-level studies because of differences in how input–output tables are constructed and how health expenditures are classified in each country. Nonetheless, several studies have used Multiregional Input–Output (MRIO) models that cover most countries and regions in order to compare performance across countries and to determine the magnitude of healthcare-related emissions on a global level. In addition, MRIO models can consider the influence of trade among countries, both of items like pharmaceuticals that are used directly in healthcare and of goods and services further upstream in healthcare supply chains.

A series of studies over the past several years have consistently found that healthcare is responsible for between 4.4% and 4.9% of global GHG emissions. The work by Lenzen *et al.* (2020) has been the most detailed, considering 189 countries and multiple environmental indicators. An example of a result, shown in Figure 31.3,

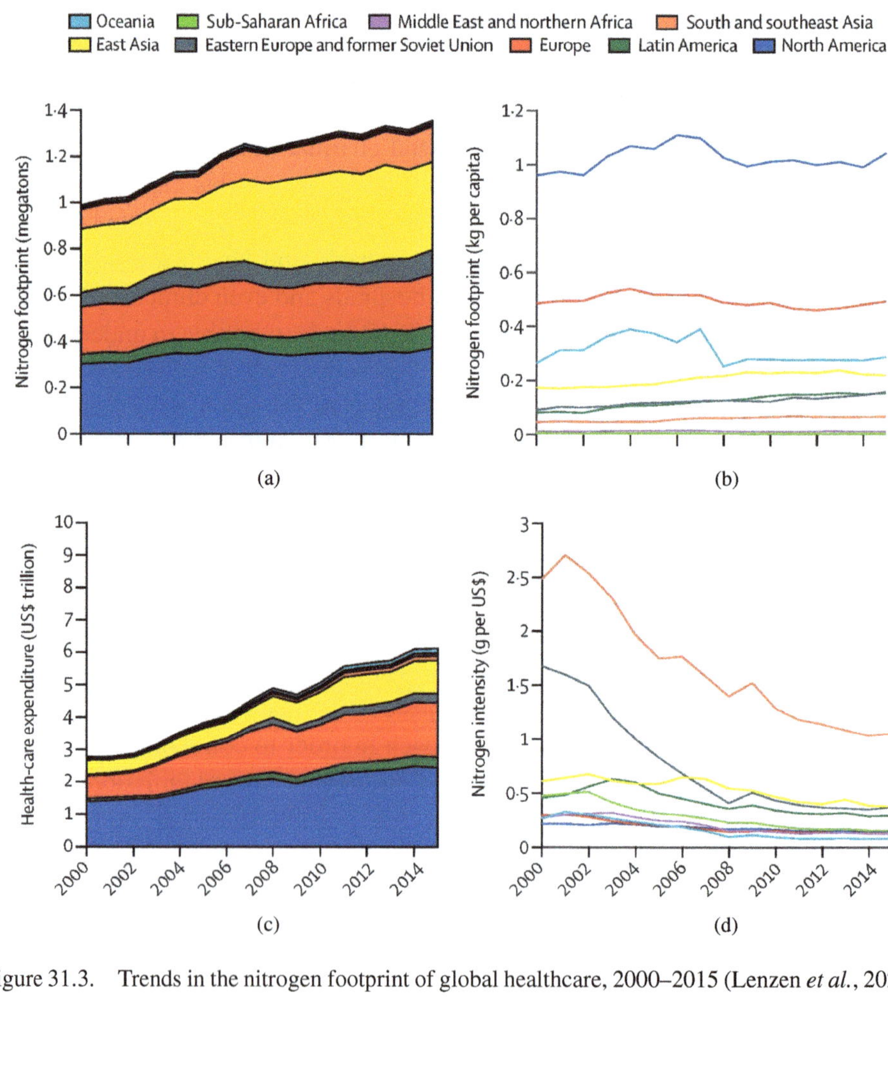

Figure 31.3. Trends in the nitrogen footprint of global healthcare, 2000–2015 (Lenzen *et al.*, 2020).

details the nitrogen footprint of global healthcare, broken down by region. Globally, the nitrogen footprint has increased nearly 40% from 2000–2015 (Panel A), driven by increases in global health expenditures (Panel C), even as the nitrogen intensity in the economy as decreased markedly over that time (Panel D).

Most of the international comparisons of healthcare environmental footprints have compared countries on a per-capita basis. But, as we learned in the chapters on LCA, when making comparisons, it is critical to consider the function being provided by each of the options. This principle also applies to healthcare systems, which provide different levels of service in terms of equitable access and quality of care. The Lancet Countdown on Health and Climate Change has tracked global healthcare GHG emissions over several annual reports, both on a national and per capita level, but also adjusted by measures such as the human development index (HDI) and the Healthcare Access and Quality (HAQ) index. Figure 31.4 shows the 2018 per capita results versus HDI, revealing a familiar trend of saturation, where additional emissions (driven by expenditures) do not necessarily lead to better human development outcomes. Conversely, the data shows that for the highest per-capita emitting countries, such as the U.S., significant decarbonization is possible while preserving (or even improving) health outcomes for the populace.

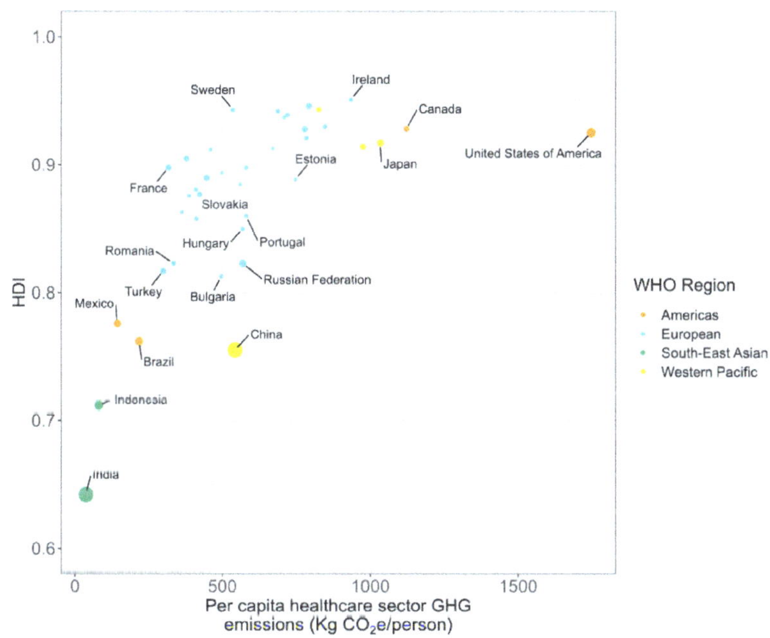

Figure 31.4. Per capita healthcare greenhouse gas emissions versus national Human Development Index in 2018 (Tennison *et al.*, 2021).

31.3 Life Cycle Assessment in Healthcare

Environmental footprinting using EEIO models is extremely useful in determining environmental impacts of healthcare systems in aggregate. Top-down results can be useful for setting national agendas, informing policy, and motivating action. But much of the actual decision-making takes place at a grassroots level, by individual clinicians and staff regarding delivery of care, by procurement professionals, or by facility managers involved in the engineering of healthcare buildings and infrastructure. Environmentally focused decisions at this bottom-up level involve specific, practical questions around which medical products or operational procedures are environmentally preferable, questions that are best addressed using life cycle assessment (LCA).

LCA has been applied to a wide variety of healthcare sustainability questions. These can generally be grouped into three major areas:

(1) comparing drugs or medical products that deliver equivalent outcomes to inform clinical and procurement choice;
(2) investigating clinical practices, procedures, settings, or modes of care to find hotspots and inform design and performance improvements; and
(3) large-scale reporting at the facility or organization level to inform strategic planning efforts.

The case studies in the following sections demonstrate how LCA can be used to provide insights and support decision-making, especially for clinicians who have the greatest direct responsibility patient health.

31.3.1 *Lessons for clinical practice*

31.3.1.1 *Anesthetic gas management*

Several LCA studies have looked at medical procedures and services, such as birthing, hysterectomy, dialysis, plastic, cataract, and minimally invasive surgeries, as well as ambulance, intensive care, dental, and telehealth services. Currently, one of the most carbon-intensive medical services is anesthesia. This is because the inhaled anesthetic drugs (sevoflurane, desflurane, isoflurane, and nitrous oxide) that are commonly used to provide general anesthesia are themselves potent greenhouse gases. They undergo very little *in vivo* metabolism in clinical use. They are exhaled and then scavenged by anesthesia machines to protect occupational exposure in the indoor environment and are then vented directly off hospital rooftops to the outdoor atmosphere. Their rates of release are largely unregulated. The different atmospheric lifetimes and infrared radiation absorption patterns determine the global warming potentials (GWPs) for desflurane (2,640 kg CO_2e/kg), isoflurane (510 kg CO_2e/kg), and sevoflurane (144 kg CO_2e/kg).

As you should recall from the chapters on LCA, any comparison among options must be made using a common functional unit. In the case of anesthesia, the functional unit is well-defined, called a minimum alveolar concentration (MAC) needed to sedate an average patient for one hour (1 MAC-hr). Different anesthetics have different medical potencies, and so different reference flows are needed to deliver the 1 MAC-hr, and the active agents are diluted with inert carrier gases. At 2 liters per minute of fresh gas flow, direct emissions of desflurane have been found to have 26 times the global warming impact of sevoflurane, whereas isoflurane has only twice the impact of sevoflurane. The addition of 60% nitrous oxide to sevoflurane or isoflurane (resulting in 1 MAC) worsens the impact of sevoflurane 6-fold, and isoflurane by 3-fold. These results clearly indicate that avoiding the use of desflurane in anesthesia would significantly reduce the carbon footprint of anesthesia while not affecting patient care, because other equivalent options are available. In fact, numerous hospitals and healthcare systems have acted to eliminate desflurane from their formularies.

What about the anesthetic gas life cycle? The earlier analysis focused just on waste anesthetic gases, but not on their manufacture, transportation, or clinical delivery. Sherman *et al.* (2012) investigated just this question by mapping out the chemical reaction schema for the drugs and using LCA to assess GHG emissions associated with drug synthesis, distribution, and the supplies and equipment (such as plastic tubing) needed in the operating room. Their analysis found that, in all cases, these other life cycle stages made only a minor contribution to overall emissions.

In addition to its impact on the climate, nitrous oxide (unlike sevoflurane, desflurane, or isoflurane) is also destructive to the ozone layer, providing an additional reason to reconsider its use. Estimates of total medical emissions of nitrous oxide range anywhere from 3% to 5% of the total, a substantial amount and therefore an important opportunity for reduction strategies.

Overall, inhaled anesthetics and nitrous oxide directly impact the environment and should be used thoughtfully. Avoiding desflurane and nitrous oxide use is one means of reducing these impacts, but there are other solutions as well. One behavioral change is for clinicians to reduce fresh gas flows once the initiation of anesthesia has occurred, rather than leaving them on at a consistently high level during an entire procedure. This ‘low-flow anesthesia’ can avoid emissions (and save drug costs) through simple training. For those emissions that cannot be avoided, gas capture systems are being developed to ensure that waste anesthetics are not simply vented to the atmosphere.

31.3.1.2 *Single-use versus reusable medical products*

One of the most common LCA topics in healthcare has been comparing single-use versus reusable medical devices. On the one hand, single-use devices must be replaced over and over, requiring new materials and all of the supply chain emissions that occur over the life cycle of the single-use product, including emissions from waste management of the used devices. The more times the reusable devices can be cycled through use, the more single-use device environmental burdens will be avoided. On the other hand, reusable devices are typically made of more durable materials, either heavier gage or metallic options that are more carbon-intensive to produce. Reusable items must also be collected, cleaned, and repackaged, which also require materials and energy, particularly if sterilization is required. It is not immediately apparent which option is preferable, so careful LCA analysis is required.

LCA studies have been published comparing different types of procedure kits, blood pressure cuffs, medical textiles, anesthetic drugs, and many other medical product and equipment alternatives. Most studies have compared only carbon emissions results, but a few have considered other environmental impacts such as air and water pollution or life cycle toxicity. For nearly every scenario, product type, and location, studies have shown that the reusable options are environmentally superior. A representative study on laryngoscopes by Sherman *et al.* (2018) is typical of the patterns of results. This study compared laryngoscope handles and blades between single-use (either plastic or metallic) versus reusable (metallic) versions, and several options for cleaning including low-level disinfection (LLD) or high-level disinfection (HLD) and sterilization. Figure 31.5 shows the main LCA results. In all cases, the emissions associated with reprocessing and repackaging were not negligible, but they were lower (sometimes much lower) than the emissions associated with replacing the single-use items. To support decision-making, these results were then scaled up to the number of laryngoscopes used at the study hospital in order to estimate potential emissions and cost savings of switching to reusable laryngoscopes of approximately $500,000 annually.

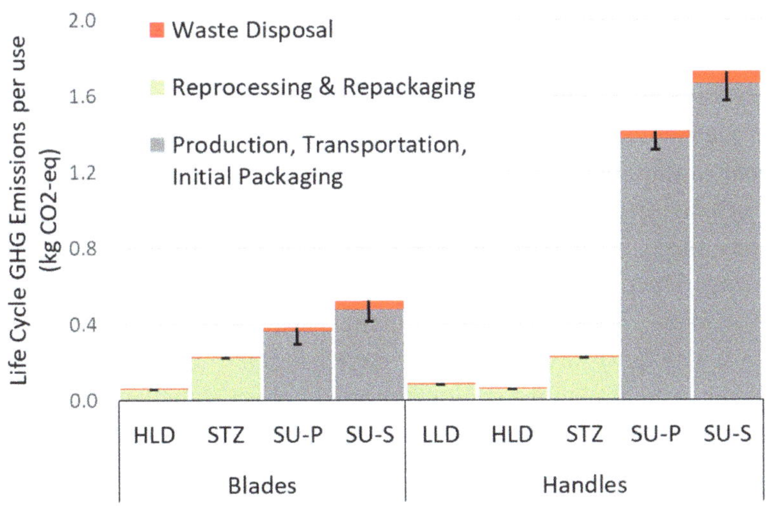

Figure 31.5. Carbon footprint comparison of single-use versus disposable laryngoscope handles and blades (Sherman *et al.*, 2018).

Note: low-level disinfection (LLD), high-level disinfection (HLD), sterilization (STZ), single-use plastic/metal (SU-P/M).

31.4 Critical Medicines and Supplies

Criticality is another industrial ecology lens that is proving useful in the healthcare arena. As with most areas of the economy, healthcare relies on global supply chains whose disruption can have dire consequences for patient care. There have been several recent catastrophic events that have elevated these concerns and led to a rethinking of healthcare supply chains. In 2017, the category IV Hurricane Maria made landfall on the U.S. Territory of Puerto Rico, causing widespread destruction and a collapse of the island's power grid. Puerto Rico was a major pharmaceutical manufacturing hub, supplying nearly 10% of all drugs. The closure of dozens of plants led to immediate shortages of pharmaceuticals, including basic items such as IV saline.

Other disruptions are economic or geopolitical. Medical isotopes used for diagnostic procedures and scans are produced at a small number of nuclear reactors worldwide. When one of these facilities halts production or trade of radioactive material is blocked by economic sanctions major supply shortages can result, as occurred in 2009–2010 for the important radioisotope molybdenum-99. Critical materials are also a concern for certain types of medical equipment. Magnetic resonance imaging (MRI) machines rely on high-strength magnets derived from rare earth elements, the great majority of which are produced in China. Concerns about potential supply disruptions have led to major investments in alternate sources of supply as well as materials science research to develop novel, earth-abundant magnetic materials.

A final example has been the major disruptions to healthcare supply chains and services due to the COVID-19 pandemic. Hospitals and clinics experienced a tremendous surge in the need for personal protective equipment (PPE). Some facilities experienced shortages that in some cases led to tragic consequences for frontline workers. Medical oxygen supplies were also strained, particularly in India, again with devastating results for patients. Such shortages raised important lessons about criticality in the healthcare supply chain and how to mitigate risk by investing in on-site or decentralized supplies. Those facilities that had the inventory and capacity to reprocess were able to reuse medical supplies and materials had alternate means of supplying PPE to their staffs. Similarly, those facilities that invested in on-site oxygen generators or concentrators were able to maintain care even if external supplies of bulk liquid oxygen were insufficient to meet demand.

31.5 Greening the Industrial Ecosystem of HealthCare

Healthcare has a critical role to play in helping societies adapt to climate change and other global environmental challenges, but also in mitigating its own environmental impacts. Healthcare is much more than just what happens at a hospital; rather, the industrial ecosystem of healthcare includes how we make drugs, medical supplies, and everything else on which healthcare delivery depends. Tools like EEIO and LCA help to uncover the largest sources of emissions and environmental impacts, so that healthcare organizations can take action to decarbonize and improve their own operations and work with suppliers to procure low-carbon goods and services.

Importantly, the ecosystem also includes people: patients, their supporters, clinicians, and staff. Holistic approaches to greening healthcare focus on keeping people healthy and avoiding hospital visits altogether. Investing in walking, biking, and public transit infrastructure allows equitable access to healthcare facilities and lowers transportation emissions, but also encourages exercise and improves local air quality. Telemedicine can also widen access to care and enable more regular check-ups that reduce the chance of complications that require a hospital visit, with its attendant resource-intensive consumption of goods and services. Public health campaigns that invest in training and encourage healthy behaviors are just as important to decarbonization, if not more so, than investments in hospital energy efficiency and renewable energy.

Further Reading

Chung, J. W., and Meltzer, D. O. (2009). Estimate of the carbon footprint of the U.S. health care sector. *Journal of the American Medical Association*, *302* (18), 1970–1972.

Eckelman, M. J., *et al.* (2020). Health care pollution and public health damage in the United States: An update: Study examines health care pollution and public health damage in the United States. *Health Affairs*, *39* (12), 2071–2079.

Lenzen, M., *et al.* (2020). The environmental footprint of health care: A global assessment. *The Lancet Planetary Health*, *4* (7), e271–e279.

MacNeill, A. J., *et al.* (2020). Transforming the medical device industry: Road map to a circular economy: Study examines a medical device industry transformation. *Health Affairs*, *39* (12), 2088–2097.

Sherman, J., Le, C., Lamers, V., and Eckelman, M. J. (2012). Life cycle greenhouse gas emissions of anesthetic drugs. *Anesthesia & Analgesia*, *114* (5), 1086–1090.

Sherman, J. D., Raibley IV, L. A., and Eckelman, M. J. (2018). Life cycle assessment and costing methods for device procurement: Comparing reusable and single-use disposable laryngoscopes. *Anesthesia & Analgesia*, *127* (2), 434–443.

Sherman, J. D., *et al.* (2020). The green print: Advancement of environmental sustainability in healthcare. *Resources, Conservation and Recycling*, *161*, 104882.

Tennison, I., *et al.* (2021). Health care's response to climate change: A carbon footprint assessment of the NHS in England. *The Lancet Planetary Health*, *5* (2), e84–e92.

Watts, N., *et al.* (2019). The 2019 report of The Lancet Countdown on health and climate change: Ensuring that the health of a child born today is not defined by a changing climate. *The Lancet*, *394* (10211), 1836–1878.

Weisz, U., *et al.* (2020). Carbon emission trends and sustainability options in Austrian health care. *Resources, Conservation and Recycling*, *160*, 104862.

Woodcock, J., *et al.* (2009). Public health benefits of strategies to reduce greenhouse-gas emissions: Urban land transport. *The Lancet*, *374* (9705), 1930–1943.

Chapter 32

Infrastructure

Chapter Concepts

- Infrastructure represents a large portion of society's material stocks, primarily in the form of construction minerals and steel
- Environmental impacts of infrastructure construction and maintenance of infrastructure are important in absolute terms, and should ideally be included when conducting life cycle assessment
- Infrastructure is closely associated with land use decisions that have long-term impacts on mobility and human geography
- Because infrastructure undergoes maintenance, refurbishment, and perhaps replacement, it may have no effective end of life and so requires rethinking the life cycle assessment approaches originally devised for products with finite lifetimes

32.1 Defining Infrastructure and Infrastructure Services

Infrastructure plays a central role in the modern economy and in industrial ecology (IE). It is almost always publicly owned and is thereby shared and funded by the general populace. One of the most common ways that infrastructure is shared is by connecting different locations. Traditional examples are roads and railways that connect cities and towns, bridges and tunnels that connect previously hard-to-access locations, or an electric grid that connects power plants with factories and homes. Even though almost everyone on Earth uses some type of infrastructure daily, it is so commonplace that we often take it for granted. The engineer Henry Petroski laments that, to the general public, "infrastructure [is] merely a fancy label for the stuff driven over, buried, and otherwise ignored compared to more glamorous aspects of the economy, society, and civilization that it makes possible" (Petroski, 2017).

For industrial ecologists, the sheer scale of physical infrastructure means that it represents a major driver of resource use and environmental impacts. Indeed, infrastructure projects are the largest physical things that we build as humans. The quantity of materials required is manifested not just in the physical roads or bridges that we see but also in all of the earthworks involved. (Recall that the largest material flows by mass, after water, are sand and gravel.) One of the world's largest infrastructure installations, the Three Gorges Hydroelectric Dam in China, used 28 million cubic meters of concrete in its construction, more than 80 times as much as the world's current largest building, the Burj Khalifa!

Many types of infrastructure have historically fallen under the field of civil and environmental engineering, and the American Society of Civil Engineers (ASCE) classifies infrastructure into 16 different types (Table 32.1). These types can be further grouped into major infrastructure areas: transportation, energy, water, waste, and recreation/education.

Traditionally, infrastructure has been defined in physical terms, but there is growing recognition that infrastructure should be defined in terms of the services it provides, rather than what it physically is. This is similar to the role that a functional unit plays in life cycle assessment (LCA). A particularly broad definition was set forth by the U.S. Department of Homeland Security, wherein infrastructure is a "framework of interdependent networks and systems comprising identifiable industries, institutions (including people and procedures), and distributing capabilities that provide a reliable flow of products and services essential to the defense and economic security of the United States, the smooth functioning of government at all levels, and society as a whole, including physical, cyber, and/or

Table 32.1. Types of Infrastructure as Devised by ASCE.

Aviation	Ports
Bridges	Public parks
Dams	Rail
Drinking Water	Roads
Energy	Schools
Hazardous Waste	Transit
Inland Waterways	Solid waste
Levees	Wastewater

human elements." In other words, infrastructure is a system of systems (a holarchy; see Chapter 29), with complexities, interdependencies, and dynamics that we are just beginning to understand. This definition is aligned with the principles of ecological science that underpin industrial ecology.

Some infrastructure connections are obvious, such as the need for electricity to run water pumps, but others are less so, such as the same need for electricity to run pumps at fuel-filling stations. In the event of a long power outage, this dependency can prevent residents from being able to obtain fuel for vehicles in order to evacuate or run backup diesel generators. Better understanding of interdependencies among infrastructure systems, including human linkages, will ideally allow us to design the next generation of infrastructure to the goals of equity, environmental sustainability, and resilience.

32.2 Estimating In-Use Stocks of Infrastructure Materials

We typically know how much money is spent on infrastructure projects, but we have only recently begun to track the physical requirements of infrastructure in terms of overall resource requirements. The techniques for studying the physical requirements of infrastructure are in many ways the same as those described in the chapters on material flow analysis (MFA). For example, using a bottom-up approach, information on the total length, size, or number of infrastructure elements is collected from statistical agencies, project records, or even remotely sensed from aerial or satellite images. These quantities are then multiplied by the material intensity of that infrastructure per unit, which can be derived from engineering designs, to find the total infrastructure material stocks.

Example 32.1.

An at-grade road is made up of several layers. The subgrade is typically soil from on-site, but the aggregate base and pavement layers use materials that must be mobilized and therefore count as material stocks. In this example, the subgrade aggregate is typically eight inches thick and the compacted asphalt six inches thick (including both base and surface layers). A typical lane width in the U.S. is 12 feet. Including voids, the aggregate layer has a density of approximately 115 lb/ft^3 and compacted asphalt has a density of 145 lb/ft^3.

(a) What is the material required for 1 lane-mile of road? Convert your answer to per lane-km of road and express in SI units.
(b) Texas is the U.S. state with the most roads, approximately 680,000 lane-miles. Extrapolate from your answer in part a to estimate the total material stock of road infrastructure in Texas.
Note: In practice, road designs and thicknesses change depending on expected traffic and speed, so it is important to do this calculation separately for different road classes. We are also ignoring other material elements like signs, lighting, barriers, and drainage.

ANSWER:

Part a.

The aggregate layer has a volume per lane-mile of:

$$V = L \times W \times H$$

$$= \left(1\,\text{lane}\cdot\text{mile} \times \frac{5280\,\text{feet}}{1\,\text{mile}}\right) \times \left(\frac{12\,\text{feet}}{\text{lane}}\right) \times \left(8\,\text{inches} \times \frac{1\,\text{foot}}{12\,\text{inches}}\right)$$

$$= 42{,}240\,\text{ft}^3$$

So, the mass of aggregate per lane-mile is

$$M = V \times \rho = 42{,}240\,\text{ft}^3 \times 115\frac{\text{lb}}{\text{ft}^3} = 4{,}857{,}600\,\text{lbs}$$

The asphalt layer has a volume per lane-mile of

$$V = L \times W \times H$$

$$= \left(1\,\text{lane}\cdot\text{mile} \times \frac{5280\,\text{feet}}{1\,\text{mile}}\right) \times \left(\frac{12\,\text{feet}}{\text{lane}}\right) \times \left(6\,\text{inches} \times \frac{1\,\text{foot}}{12\,\text{inches}}\right)$$

$$= 31{,}680\,\text{ft}^3$$

So, the mass of asphalt per lane-mile is

$$M = V \times \rho = 31{,}680\,\text{ft}^3 \times 145\frac{\text{lb}}{\text{ft}^3} = 4{,}593{,}600\,\text{lbs}$$

And the total mass of material per lane-mile is

$$M_{\text{1-mi}} = 4{,}857{,}600 \text{ lbs} + 4{,}593{,}600 \text{ lbs} = 9{,}451{,}200 \text{ lbs}$$

Considering 1 km of road with the same lane design, the total mass would be

$$M_{\text{1-km}} = 9{,}451{,}200\,\text{lbs} \times \frac{1\,\text{kg}}{2.2\,\text{lbs}} \times \frac{1\,\text{ton}}{1000\,\text{kg}} = 4296\,\text{tons}$$

Part b

Extrapolating to the entire U.S. state of Texas, the total mass of material stock in roads is

$$M_{TOT} = \frac{9{,}451{,}200\,\text{lbs}}{\text{lane} \cdot \text{mile}} \times 680{,}000\,\text{lane} \cdot \text{miles} = 6{,}427\,\text{billion lbs}$$

or 2915 million metric tons (Tg) of material. The population of Texas is currently 29 million people, so the roads represent 100 metric tons per capita of material stocks.

Industrial ecologists have used this bottom-up approach to understand material requirements to support infrastructure. Building off Example 32.1, a full examination of stocks of construction minerals in U.S. roads was conducted by Miatto *et al.* (2017), who found a total of 15.1 billion metric tons in 2015, or 47 metric tons per capita. This represents 13% of all in-use material stocks in the country!

A two-dimensional (2-D) approach works well for relatively standardized construction like at-grade highways, but cannot capture the high variability of design and heights for infrastructure like bridges. More sophisticated bottom-up estimates use three-dimensional (3-D) maps that capture the full volume of materials used, with data from design drawings, remote sensing, or even laser scans from drones. Importantly, below-ground infrastructure must also be counted, including tunnels and foundations.

Numerous studies have been conducted in China, where so much of the world's infrastructure construction has been concentrated in recent years. This has had major implications for material use. For example, Wang *et al.* (2016) analyzed the material requirements of China's high-speed rail network, which they estimated to require nearly 100 million metric tons of steel. When considered on a per unit length basis and compared with results from other countries, this type of analysis can reveal the material implications of different engineering designs. In the case of China, much of its high-speed rail network is built on elevated or underground track and so has a higher material intensity than European rail that is largely at-grade. In general, bottom-up dynamic stock accounting of infrastructure such as this allows for the comparison among different systems or projects to determine which designs are most resource efficient.

Top-down input–output or trade-based models can also be used in MFA to estimate in-use material stocks in infrastructure by assigning certain flows to infrastructure categories or uses, such as steel in structural elements for bridges or copper for electricity transmission and distribution. In a global top-down MFA study spanning 80 years of cement use, Cao *et al.* (2007) found that just in the last decade, China consumed three times more cement than the U.S. had over the past

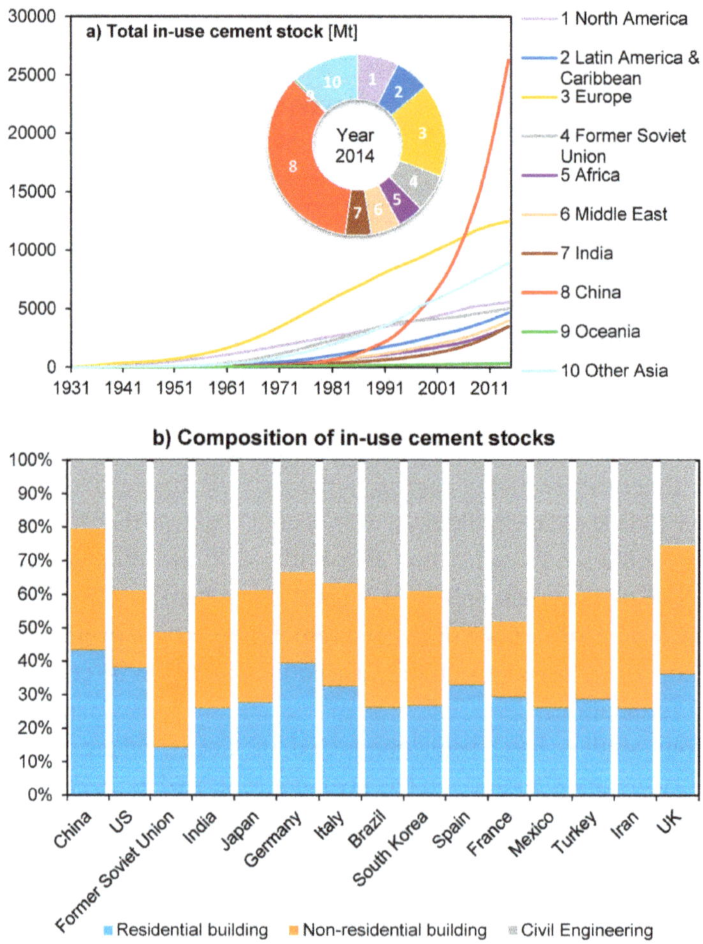

Figure 32.1. (a) Global in-use stocks of cement from 1931–2014. (b) Sectoral shares of in-use stocks of cement in major economies (Cao *et al.*, 2017).

eight decades! Using input–output sector categories, the authors were also able to determine the proportion of cement used in buildings versus civil engineering infrastructure (Figure 32.1).

32.3 Mapping Infrastructure Materials over Time

Infrastructure generally grows over time, especially as populations increase and where sufficient funding can be made available. A visual example is provided by Figure 32.2, showing urban and semi-urban growth as related to the availability of regional rail service in the Netherlands from 1850–2010.

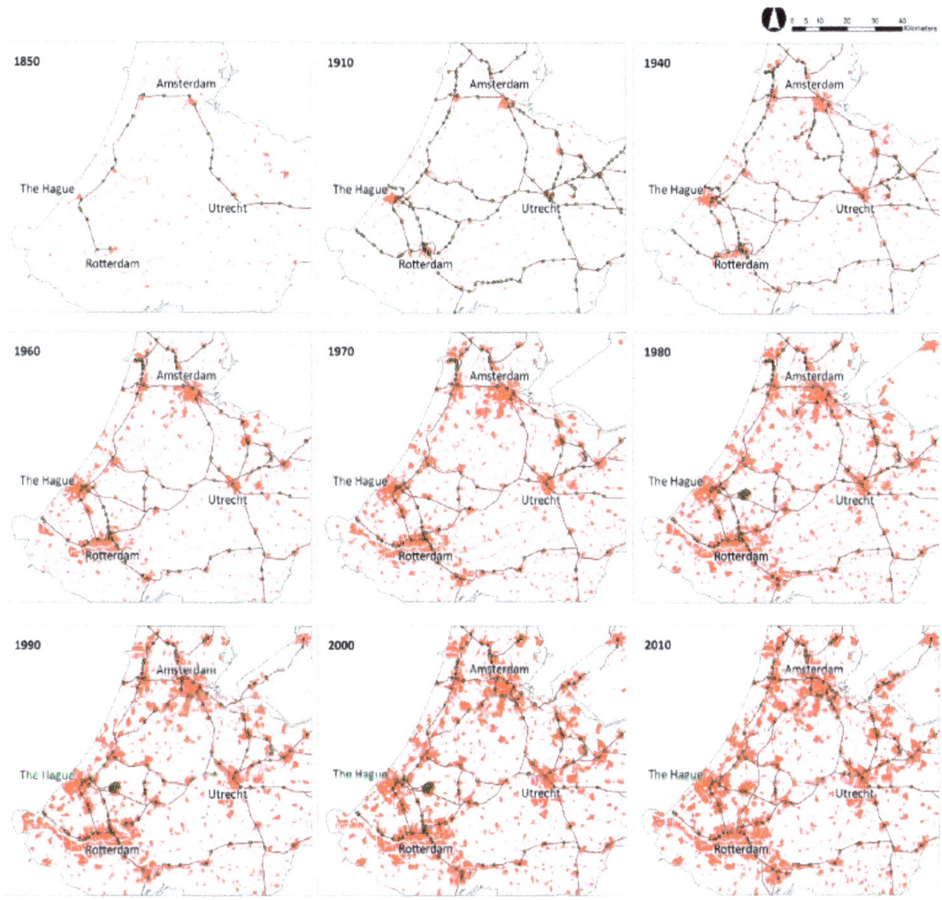

Figure 32.2. The growth in built-up areas (red) and rail systems (black) in the Netherlands over a 160-year period (Kasraian *et al.*, 2017).

IE researchers have used maps like these to map infrastructure material stocks in four-dimension (4-D). Probably the most extensive efforts to track material stocks of infrastructure have taken place in Japan, starting with individual cities and eventually applying bottom-up methods to the entire country, taking advantage of Japan's excellent statistical agencies. Tanikawa *et al.* (2015) tracked infrastructure materials used in Japan from the mid-1940s to the present, including roads, railways, underground water infrastructure, airports, seaports, and dams. A high-resolution map of the material intensity of Japanese infrastructure is shown in Figure 32.3, where urban areas and major travel corridors can be seen.

Long-run, spatially explicit MFA is useful in infrastructure planning because it can reveal where and when infrastructure materials may need to be updated or replaced. It is also possible to study the effects of improvements in infrastructure materials, technology, or design practice. For example, advances in civil engineering analysis over the past 50 years, especially with the advent of computer simulation, have allowed engineers to use lower factors of safety (multiplicative factors

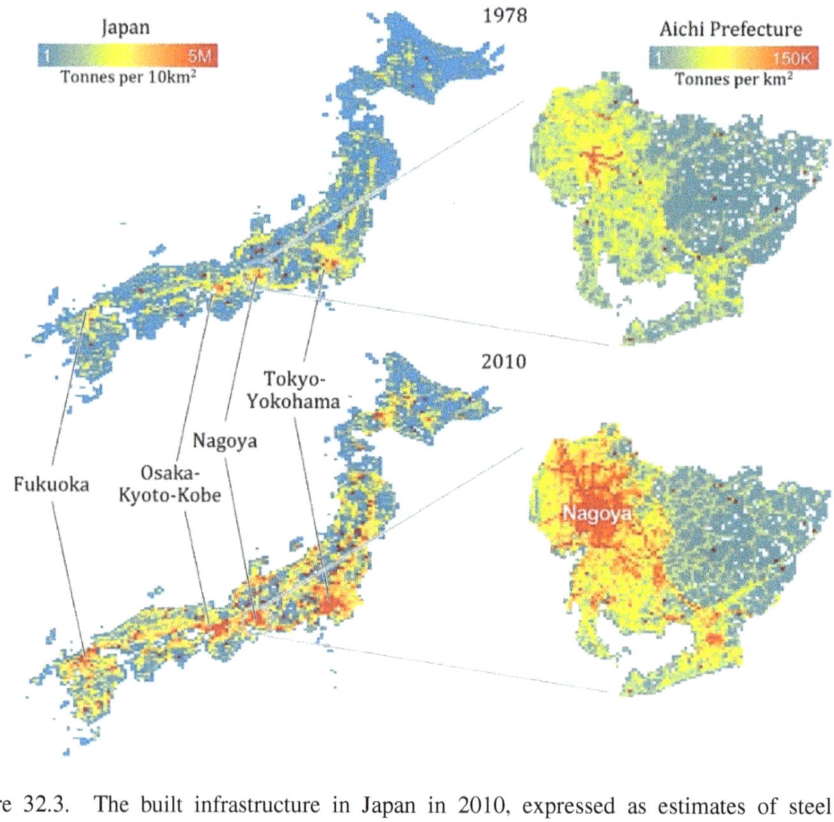

Figure 32.3. The built infrastructure in Japan in 2010, expressed as estimates of steel stock (Tanikawa *et al.*, 2015).

for overdesigning projects), thus reducing the amount of construction materials required.

32.4 Infrastructure and Environmental Impacts

Material stock accounting indicates total resource requirements of infrastructure between regions or across space and time. Often, however, infrastructure designers and managers are concerned with quantifying environmental impacts of specific projects. In many countries, large projects must obtain environmental permits before construction can begin, often through submitting an environmental impact assessment (EIA) or statement. In the U.S., these requirements and procedures fall under the National Environmental Policy Act (NEPA). An EIA typically includes direct impacts on habitat, soil, water, and air quality, and impacts on surrounding communities such as noise.

In industrial ecology, environmental impacts are considered at a systems level using LCA. So, while EIA considers only direct impacts from infrastructure construction and use, LCA considers additional environmental burdens stemming from infrastructure site preparation, materials production and transport, maintenance, and eventual decommissioning. Crucially, LCA can be used to evaluate trade-offs among life cycle stages that EIA cannot, and is therefore useful in evaluating alternative designs, especially those that take very different approaches in providing services to the public.

Thousands of LCA studies have been done specifically on infrastructure, covering all types listed by ASCE and across many countries. Many studies evaluate specific projects or case studies, but some also examine broad sustainable design and technology innovations that are widely applicable and at the core of our changing conception of infrastructure. The following are some examples of large infrastructure research questions that have been evaluated using LCA in recent years:

Transportation
- Is asphalt or concrete a better material for roads?
- Which public transit option is most environmentally preferable?
- How can streets be redesigned to provide multiple services at the lowest environmental costs?

Energy
- Is a decentralized energy system preferable to a centralized system?
- What are the carbon and water implications of different methods for generating electricity?

- What are the energy costs associated with the internet, and is moving to teleworking and cloud-based computing preferable?

Water

- What is the most environmentally preferable method for supplying safe drinking water?
- What are the life cycle benefits of recovering and utilizing nutrients from wastewater, and which treatment technologies are best suited for this?

Waste

- What should we do with different types of solid waste — is it better to recycle, compost, incinerate, or landfill?
- How important is collection and transport, and what distances make one option preferable over another? Or, should people be responsible for transporting their own waste?

These are big-picture questions that require a life cycle perspective and are often context- or location-dependent, and so LCA has proven to be an important tool for policy makers and industry in guiding infrastructure investments, to the point where each of the topical areas indicated earlier has dedicated conferences, technical standards, and communities of practitioners.

Of course, with infrastructure being so central to the economy, various infrastructure processes are a factor in almost *any* LCA. Consider the life cycle of a tomato: transportation infrastructure is used for shipping, energy infrastructure is needed to deliver fuels and electricity required for processing and storage (and perhaps cooking), water infrastructure is used for cleaning and wastewater treatment, and waste infrastructure is used for processing food scraps, and there are many other possible infrastructure inputs as well. In general, LCA does not consistently include infrastructure inputs and LCA practitioners can choose. For example, power transmission and distribution infrastructure is typically included as this has been factored into many existing electricity life cycle impact (LCI) data sets, but transportation infrastructure is often excluded. How important is this choice?

A seminal piece by Chester and Horvath in 2009 revealed part of the answer. In this work, the authors compared different personal transportation options, a topic that had been well-covered in previous analyses, but this time, infrastructure materials and processes were included (Figure 32.4).

First, notice the extremes. An urban diesel bus is both the most carbon-intensive way to travel (when mostly empty, off peak) and the least (when mostly full, peak). This clearly shows that social and management aspects of

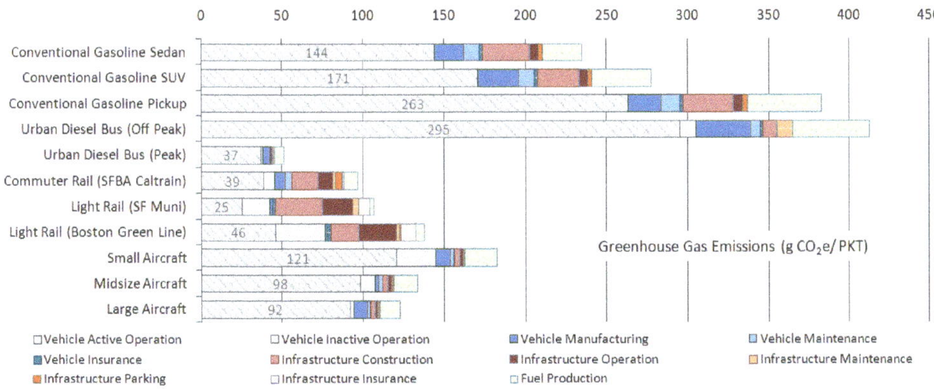

Figure 32.4. Life cycle greenhouse gas (GHG) emissions of different travel modes, per passenger kilometer traveled (PKT) (Chester and Horvath, 2009).

infrastructure, such as maintaining high ridership, can substantially influence LCA results. Second, look at the breakdown between use phase (white shades), vehicles (blue shades), and infrastructure (orange/red shades). For most travel modes, greenhouse gas (GHG) emissions associated with infrastructure (including construction, operation, maintenance, parking, and insurance) are non-negligible, especially for light rail in San Francisco and Boston, where they approached half of the total. One exception was air travel, where the contribution of infrastructure to overall emissions is small. For air travel, this is the case only for the origin and destination which require physical infrastructure (the airports), but not the path connecting them, unlike for roads or rail. This and related studies brought a life cycle approach to understanding the environmental impacts of transportation infrastructure and emphasized that ignoring infrastructure can lead to underestimation.

A life cycle perspective has also seeped into infrastructure design. Infrastructure engineers and owners also increasingly want to build infrastructure that is sustainable and resilient. In all areas of engineering, published standards are critical in ensuring that materials have the correct specifications, testing has been carried out, or generally that agreed-upon procedures have been followed to ensure the desired performance. We have already learned about the ISO standards for LCA, carbon and water footprinting, and environmental product declarations, as well as the Leadership in Energy and Environmental Design (LEED) certifications for green buildings. The infrastructure community has also developed certification standards for infrastructure, the most prevalent of which is called Envision, published by the Institute for Sustainable Infrastructure. Envision ranks infrastructure projects on their effects in five areas: (1) quality of life, (2) leadership, (3) resource allocation, (4) natural world, and (5) climate and risk. Many of the metrics included are based on industrial ecology ideas of resource efficiency, and the climate and risk category has a metric explicitly looking at life cycle GHG emissions.

32.5 Infrastructure and Land Use

Perhaps the most important environmental impact to consider is the influence of infrastructure on land use. Many infrastructure decisions last for millennia, through a process called technological or spatial "lock-in". In short, once infrastructure has been placed, it becomes so expensive, bureaucratic, or time-consuming to move them that things tend to stay where they are. For example, where roads are placed is extremely difficult to change once surrounding privately-owned properties have been developed. Even after the Great London Fire in 1666, where there was a chance to rationalize the street system in the core of the city, residents quickly returned to stake out their plots, essentially keeping the roads where they had been in medieval times.

Our modern era has witnessed an incredible expansion in urban areas around the world, including low-density urban sprawl in the U.S. Decisions made now on how road networks are laid out (with accompanying water and energy infrastructure) will largely govern which transportation options are available, where people live and work, and how urban land is used for centuries to come. An important implication of infrastructure lock-in and land use is that it limits our options when new environmental issues are realized. The relationship between density and environmental impacts is complex, but it is generally accepted that density enables walking, biking, and public transportation options, with benefits for emissions and public health. For communities that have been designed to be far from urban centers, largely residential, with disconnected road patterns (describing much of the past 50 years of U.S. land development), it will be difficult to achieve a low-carbon, resource-efficient lifestyle.

Infrastructure's effect on land use also has severe implications for other species. While low-density development preserves more green space than an urban core, what remains is fragmented, making it unsuitable for species that require contiguous corridors for migration or habitat. The same can be true for industrial infrastructure in rural areas. In analyzing land use changes by oil and gas infrastructure in Canada, Jordaan *et al.* (2009) showed that although the individual land occupied by a production site was modest, the land between them also became unusable as habitat, and the effective land footprint was therefore many times larger than previously reported.

Saxe and Kasraian have suggested that the life cycle characteristics of infrastructure are so different in scale and influence from those for industrially-manufactured products with discrete lifetimes that existing LCA approaches need

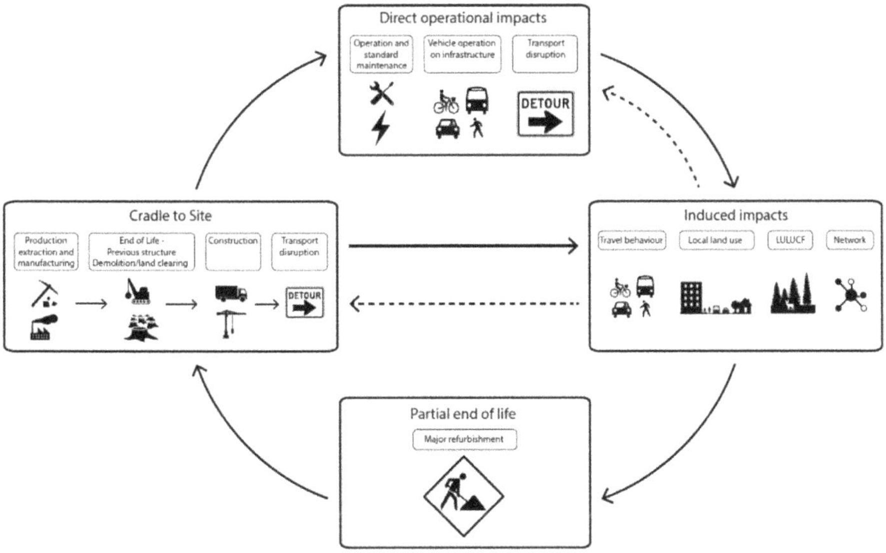

Figure 32.5. The life stages of transport infrastructure (Saxe and Kasraian, 2020).

significant revision to make them suitable for application to infrastructure. They recommend a system diagram of the sort pictured in Figure 32.5, with specific recognition of the role of infrastructure in inducing long-term changes in mobility, urban development patterns, and land use.

The environmental impacts of infrastructure are wide-reaching, and industrial ecology is in many ways an ideal lens through which to analyze infrastructure using a systems-level approach. This is an area of cutting-edge research in industrial ecology, as new analytical tools, methods, and metrics are being developed. Close collaboration with other disciplines is a key feature of the next generation of IE work on infrastructure, allowing us to add sustainability considerations into existing engineering design tools, incorporate social science approaches to evaluating equity in infrastructure access and effects, and utilize evolving understanding of natural ecology in urban areas to represent infrastructure's impacts on other species and ecosystems.

Further Reading

Cao, Z., Shen, L., Løvik, A. N., Müller, D. B., and Liu, G. (2017). Elaborating the history of our cementing societies: An in-use stock perspective. *Environmental Science & Technology*, *51* (19), 11468–11475.

Chester, M.V. and Horvath, A. (2009). Environmental assessment of passenger transportation should include infrastructure and supply chains. *Environmental Research Letters*, *4* (2), 024008.

Jordaan, S. M., Keith, D. W., and Stelfox, B. (2009). Quantifying land use of oil sands production: A life cycle perspective. *Environmental Research Letters*, *4* (2), 024004.

Kasraian, D., Maat, K., and Van Wee, B. (2017). Development of rail infrastructure and its impact on urbanization in the Randstad, the Netherlands. *Journal of Transport and Land Use*, *9*(1).

Miatto, A., Schandl, H., Wiedenhofer, D., Krausmann, F., and Tanikawa, H. (2017). Modeling material flows and stocks of the road network in the United States 1905–2015. *Resources, Conservation and Recycling*, *127*, 168–178.

Petroski, H. (2017). *The Road Taken: The History and Future of America's Infrastructure*. USA: Bloomsbury Publishing.

Saxe, S. and Kasraian, D. (2020). Rethinking environmental LCA life stages for transport infrastructure to facilitate holistic assessment. *Journal of Industrial Ecology*, *24*, 1031–1046.

Tanikawa, H., Fishman, T., Okuda, K., and Sugimoto, K. (2015). The weight of society over time and space: A comprehensive account of the construction material stock of Japan, 1945–2010. *Journal of Industrial Ecology*, *19* (5), 778–791.

Wang, T., Zhou, J., Yue, Y., Yang, J., and Hashimoto, S. (2016). Weight under steel wheels: Material stock and flow analysis of high-speed rail in China. *Journal of Industrial Ecology*, *20* (6), 1349–1359.

Chapter 33

Urban Industrial Ecology

Chapter Concepts

- Urban areas concentrate on people as well as material stocks and represent the majority of human demand for energy and materials
- Urban metabolism uses the analogy of the city as an organism, tracking inputs of nutrients, growth of physical stocks, and waste outputs
- Urban industrial ecology studies map material stocks and flows in space and time, enabling better infrastructure planning, materials management, and understanding of material inequities

33.1 The City as an Organism

Cities have often been compared to organisms. They are born, they grow, and they pulsate. As with other organisms, they utilize resources. The resource spectrum of a city is spectacular: apples from New Zealand, electronics from Japan, automobiles from Germany, leather from Argentina, and lobsters from the U.S. Cities also dissipate resources after use, sometimes to the local air and water, often to land fills, generally close to home. Cities are accomplished attracters and poor dispersers.

If a city can be regarded as an industrial ecology organism, what might we learn about cities from traditional methods of biology? What does a biologist learn from studying, for example, the metabolism of a squirrel? It turns out that a number of useful facts emerge — the amount and types of nutrients the animal requires, how the nutrients are utilized, what proportion is stored, how rapidly nutrients pass through the body, what fraction is devoted to growth, and how all of

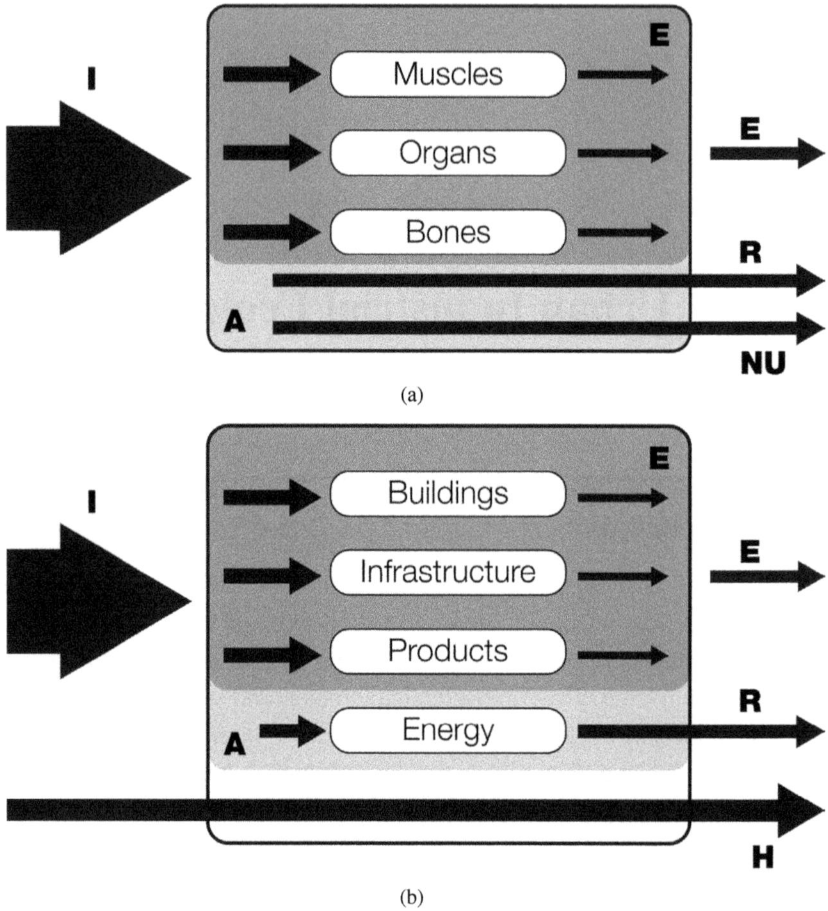

Figure 33.1. Metabolic flows of organisms (a) and of urban areas (b). I = ingestion, A = assimilation, E = egestion, R = respiration, NU = not used, H = hidden flows.

these change with time. The metabolic cycle may be pictured as shown in Figure 33.1(a). The ingested material (I), shown at the left of the diagram) has three fates. Some is not utilized at all (NU), as with the bones of small mammals eaten by owls. The remainder, the assimilated portion, consists of two parts: that devoted to respiration (R), thus providing energy for the organism, and that devoted to the maintenance and growth of the organism's parts (E). The change in mass of the organism (its growth) is given by

$$\Delta M = I - E - R - NU \tag{33.1}$$

Analogously, the urban ecologist can, at least in principle, use the information in an urban metabolic analysis to gain perspective about steel beams, gasoline, water, food, and so on. As with the squirrel, we wish to know the uses and fates of these "urban nutrients" — their rates of flow, their degree of spatial concentration, their modes of loss, and their potential for reuse. Figure 33.1(b) is the urban version of Figure 33.1(a). As before, there is ingestion (of crushed stone, diesel fuel, coffee makers, etc.), and egestion (of garbage, discarded cellular telephones, etc.). The assimilated material, as with the organismal analog, is of two parts. One is the energy sources — fuel oil and natural gas that provide energy to the urban system and generate respiratory products such as carbon dioxide. The second consists of the materials and products that enable our modern technological world to function. The urban organism changes mass with time according to

$$\Delta M = I - E - R \tag{33.2}$$

A difference between this formula and that for natural organisms is that there is no "not used" portion, at least if everything occurs as planned. A second difference is the arrow termed "hidden flows" (H); these represent the mobilization of waste rock and soil and the use of energy in the generation of the urban nutrients. These flows occur away from the city, and so are hidden from it, but are attributable to it.

A biological organism's metabolic flow consists largely of organic matter (which is often expressed in energy terms) and water. The flows connected to a city are much more complex. They include five components at minimum: (1) nutrients, (2) materials, (3) water, (4) energy, and (5) dissipative flows, but these categories can be disaggregated considerably for various purposes. Nutrients can be subdivided into many food types, for example, and materials into a diversity of product groupings. Combining this broad scope with the challenge of assembling accurate and complete data renders an urban metabolic study a major challenge. Nonetheless, given that resources follow people, and people increasingly congregate in cities, the long-term sustainability of resource supplies, the ability to reuse them, and the environmental consequence of discarding them are all stories that will largely be told in cities.

More than half a century ago, a sanitary engineer named Abel Wolman published what can be regarded as the first *urban metabolism* study, and probably the earliest industrial ecology scientific study of any kind. In the paper, he characterized urban metabolism as follows: "The metabolic requirements of a city can be defined as all the materials and commodities needed to sustain the city's inhabitants at home, at work, and at play". This definition remains valid many decades later, as will be seen in this chapter. Wolman's focus was on urban water acquisition, treatment, and losses, but in his paper, he managed to estimate the overall metabolism of a typical American city of one million people. Fundamentally, not much has changed, and with a few revisions and updates, the paper could probably be published today.

33.2 Urban Metabolic Flows

A number of material flows are routinely monitored at the national level, largely because taxes and import/export duties are collected at that level. No comparable information usually exists for cities because there is generally little incentive for governments to collect it. However, useful information sometimes can be acquired because water companies monitor water flow, electrical suppliers bill customers for the amount of electricity that is used, and so forth. As a result, at least an approximate, likely partial, urban metabolism can be developed by merging these sources of information. This can be done for single nutrients or discards or by combining information into a more comprehensive urban metabolism.

An example of the results generated by an urban metabolic study is that for Bogotá, Colombia, shown in Figure 33.2. In this relatively simple picture, the "nutrients" to the city are grouped into four categories: (1) energy, (2) water, (3) food, and (4) construction minerals. The related outflows are also quantified as emissions, sewage, solid wastes, and construction wastes. In tables within the research report, the categories are further divided into fuel use by type, food

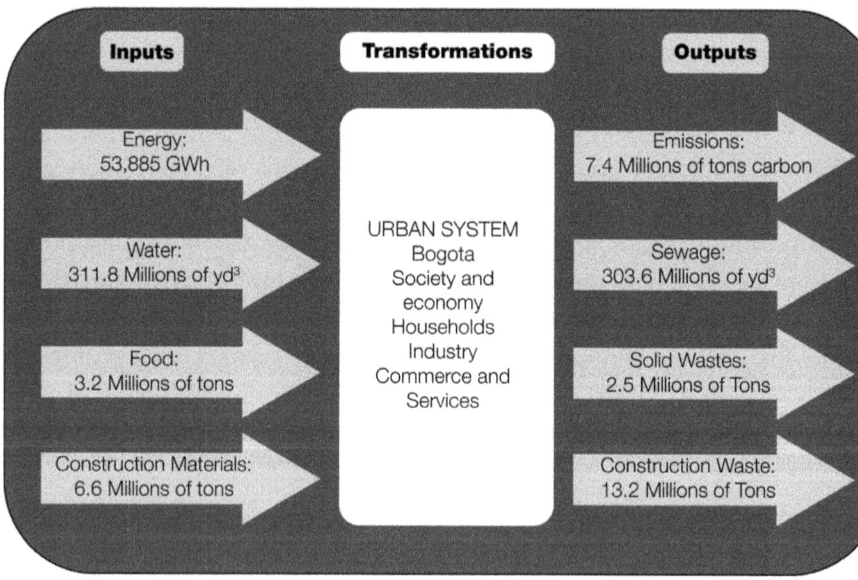

Figure 33.2. A material flow analysis for Bogotá, Colombia, 2010 (adapted from Piña and Martinez, 2014).

amount by household wealth, etc. The results are of significant usefulness in understanding Bogotá's materials flows, and perhaps for the purposes of policies designed to respond to that information (to reduce emissions, for example).

A more detailed but still grouped evaluation is shown in Figure 33.3, comprised of urban metabolism results for two cities, a metropolitan region, and two agricultural regions in France. The two medium-sized cities (about 400,000 in

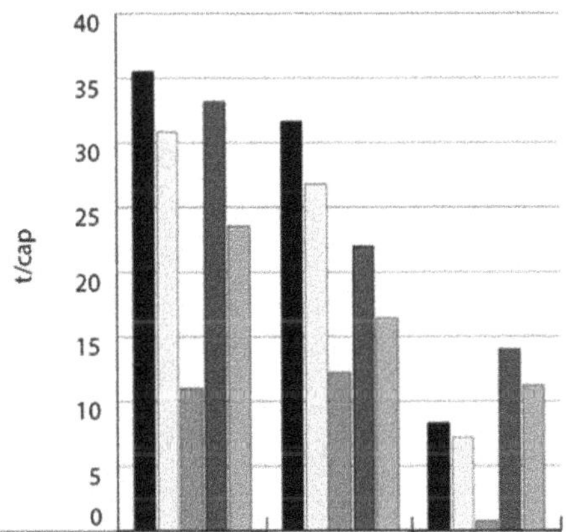

	DMI	DMO	NAS
Le Mans	35.5	31.7	8.3
Rennes	30.8	26.8	7.2
Paris (megacity)	11	12.2	0.7
Ariège (rural-industrial)	33.2	22	14
H-Garonne (rural-urban)	23.6	16.4	11.2

(a)

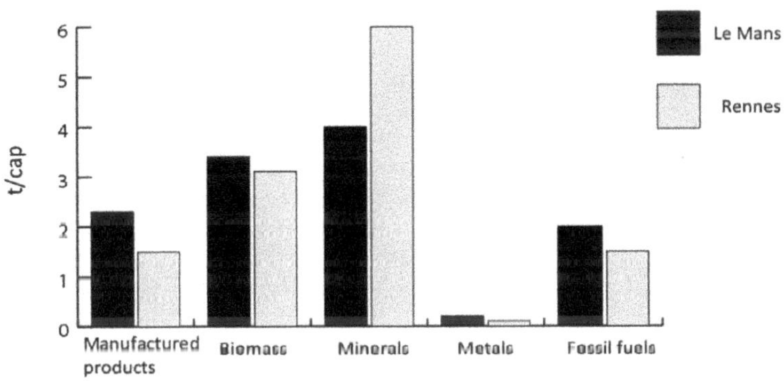

(b)

Figure 33.3. (a) A comparison of the principal material flow analysis indicators for the cities of Paris, Le Mans, and Rennes, and the regions of Ariège and Haute-Goronne, France: DMI = Direct material Input, DMO = Direct Material Output, NAS = Net addition to stocks. (b) Direct Material Consumption (DMI-DMO) for Rennes and Le Mans by materials category (Bahers *et al.*, 2018).

Rennes, about 200,000 in Le Mans) show similar pictures, the per capita flows in Le Mans being somewhat higher. The Paris region, as a large, compact urban area without significant industry and agriculture, has much lower per capita material flows. The two agriculturally focused regional areas have the highest flows. The section of the figure at the right shows the Net Additions to Stock (NAS), again on a per capita basis: quite high for the agricultural regions, nearly as high for the two cities, and quite low for the Paris region.

Despite the challenges of conducting an urban metabolism study, interesting and useful results are increasingly being generated. Studies for multiple sites permit the derivation of important information, as Figure 33.4 demonstrates for studies of 27 megacities (populations greater than 10 million) in 2011. The graph shows that the megacities, with some 6.7% of the global population, have combined gross domestic products (GDP) of nearly 15% of the global amount, and generate nearly 13% of global waste. The tenor of the results is probably not surprising, but the quantifications are quite valuable in taking global results to higher resolution.

An innovation in urban metabolism studies is not only to track material flows across the city boundaries but also to spatially locate where flows are utilized. This is done by taking advantage of the geo-spatial data gathered by global navigation satellite systems to manage vehicle fleet operations. In this work, the geo-spatial data on movements is enhanced by interviews with truck drivers, who validate the tracking data and also provide information on the types and volumes of commodities that are transported. In the example shown in Figure 33.5, the arrows indicate the paths of significant flows of construction materials in Singapore in 2017–2018,

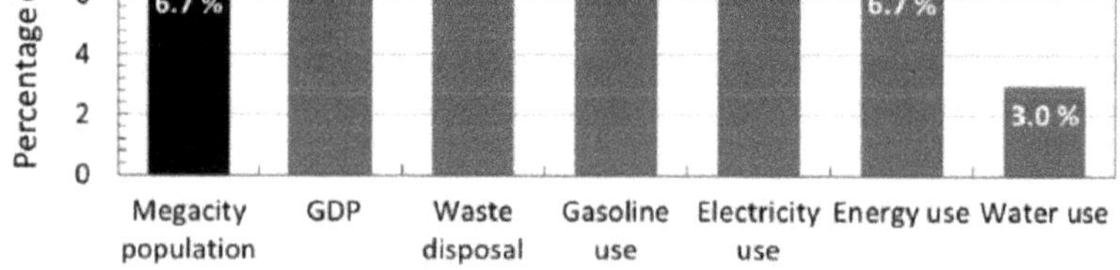

Figure 33.4. Megacity resource and waste flows as percentages of their global values (Kennedy *et al.*, 2015).

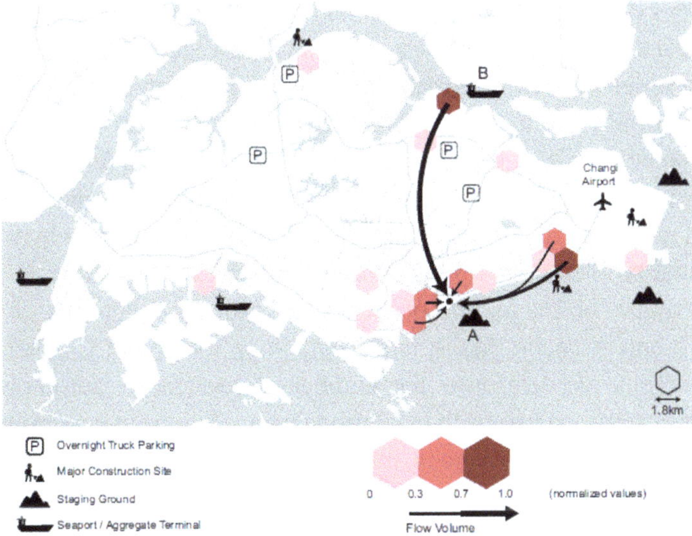

Figure 33.5. The origins and amounts of construction material flows to Location A in Singapore for five days in 2017–2018 (Yeow and Cheah, 2019).

and the hexagons indicate flow magnitude to Point A. The largest flows are from the port at Point B, with smaller amounts from a construction site to the east and from areas near Point A. This example indicates the extent of quantitative spatial information that can be derived for urban material flows, given the utilization of non-traditional data sources.

33.3 Urban Metabolic Stocks

Many measurements or estimates in science and technology are derivative; that is, they measure something related to what one wishes to determine rather than measuring the actual entity of interest. Perhaps the most famous examples of derivative measurements are those related to the structure of the atom — quarks and gluons are well understood because of analyses of the signatures of various types of collision fragments in atom accelerators, not because the entities have been weighed on a balance.

From the perspective of the atomic derivative measurements, we can address our desire to know the stocks of urban nutrients, especially materials of interest like copper or lead that exist in long-lived uses like building wiring and batteries. If we have good information on inputs and outputs over time, we can take the difference to compute stocks in use. (This is the *top-down* method of Chapter 7 the analogous problem for the atmosphere is computing carbon dioxide concentrations by measuring or estimating the CO_2 emissions to the atmosphere and the rate at which CO_2 is removed by such processes as the growth of vegetation.) Only rarely do we have a time series of such information for cities, however, either long or short. In addition, we would like to know more than just the amounts — we would like to know with some certainty the location of the material within the city. Such knowledge serves two purposes — it could make it easier to collect and reuse the material, and if the material is environmentally toxic, it could help to anticipate environmental problems should some of the material be dissipated.

To determine the magnitude of stocks, we turn to inventories of products in which the resources occur, that is, to the *bottom-up* method of Chapter 7. For copper, for example, we determine the number of residences, automobiles, power distribution systems, and so forth in an urban area, often from geographic information system (GIS) maps. We then multiply those reservoir numbers by typical copper concentrations within them to arrive at an estimate of both total copper stock and its location. (This is analogous to determining atmospheric CO_2 stock by direct measurement, as is done commonly.) If desired, spatial techniques can be used to distribute the stock determinations over the urban area. The characterization of large spatial variations enables government officials to plan more wisely for the eventual recovery of those resources by knowing where they are located, when it is likely to be discarded, and how much can be expected to be available.

33.4 Urban Metabolism and Society

The study of urban metabolism is perhaps the most obvious link between industrial ecology and human populations. Ultimately, the flows of material resources that constitute that metabolism exist to provide services to city dwellers, and those people also value attributes of society and of the local environments.

How can human/material interactions be best quantified and then made vivid for planners and citizens alike? An example of one tool for communication is the expanded urban material flow analysis shown in Figure 33.6, an "edge bundling diagram" for the city of Bangalore, India. In this complicated but strangely engaging diagram, 28 supplying sectors (left side of the diagram) are linked with 36 receiving sectors (the right side). The colored arcs are proportional to the flows (if such exist) from each of the former to each of the latter. For example, a red arc connects sector BM1 (agricultural biomass, at top of diagram) to sector S24 (hotels and restaurants, at center right of diagram). A light green arc depicts the flow of stone from sector NM1 (sand, lower left) to sector S28 (real estate activities, upper right). The widths of the arcs are proportional to the flow magnitudes

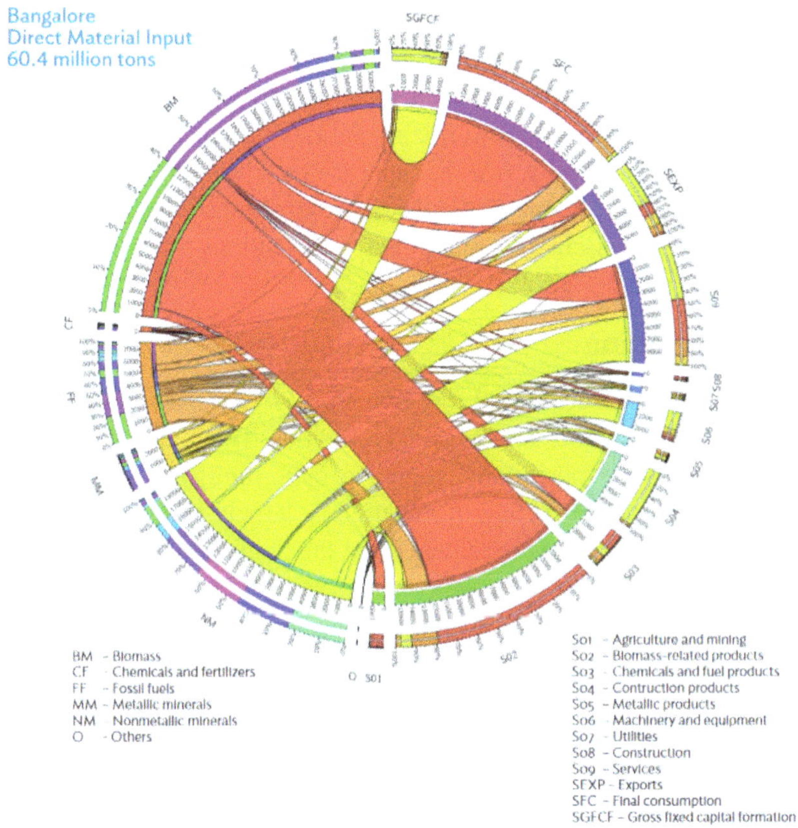

Figure 33.6. The distribution of resources across industrial and social sectors in Bangalore, India (Asian Development Bank, 2014).

from the supplying sector (the solid base of the arc) to the receiving sector (white gap at the end of the arc). Narrow white wedges and black connecting lines indicate where no material flows exist, as from the nuclear fuels sector (MM5) to the education sector (S34). This diagram and its supporting documentation obviously provide material for intensive study by industry, government, and society, but this example provides an introduction to the potential benefits that can be derived by quantifying this network of materials interchanges.

How do detailed diagrams like Figure 33.6 come about? The approach was originated in 2014 by Rosato *et al.* (2014). They realized that urban metabolism studies generally possessed several serious liabilities, including the lack of a unified methodology, the lack of material flows at the urban level, and limited categorization of material types. They drew upon extensive databases from the

European Union (EU), the Food and Agricultural Organization (FAO), and others, to remedy most of the defects and generate the needed information. Other organizations such as the Asian Development Bank (ADB) repeated the approach with somewhat less starting data but by utilizing proxies such as the number of workers per specific economic activity. In either approach, they recognized that high levels of accuracy were unlikely, but that good approximations were quite sufficient to guide contemplation, discussion, and policy interventions.

33.5 Urban Mining

A key concept in the discussion of resource reuse is that materials are used where people are located, and this is increasingly in cities. This means that materials in use in cities present a potentially attractive opportunity for eventual resource recovery. Currently, the recycling of materials is largely opportunistic and short-term. However, the increasing urbanization of the global population is concentrating the materials in use, or, we might say, enriching the grade of the "urban ore bodies", at the same time as ore grades in conventional mines are decreasing.

Urban mining, or what might be termed spatially oriented recycling, has a number of specific benefits for society. These include the following:

- mitigating problems of resource depletion by providing an alternative source;
- avoiding (for recycled materials) the environmental impacts of mining;
- saving energy;
- saving water;

What are the practical requirements for implementing the urban mining? There are at least five:

1. an accurate determination of in-use magnitudes or stocks of urban materials, good estimates of the discard flow rates as a function of time, and specification of the physical and chemical forms of the discards;
2. an accurate determination of the spatial distributions of the urban stocks;
3. the design of buildings, equipment, and products of all kinds so as to facilitate disassembly and reuse;
4. a system of efficient collection and sorting of discards;
5. the provision of incentives, monetary or otherwise, for materials reuse.

Can urban mines replace those of nature? This cannot happen so long as the total usage of materials is increasing because even if 100% of discarded material is captured for reuse, this material, re-entering the system after some years of service, can at best provide only a fraction of the total requirements. This situation can approach sustainability only as the total annual requirements for materials approach constancy.

Lastly, if we want to mine cities, should we mine *all* cities? We do not know very accurately whether stock in cities scales linearly with population, how stock differs as a function of city wealth or location, or whether large cities tend to have more urban ore per capita or less. One could imagine unique situations as well; for example, that large commercial cities might be rich in copper, germanium, and other materials as a result of a large concentration of computers and other industrial electronics. A local manufacturing industry might constitute a high density of steel. A community centered around agriculture might have lower densities of some of these technologically advanced materials. There is clearly a research challenge in characterizing these diverse urban stocks.

To have urban mines, we need urban miners and urban mining companies. Such entities exist now, but they are smaller than they might be and less efficient. It is clear, however, that there is significant potential for urban mining as a contribution toward sustainability. It will take initiative, effort, and political will to make extensive urban mining a reality.

33.6 Potential Benefits of Urban Metabolic Studies

Plants and animals are understood in part because their metabolic information allows us to understand the functioning of those organisms. The same potential holds true for studies of urban physical metabolisms. The lure of such investigations is immediately apparent: allowing researchers and policy makers alike to examine what (and how much) enters a city, how long it is retained, and when and how it is discarded. This information has potential for informing discussions of resource availability, recycling, energy use, and environmental impact. It could enable the comparison of one city's performance with that of another, or with itself over time.

A comprehensive characterization of urban metabolism includes the need to measure not only the combined flows of all forms of materials, but the flows of *each* form. This is crucial because for recycling and reuse purposes, the viability of the stock of a material generally depends on the efficiency with which it can be recovered. A material discarded in pure form, such as copper pipe, is reused readily. The same amount of material in complex manufactured products — cellular telephones, for example — may be much more likely to be consigned to the landfill. Impacts to the environment are also highly dependent on its chemical form — a chromium-plated bumper is environmentally benign, but the Cr^{6+} ion is not.

As with any organism, a city's metabolism is characterized by the quantities of flows of resources, their distribution throughout the organism, and the dynamic nature of the flows. The sustainability of a city cannot be evaluated unless those factors are quantified. If they are, they define the urban industrial ecology of the city and permit decisions to be made on an informed basis.

Further Reading

Asian Development Bank (2014). *Urban Metabolism of Six Asian Cities*. Retrieved from: https://www.adb.org/sites/default/files/publication/59693/urban-metabolism-six-asian-cities.pdf.

Bahers, J.-B., Barles, S., and Durand, M. (2018). Urban metabolism of intermediate cities. *Journal of Industrial Ecology*, *23*, 686–698.

Bristow, D. N. and Mohareb, E. A. (2020). From the urban metabolism to the urban immune system. *Journal of Industrial Ecology*, *24* (2), 300–312.

Kennedy, C. A. *et al.* (2015). Energy and material flows of megacities. *Proceedings of the National Academy of Sciences*, *112*, 5985–5990.

Piña, W. H. A., and Martinez, C. I. P. (2014). Urban material flow analysis: An approach for Bogotá, Colombia. *Ecological Indicators*, *42*, 32–42.

Rosato, L., Niza, S., and Ferrão, P. (2014). A material flow accounting case study of the Lisbon metropolitan area using the urban metabolism analyst model. *Journal of Industrial Ecology*, *18* (1), 84–101.

Wolman, A. (1965). The metabolism of cities. *Scientific American*, *213* (3), 178–190.

Yeow, L. W. and Cheah, L. (2019). Mapping urban material flows with freight data. *Journal of Industrial Ecology*, *23*, 1121–1132.

Zhang, Y., Yang, Z., and Yu, X. (2015). Urban metabolism: A review of current knowledge and directions for future study. *Environmental Science & Technology*, *49*, 11247–11263.

Prospective and Systemic Industrial Ecology

Chapter 34

Scenarios in Industrial Ecology

Chapter Concepts

- Unlike material flow analysis, which examines what occurred in the past, and in-use stock analysis, which looks at materials presently in use, industrial ecology scenarios enable and encourage thinking and planning about the future
- Scenarios are not predictions but are well-grounded stories about how the future might unfold under different constraints and opportunities
- Scenarios are policy-relevant and link industrial ecology to societal decision-making

34.1 Purposes of Scenarios

The chapters in this book thus far have been directed at a single objective: to understand how modern societies employ natural and manufactured materials, how these materials are exchanged, utilized, and lost, what the environmental consequences of those uses might be, and how we can do better. This is an important common thread in these topics that is worth pointing out: they are all addressed to understanding what is occurring *at present*. This chapter presents a complementary approach: devising, generating, and analyzing depictions of how the future *might* unfold. Such depictions are termed *scenarios*.

Scenarios were a tool originally developed to serve national defense organizations but were adopted by major industrial corporations in the 1970s. The most famous example of corporate benefit from developing and utilizing scenarios is that of Ted Newland and Pierre Wack's group at Royal Dutch Shell. That team developed alternative pictures of global political and resource-related actions and their consequences. One of these pictures (i.e., scenarios) asked what would happen to Shell's business if Middle East oil producers suddenly decided to halt production for political purposes. Although Shell's management thought such action unlikely, they developed a corporate action plan should it occur. When it did, Shell's prompt actions saved a great deal of money and positioned the corporation for leadership in the oil industry for many years to come.

The consumption of oil and its products in our modern technological society is just one example of society's use of the material resources of Earth. It is possible to imagine different material-related futures, oil or otherwise, without knowing which (or if any) might become true. Such stories are *not* predictions, but rather are alternative depictions (or what social scientists call cultural constructs). Figure 34.1 illustrates the difference. A prediction assesses the present, sketches out the transitions, and presents its presumption of the future. In contrast, a scenario activity looks at the present situation from different perspectives and imagines several possible transitions to a future state. The goal of a set of scenarios is to provide diverse but believable pictures of situations that could occur over time and to use the development of those situations over time to stimulate thinking about informed responses that might prove to be useful.

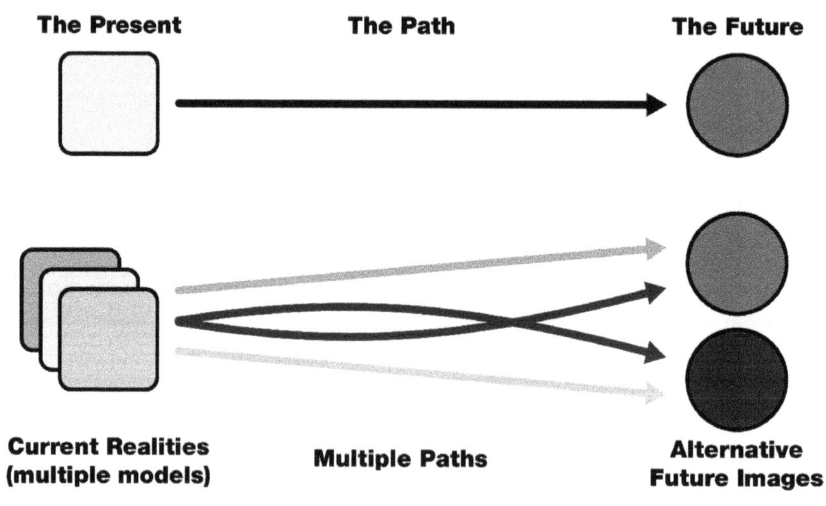

Figure 34.1. Distinguishing predictions from scenarios.

34.2 The Process of Building Scenarios

Because a set of scenarios provides alternative pictures of the future, the set needs to have several components: stories that communicate to users the thrusts of the various scenario components and of the individual scenarios. Although scenarios are sometimes limited to narrative depictions, they generally are most useful if told in both narrative and quantitative form. An initial decision in formulating a scenario set is to determine how many stories to generate. Choices have varied widely over the past half-century or so, the most common advice being to choose a relatively small number of scenarios and to choose an even number. The latter guidance exists because experience has shown that given an odd number, many people will instinctively choose the scenario in the middle as the "most likely", and thus not explore other depictions that may be of equal or greater value. In fact, it is well known among scenario planners that most government and corporate planning assumes that the future will look much like the present. However, as futurist Alan Toffler said, "It does little good to forecast the future...if the forecast springs from the premise that everything else will remain unchanged".

In a 2013 publication by Wilkinson and Kupers (see References), several attributes of scenario construction were identified. Among them were the following:

- Make scenarios plausible, not probable.
- Avoid expressing a preference for one scenario over another.
- Make scenarios relevant to the current situation but challenging to its assumptions.
- Give scenarios names that are short and memorable. One- or two-word names such as Relapse or Regional Rivalry can provide recognition and then be enhanced by the storylines.
- Add numbers to narrative. Corporate and governmental organizations often find resonance in quantitation. As an old saying has it: "Numbers beats no numbers every time".

A frequent scenario development approach is to select and populate four scenarios, as shown in Figure 34.2, with the vertical axis separating the most uncertain of the anticipated influencing variables from the least uncertain. The horizontal axis might be used to separate the most important influencing variable from the least important. The result is a set of guidelines that forms the basis for four scenarios. Each scenario can then be given a tentative title, the shorter and more memorable the better.

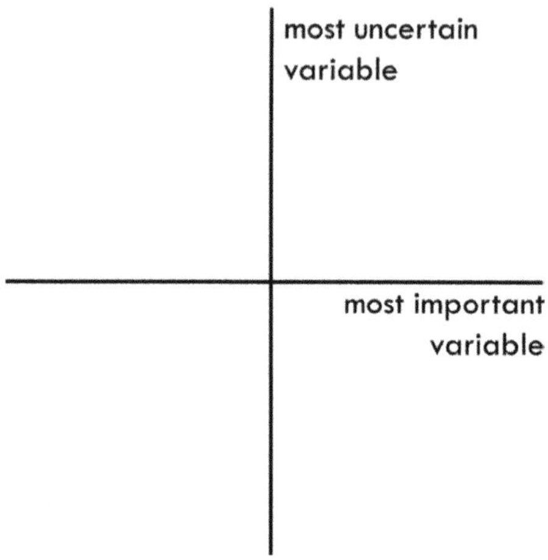

Figure 34.2. The two-axis approach to scenario design.

The team then follows a sequence of steps:

1. Develop a set of storylines.
2. Ask an interesting question that might be addressed with scenarios (examples might be as follows: "Are any resource supplies in peril?" "How much influence could recycling have?" "What would be the material implications of a major civil war in Country X?")
3. Incorporate essential components (variables) into the scoping of the scenarios.
4. Identify and quantify the most important motivating forces for the storylines, which are as follows:
    - Driving forces — well-known factors that will help shape the future
    - Prime movers — powerful individuals, organizations, and governments
    - Predetermined elements — factors that are highly likely to occur
    - Uncertainties — factors that could change the future in major ways but have moderate or low probabilities.

Examples of these motivating forces are likely to include such factors as growth in human populations and changes in per capita income over time.

A "storyline" is now constructed for each scenario. Storylines are depictions of plausible alternative futures, describing in human terms (rather than lines on a graph) how those future could unfold. They will ultimately be several pages long, and probably include displays and tables that will help a reader understand in some detail what the story is about. Figure 34.3 shows a graphical storyline of the

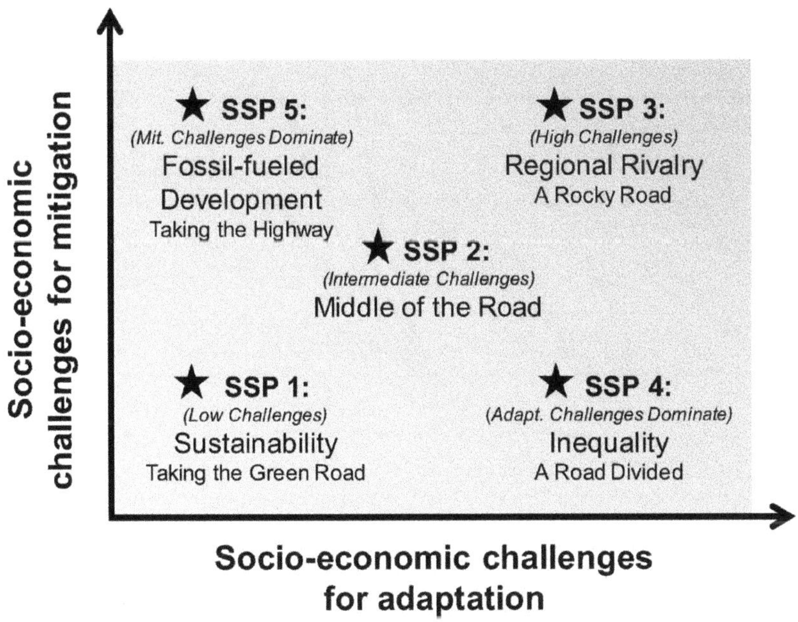

Figure 34.3. Distinguishing the shared socioeconomic pathways (SSPs) trajectories (O'Neill *et al.*, 2017).

widely employed and extensively characterized Shared Socioeconomic Pathways (SSPs) (see Riahi *et al.*, 2017).

34.3 Forming Industrial Ecology Scenarios

Why are scenarios useful in industrial ecology? Three reasons can be postulated as follows:

1. No science ever truly has impact until it can make predictions based on a possible course of action. (The success of scenarios in communicating the poten tial impacts of climate change is a good example of this impact.)
2. Industrial ecology can think on time scales that are typically longer than those of businesses or governments.
3. Industrial ecology-related scenarios have the potential to become the operational strategies for sustainable development.

Scenarios are usually directed at a specific topic or question. In industrial ecology, the issues most frequently focus on addressing the future magnitudes

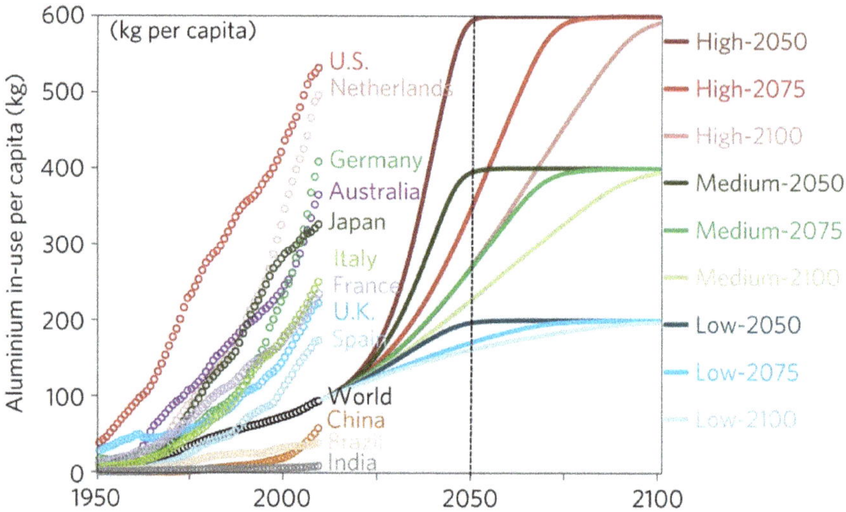

(a)

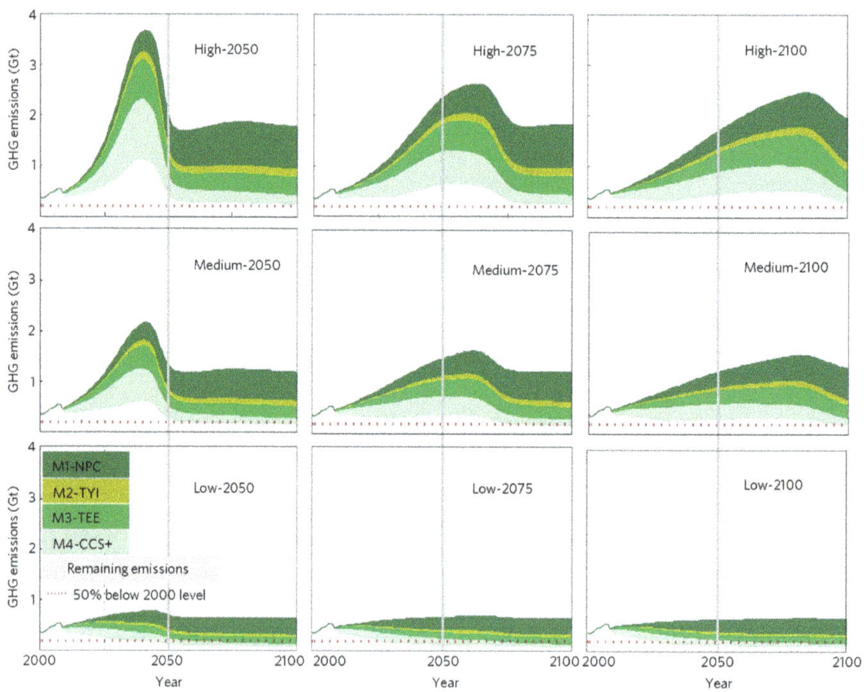

(b)

Figure 34.4. (a) Stock dynamics and (b) emission pathways of the global aluminum cycle (Liu *et al.*, 2012).

of the use of material and energy in industrial and consumer activities, of the related effects on the environment, potential limitations in resource supply, and of the influences of economic, political, regulatory, and social factors on future flows, uses, stocks, and transformations of resources. A common starting point in recent years is to employ a variation of the SSPs and then superimpose industrial ecology-related parameters such as technology uptake or recycling regulations.

34.4 Example Results and Implications of Industrial Ecology Scenarios

Primary aluminum processing is one of the major industrial generators of greenhouse gases (GHG), and those working in the aluminum industry and related to it need guidance on how best to enable the transition to aluminum use over the next several decades and the implications for greenhouse gas emissions. A crucial variable in this connection is the average amount of aluminum anticipated to be in use per capita. The data available do not yet show that any country has reached saturation for this variable. Consequently, Liu *et al.* assumed that three different values were possible, as were three dates on which saturation would occur. The result is nine development curves that need to be followed out to 2100 (Figure 34.4(a)). The scenario study was thereby able to generate nine

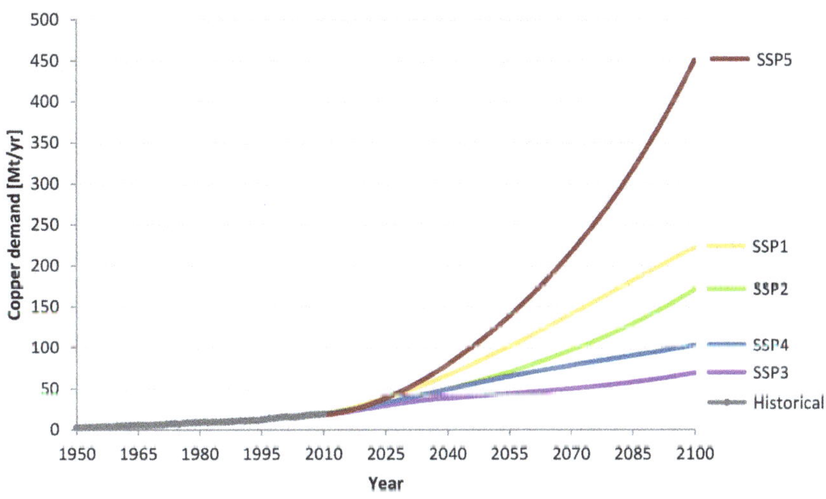

Figure 34.5. Copper demand, supply, and associated energy use in the 21st century from the basis of the Shared Scenario Platform (SSP) scenarios (Schipper *et al.*, 2018).

stories of possible relationships between aluminum development and greenhouse gas generation (Figure 34.4(b)) and thus provide extensive food for thought by corporate and government executives who wish simultaneously to develop modern technological benefits for citizens while at the same time minimizing climate impacts.

The SSP development framework has also provided the basis for a set of resource scenarios related to future copper demand. Figure 34.5 pictures five alternative copper demand patterns, with SSP5, the scenario with the largest global GDP growth generating by far the largest copper demand, most of it for infrastructure. The difference of some six times the demand, depending on which scenario occurs would surely get the attention of resource planners in the copper industry and of governments in countries anticipating extensive capital development in the 21st century.

Further Reading

Fishman, T., Myers, R. J., Rios, O., and Graedel, T. E. (2018). Implications of emerging vehicle technologies on rare earth supply and demand in the United States. *Resources, 7*, 7010009.

Liu, G. *et al.* (2012). Stock dynamics and emission pathways of the global aluminum cycle. *Nature Climate Change*, *3*, 338–342.

O'Neill, B. C. *et al.* (2017). The roads ahead: Narratives for shared socioeconomic pathways describing world futures in the 21st century. *Global Environmental Change*, *42*, 169–180.

Pauliuk, S., Arvesen, A., Stadler, K., and Hertwich, E. G. (2017). Industrial ecology in integrated assessment models. *Nature Climate Change*, *7* (1), 13–20.

Riahi, K. *et al.* (2017). The shared socioeconomic pathways and their energy, land use, and greenhouse gas emissions implications: An overview. *Global Environmental Change*, *42*, 153–168.

Rounsevell, M. D. A. and Metzger, M. J. (2010). Developing qualitative scenario storylines for environmental change assessment. *Wiley Interdisciplinary Reviews — Climate Change*, *1* (4), 606–619.

Schandl, H. *et al.* (2020). Shared socioeconomic pathways and their implications for global material use. *Resources, Conservation and Recycling*, *160*, 104866.

Schipper, B. W., Lin, H.-C., Meloni, M. A., Wansleeben, K., Heijungs, R., and van der Voet, E. (2018). Estimating global copper demand until 2100 with regression and stock dynamics. *Resources, Conservation and Recycling*, *132*, 28–36.

Van der Voet, E., Van Oers, L., Verboon, M., and Kuipers, K. (2018). Environmental implications of future demand scenarios for metals: Methodology and application to the case of seven major metals. *Journal of Industrial Ecology*, *23*, 141–155.

Wilkinson, A. and Kupers, R. (2013). Living in the futures. *Harvard Business Review*, *91* (5), 119–127.

Chapter 35

Industrial Ecology and the Social Sciences

Chapter Concepts

- An important theme within industrial ecology studies socio-technical systems, recognizing that our industrial system is not only a network of physical stocks and flows but also of information and controls, influenced by social, cultural, and governance factors
- Human behavior is a critical driver of how technologies are designed and used and must be accounted for in accounting of stocks and flows
- Servicizing responds to increased demand in environmentally beneficial ways: it can save materials, encourage reuse, and generate fewer and less problematic environmental impacts
- Governments at all levels can take actions to decouple between consumer demand and environmental impacts by adopting policies to encourage servicizing, low-carbon modes of transportation and recreation, and sharing economy

35.1 Industrial Ecology: A Science Embedded in Society

Industrial ecology (IE) has traditionally focused most of its attention on the study of the acquisition of materials from their natural deposits, the transformation of those materials, material uses, recycling, losses, and environmental implications. The field has paid much less attention to why these processes occur, but the answer to that question ultimately provides the motivation for all that industrial ecology does.

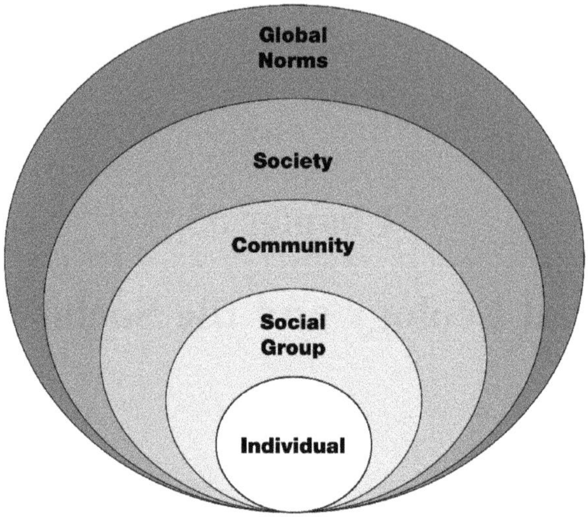

Figure 35.1. The hierarchy of social connectivity.

An insight into the motivation is provided by Figure 35.1, which suggests the nested social relationships of a typical human being. Each person is a member of a social group, and the group's habits and preferences influence the decisions of each group member. At the next higher level, the norms and requirements of the community also influence the individual, but usually to a lesser degree than does the social group. The process continues at still higher levels, typically with even less effect on the motivation of the individual and on the preferences of her or his own choices and those of the immediate social group.

In Chapter 4, industrial ecology was defined as "the science of the physical functioning of society". This implies that the field needs to study society's predilections and the resulting influences on materials acquisition, use, loss, and environmental implications. Following this line of reasoning then leads to the following areas of inquiry:

- How many people are there on the planet? How many are there likely to be in the future? How likely are their incomes to increase, and at what rates? These are topics typically addressed in the field of *demography*.
- As incomes increase, for an individual or for society, to what extent will those incomes result in new or growing demand for materials? These are topics typically addressed in the field of *psychology*.
- As material demand grows, how might the demand be met not with more material use but with the invention and deployment of materially minimal and

environmentally preferred alternatives? These are topics typically addressed in various industrial ecology initiatives, such as *servicizing*.
- What might be the roles of governments in promoting environmentally responsible alternatives and activities to the current approaches? These are topics typically addressed in the field of *governance*.

This chapter addresses these relevant social science issues in demonstrating that human action for environmental betterment is the foundation of the industrial ecology area of study.

35.2 Demography and Industrial Ecology

Industrial ecology has among its foci the challenges of reducing material extraction from nature, reusing materials once acquired, and minimizing environmental impacts. These challenges arise from the analyses discussed earlier in this book, but they are not realizable without the involvement of individuals, their actions related to physical resources, and the potential of these actions to make a difference. This realization leads to the consideration of several issues related to demography: the study of human populations, their properties, and their evolution over time. The first and most obvious question is "How many people now exist, and how many are likely to exist in the future?" This and other demographic issues have been and are being addressed by the Population Division of the United Nations (UN) and other organizations, which estimate the global population reaching more than nine billion people by 2060 (as shown in Figure 2.4), as well as a (anticipated) strong trend toward increasing amounts of education. It is well established that people who are more educated tend to have higher incomes and thus more buying power. It has also been shown by global surveys that women's fertility (the average number of children produced over her lifetime) is inversely proportional to level of education.

Improvements in education level are important not only in providing more rewarding lives for the more educated but also in how population evolves. Changes in overall population as a result of education access take time, of course, because young women require 15 to 20 years to move into reproductive age. By a few decades, however, the effects of educational transitions are readily seen. This occurrence can be illustrated by age pyramids (Figure 35.2), in which color represents the level of education for male and female cohorts. Note in the situation in 2010 that at all ages the level of education for females at that time was lower than that for males. Three scenarios of possible age pyramids in 2060 are also shown:

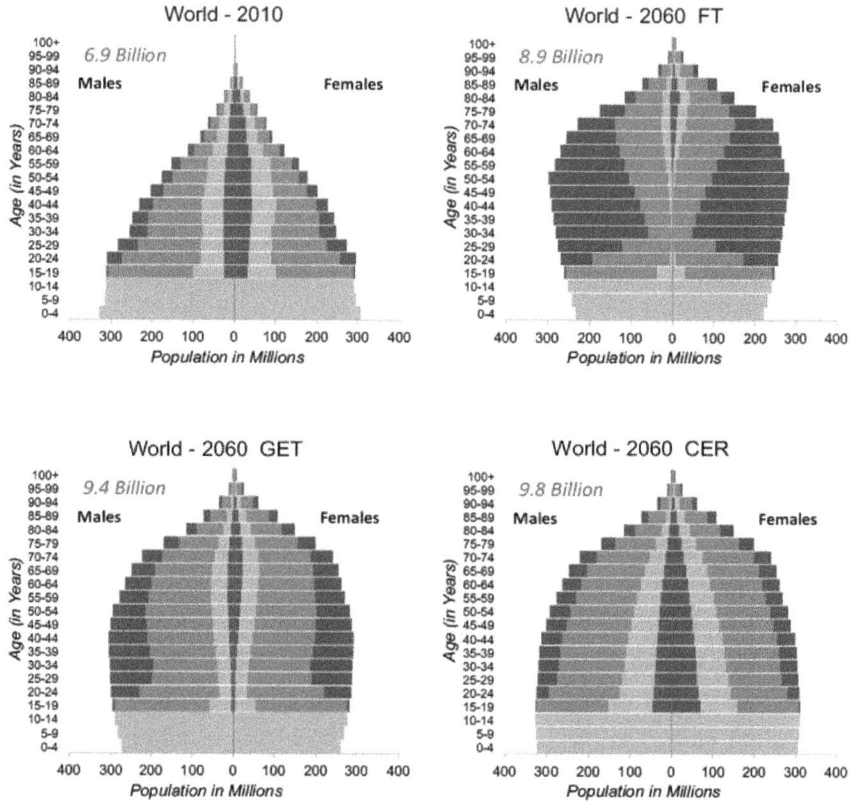

Figure 35.2. Age pyramids for year 2010 at global level (historical) and for three 2060 scenarios depicting global educational options: rapid expansion of education (FT), continuation of current educational trends (GET), and educational stability over time at 2010 levels (CER) (Lutz *et al.*, 2014).

one for rapid expansion of education (Fast Track), another for continuation of the current trend (Global Educational Trend), and a third for educational stability at the 2010 level (Current Enrollment Rate). Higher education for women in their reproductive years leads to fewer births and a narrower age pyramid (for the bottom several cohorts). In contrast, a Current Enrollment Rate (CER) future produces a higher population, and one less well educated.

Demographers generate much other useful data, such as population age distributions, which have implications for both the quantities and types of resource use and can serve to help anticipate levels of material demand in developing countries over the next few decades. Data on population, especially, has important implications for both the quantities and the types of resource use in regions with different population pyramids. Per capita income and its growth over time are also quite important in this regard. That measure is predicted to increase globally over the next several decades and in many countries and regions as well. Such information is potentially important in predicting resource demand, which by itself justifies the value of industrial ecology's link to demographic science.

35.3 Psychology and Industrial Ecology

In a provocative 1999 paper, as relevant today as it was at the end of the 20th century, Ropke (1999) describes the transition in the 1940s and 1950s from societies in which family farming was the lifestyle for perhaps half the planet, to one in which "factory farming" resulted in large numbers of people moving into urban and suburban areas. Among the results of this transformation were more informal and less traditionally constrained populations with generally higher incomes, more desire for individual transportation, and an increased ability to enjoy travel and other benefits of material wealth. With this flexibility in residential location and lifestyle, the demand for materials has increased rapidly over the last several decades.

"Consumption makes sense to people", Ropke (1999) says, and considerable research bears that out. We find that buying things makes us feel good. It fills empty time, exposes us to new and interesting products, and may help counter the stresses of life. As a result of these encouragements to consume, the average person in a stable society owns perhaps two to three times as much "stuff" as was the case a generation earlier.

Perhaps the commanding reason for the attractiveness of a consumptive lifestyle is that consumption is related to power. Achieving and maintaining a social position requires conforming to the norms of that position, be they "fast fashion", a newer automobile, or an upscale home filled with modern gadgetry.

It is widely recognized that there are significant benefits to leading less materialistic lives, including the pleasures of having more time for reading, or learning new skills, or enjoying nature. The benefits to the environment that result from decreased consumption are significant and easily measured. Nonetheless, achieving a simpler lifestyle is not easy and most do not find it an attractive goal (Osikominu and Bocken, 2020). It truly appears that humans are psychologically predisposed to favor consumptive lifestyles if such lifestyles are available to them. If so, a substantially more thoughtful and greener technology will be needed as a counterweight to human psychological instincts.

IE tools such as urban metabolism, material flow analysis (MFA), and Environmentally Extended Input–Output (EEIO) modeling analyze physical stocks and flows in aggregate, where different patterns represent the collective action of an entire city or country. But for LCA, when conducted at the level of the individual, personal psychology is a major factor in driving decisions and behavior. Different people can interact with the same technological system and achieve dramatically different results in terms of resource use or emissions. For example, people have different driving styles that lead to different levels of fuel economy for the same vehicle. Faced with an overly warm room, one person might remove a layer of clothing, another might turn down a thermostat, and a third might open a window. Early risers might wake with the sun and go to bed early, while others prefer to stay up later, requiring more artificial light. In many cases, behavior can be a determining factor in LCA studies, even more important than differences in materials or processing routes. For example, whether a person prefers to wash clothes after each wearing or after multiple wearings can be a more important driver of life cycle impacts than what the clothes are made of or how they were acquired.

Rather than simply make assumptions or scenarios about behavior, IE social scientists rely on observations and surveys in order to collect information directly from people. This allows them to understand not just average behavior but also behavior at the extremes, and in turn, how behavior might be changed in order to reduce impacts or increase environmental benefits. The purpose of LCA is decision support, but the most effective places to make decisions for environmental improvements may be at the extremes. The efficient techniques of one person can be held up as the best practice to follow. Or, if a small subset of behavior is driving a large portion of emissions for an entire group, then changing decisions within just of that small subset may be the highest priority.

Goal 12 Sustainable Development Goals (SDGs) is 'Ensure sustainable production and consumption patterns'. Traditionally, we have analyzed the resource demands and waste from the 'average consumer', but in fact this doesn't actually represent real people! A cutting-edge area in IE is to consider the consumption of subgroups or even individuals in order to better understand inequities and provide directed decision support for individuals, which after all is where most consumption decisions occur. Considering socio-economic inequities, it is intuitive that higher incomes are correlated with more consumption and higher emissions, but the scale of difference among the rich and poor is surprising. The UN Emissions Gap report found that the richest 1% of people globally are responsible for more than double the carbon emissions of the poorest 50%. So, shifting consumption patterns and reducing emissions intensities for the very wealthy must be an important goal of decarbonization efforts.

Another important theme of social science research is to analyze and explain the differences among social groups, and here also, industrial ecology is opening up new capabilities. For instance, advances in spatial life cycle impact assessment (LCIA) methods have also revealed differences in who is affected by consumption in terms of air pollution, and how this relates to environmental justice and racial-ethnic differences. Analyzing consumption patterns in the U.S., Tessum *et al.* (2019) found that non-Hispanic white consumers are exposed to 17% less pollution than is caused by their consumption, whereas Blacks and Hispanics are exposed to an excess pollution burden of more than 50%, relative to their consumption.

IE researchers have developed a range of tools to help model complex decisions and relationships that lead to environmental consequences, either positive or negative. Agent-based models (ABMs) model the decision process itself by assigning probabilities of a choice based on factors like proximity, social outlook, or past behavior. For example, you are more likely to opt for solar panels on your roof or low-carbon agricultural practices on your farm if your neighbor has already done the same things successfully. IE researchers have used this approach to simulate technology adoption and life cycle implications, such as Miller *et al.* (2013) who built an ABM to model farmers' adoption of switchgrass cultivation for bioenergy, based in part on whether neighboring farms had already adopted. ABMs provide a way of simulating complex socio-technical systems made up of large numbers of individuals, based on a relatively small set of decision parameters. The collective patterns that arise are called 'emergent behavior', which would be impossible to see by modeling just one individual or system by itself.

In the same way that metabolism researchers map flows of resources, social scientists can map the relationships that give rise to flows. Tools such as Social Network Analysis model connections between agents (rather than the agents themselves as in ABM) to explore questions under what conditions is a relationship likely to form, or break. Social network analysis has been applied extensively to industrial symbiosis like the one in Kalundborg, Denmark, to explore the role of personal connections, economic conditions, or contractual details in determining the resilience and environmental performance of a symbiosis (Ashton and Bain, 2012).

35.4 The Servicizing Approach to Human Society

Servicizing is a concept in which a customer purchases a service that is enabled or provided by a product rather than purchasing the product itself. The adoption of the servicizing approach by a customer is a decision that is somewhat contrary to historical practice, in which product ownership enabled the desired service to be performed by the owner. In situations where servicizing has been accepted either by corporations or individuals, however, the process combines generally lower environmental impacts with the feeling of having made a meritorious decision.

Businesses with foresight have developed servicizing approaches since about the mid-1990s, when Xerox offered U.S. customers the option of renting a copying machine (or perhaps contracting for the ability to make a fixed quantity of copies per month) rather than buying the machine: perhaps the first clear example of industrial ecology thinking at work within a corporation. Traditionally, the customer purchases the copying machine and is responsible for purchasing the necessary consumables (paper, ink cartridges, etc.), arranging for maintenance while the machine is in service, and dealing with eventual recycling or disposal at end of product life. In the servicizing approach, the machine manufacturer selects the appropriate equipment, maintains it while it is in service, and recycles it when the contract commitments are complete (see Baines *et al.*, 2007). Xerox found that the certainty of recovering the product at end of "first life" meant that the copiers could be designed for a longer life from the start, easier component substitution and upgrading, and avoidance of most waste material flows.

A second and more recent example of corporate servicizing is that of Philips in the Netherlands, which contracted to provide the service of lighting for Schiphol Airport. Philips uses rather costly but highly dependable lighting so as to minimize bulb replacement and other high-cost maintenance, and designed rugged light fixtures that ae very long-lived. When the fixtures eventually need redesign or replacement, most of the material is readily recovered. Corporate servicizing is also attractive in countries where extended producer responsibility (EPR) is required. If appropriate end-of-life treatment is required, it is generally convenient to have a servicizing company assume that responsibility rather than have the requirement fall on the recipient of the service.

Another major example of servicizing is that of individuals rather than corporations choosing to purchase services rather than products, perhaps because they find it more convenient or perhaps because of personal environmental commitments. There are many interesting examples of this approach, including short-term rentals of bicycles or motor vehicles, or community organizations that rent tools for one-time use. An innovative version of servicing for individuals is that of renting baby prams, which are customarily needed for only six to eight months. Through careful design, the prams can be refurbished or remanufactured at the end of use, or can be responsibly recycled.

A related idea to servicizing is the 'sharing economy', where people procure goods and services from each other rather than directly from a business, mediated through some sort of market-making platform. Sharing economy businesses have grown to dominate various service sectors, such as Uber for mobility, AirBnB and

VRBO for property, and cloud platforms like Amazon Web Services for computing. There is a supposed environmental advantage to avoiding personal ownership, as products can be shared among multiple users, thus avoiding manufacturing impacts. But, as always in IE, it is critical to take a systems view, and some research has found major environmental trade-offs. In one example, the extra cruising that car-sharing drivers were doing in between rides was increasing fuel consumption and associated GHG emissions by 20%, as well as contributing to congestion, crashes, and noise. Again, behavior is a key factor in whether car sharing can realize environmental benefits.

35.5 Governance and Industrial Ecology

Environmental governance is a large and complex field of study, and in many national and international aspects, it interacts with industrial ecology only indirectly, as when emission limitations are imposed on releases of ozone-depleting substances or constraints are imposed on fishing for endangered species. Financial incentives are another common tool, as are consumption taxes that reduce purchases of targeted goods or tax credits given for investments in clean technologies.

In situations where industrial ecology-related information can be a useful input to governmental policy, however, industrial ecologists have been quite active as a result of working within international organizations. The longest-running connection of this type is through active participation in the International Resource Panel (IRP). An example is provided by Figure 35.3(a), in which a published study of historical and projected copper production has been augmented by an insertion at upper right to emphasize the relationships between metal mining and policy options. Many other reports generated since the IRP's formation in 2009 present industrial ecology-related concepts to governments in policy-relevant form.

A second important involvement of industrial ecologists in governmental policy is with the UN's Intergovernmental Panel on Climate Change (IPCC). In the most recent IPCC overview report, the section title "Mitigation of Climate Change" was largely built on industrial ecology data: emissions of carbon dioxide and other gases from industry to the atmosphere, growth trends for mining and product manufacturing, and a variety of tables, discussions, and graphics. An example where the importance of industrial ecology is evident is in the introduction to the overall report, where a graphical display of greenhouse gas emissions by economic sector (Figure 35.3(b)) introduces the entire sectional report. A much more detailed example is the figure that opens the report chapter on industry: a Sankey diagram of the quantitative linkages between industrial sectors and greenhouse gas emissions, circa 2010.

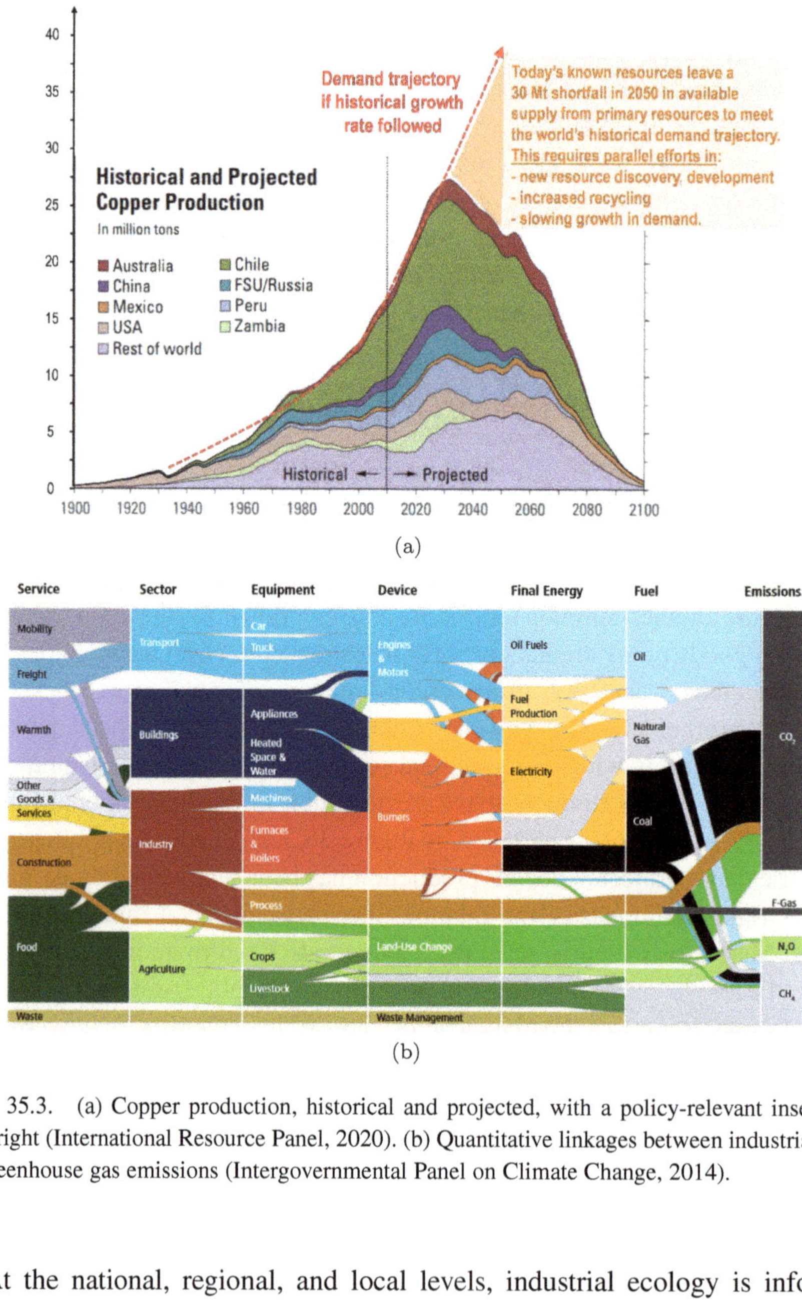

Figure 35.3. (a) Copper production, historical and projected, with a policy-relevant insertion at upper right (International Resource Panel, 2020). (b) Quantitative linkages between industrial actors and greenhouse gas emissions (Intergovernmental Panel on Climate Change, 2014).

At the national, regional, and local levels, industrial ecology is informing policy and environmental governance. The full range of possible environmental governance options is suggested by Figure 35.4, still relevant despite being created a quarter-century ago. Its useful ideas aim to correct various failures: (1) lack of information; (2) misaligned incentives; and (3) harmful behaviors.

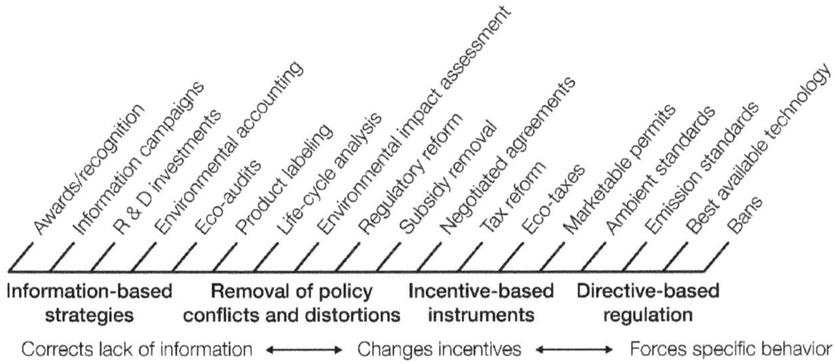

Figure 35.4. Governmental strategies to achieve environmentally responsive behavior in a market economy (courtesy of B. Long, OECD).

An example at the national level is the institution of extended producer responsibility (EPR), the requirement that manufacturers take responsibility for their products at the end of useful life. The intent behind the concept is that product reuse and recycling (and ultimately product design) will be much more likely to occur in environmentally responsible ways if the original manufacturer bears the burden. This approach has been adopted largely in Europe and has often involved industry-collaborative programs for product collection and subsequent treatment.

Several countries have enacted low-carbon standards that rely on LCA to determine whether the product is eligible, such as for low-carbon fuels under the Renewable Fuel Standard in the U.S. or the Renewable Energy Directive in the EU. Governments also set minimum water- and energy-efficiency standards for appliances and other products that reduce resource use and environmental impacts per unit of service delivered.

National governments can also take advantage of the fact that they themselves are among the largest purchasers and users of environmentally related products and services: buildings, heating and air-conditioning equipment, transportation, electronics, etc. To the extent that environmentally superior products are required in all government facilities and contracts, more responsible acquisition and use are built into the public sector by extensive use of environmental purchasing.

Industrial ecology-relevant governance can also be easily observed at the city and community levels. A common example is the refashioning of urban areas to minimize traffic and its attendant pollution, such as the bicycle-friendly streets and convenient bicycle rentals available in many locales. The minimization of pollution is also aided by convenient vehicle rental programs that tend to substitute for the large numbers of private vehicles used more reflexively and more often.

In areas served by commuter trains, urban design that includes reserved parking at commuter rail stations for rental vehicles can stimulate vehicle rental programs that substitute for less environmentally friendly alternatives.

Further Reading

Ashton, W. S. and Bain, A. C. (2012). Assessing the "short mental distance" in eco-industrial networks. *Journal of Industrial Ecology, 16* (1), 70–82.

Baines, T. S., *et al.* (2007). State-of-the-art in product service systems, *Proceedings of the Institution of Mechanical Engineers, 221B: Journal of Engineering Manufacture,* 1–11. DOI:10.1243/09544054JEM858.

Intergovernmental Panel on Climate Change (2015). *Climate Change 2014: Mitigation of Climate Change: Working Group III Contribution to the IPCC Fifth Assessment Report.* Cambridge: Cambridge University Press. DOI:10.1017/CBO9781107415416.

Leimback, M., Kriegler, E., Romig, H., and Schwanitz, J. (2017). Future growth patterns of world regions — A GDP scenario approach. *Global Environmental Change, 42,* 215–225.

Lutz, W., Butts, W. P., and Samir, K. C. (2014). *World Population and Human Capital in the Twenty-First Century.* Oxford, England: Oxford University Press.

Miller, S. A., Moysey, S., Sharp, B., and Alfaro, J. (2013). A stochastic approach to model dynamic systems in life cycle assessment. *Journal of Industrial Ecology, 17* (3), 352–362.

Mont, O. K., Dalhammar, C., and Jacobsson, N. (2006). A new business model for baby prams based on leasing and product remanufacturing. *Journal of Cleaner Production, 14,* 1509–1518.

Osikominu, J. and Bocken, N. (2020). A voluntary simplicity lifestyle: Values, adoption, practices, and effects. *Sustainability, 12,* 1903.

Plepys, A., Heiskanen, E., and Mont, O. (2015). European policy approaches to promote servicizing. *Journal of Cleaner Production, 97,* 117–123.

Ropke, I. (1999). The dynamics of willingness to consume. *Ecological Economics, 28,* 399–420.

Tessum, C. W. *et al.* (2019). Inequity in consumption of goods and services adds to racial–ethnic disparities in air pollution exposure. *Proceedings of the National Academy of Sciences, 116* (13), 6001–6006.

United Nations Environment Programme (UNEP) (2021). *Emissions Gap Report 2021: The Heat Is On — A World of Climate Promises Not Yet Delivered.* Nairobi, Kenya: United Nations Environment Programme.

United Nations Environment Programme — International Resource Panel (2020). *Mineral Resource Governance in the 21st Century: Gearing extractive industries towards sustainable development*. Paris, France: United Nations Environment Programme.

Vermeulen, W. J. V. (2006). The social dimension of industrial ecology: On the implications of the inherent nature of social phenomena. *Progress in Industrial Ecology*, *3* (6), 574–598.

Chapter 36

The Status of Resource Supply and Demand

> **Chapter Concepts**
>
> - Growing populations and increasing per capita incomes are resulting in rapidly increasing demands for resources
> - Supplies of some important resources appear unlikely to meet demand by mid-century or end-of-century
> - In the short term, innovative technology may be sufficient to achieve supply/demand balance; in the longer term, constraints on demand may be required

36.1 How Much Resource Supply Is Needed?

The anthropogenic demand for resources is generated by human needs and wants, which provide the incentive for acquiring and processing the resources necessary to meet demand. Demand is generally not for basic resources such as minerals or fibers, but for the variety of industrial products that can be produced from those resources. The links between supply and demand are suggested by the dashed arrows from Industrial Products to Demand Segments in Figure 36.1. In most cases, a failure by any of the Industrial Product activities to provide the desired flows to the Demand Segments will result in significant societal disruption. And, as can be seen, most of the Industrial Product activities serve most of the Demand Segments. In fact, the challenge is far greater than the previous statement implies. As chronicled in earlier chapters, some 60 different metals are employed in modern technology, more than a score of soil nutrients and other mineral additives, perhaps a dozen major plastics, and at least half a dozen fibers. In fact, a survey

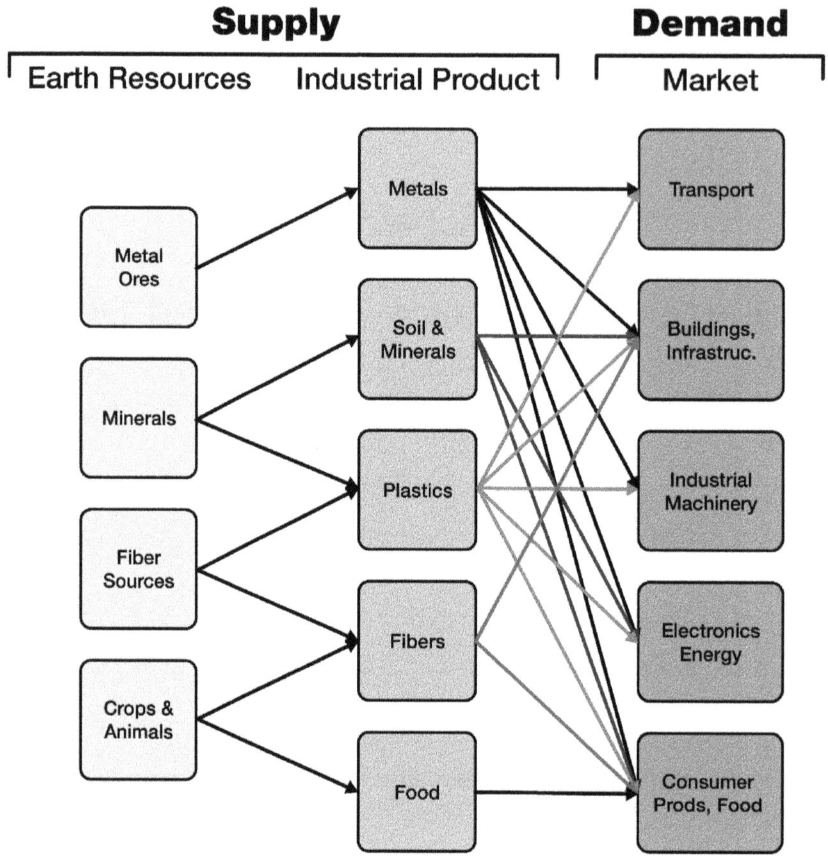

Figure 36.1. Material flows for the assessment of supply and demand.

of elemental use in automobiles lists more than 70. A supply shortage of *any* of these resources or industrial products has the potential to cause hardship. Such supply/demand mismatches may be short-lived, a situation that can often be surmounted with effort and expense, but in some cases, it may threaten to be long-lasting or perhaps permanent. The challenge is for us to look in detail at cases that appear to be in the latter category and to attempt to determine whether long-term mismatches indeed seem likely. If so, what are the possible response options?

36.2 Scenarios of Resource Supply and Demand

A classic way to imagine the future is to look to the past and present for evidence of change. From the perspective of materials of various kinds, an authoritative compilation of material flows over the period 1970–2010 clearly demonstrates a

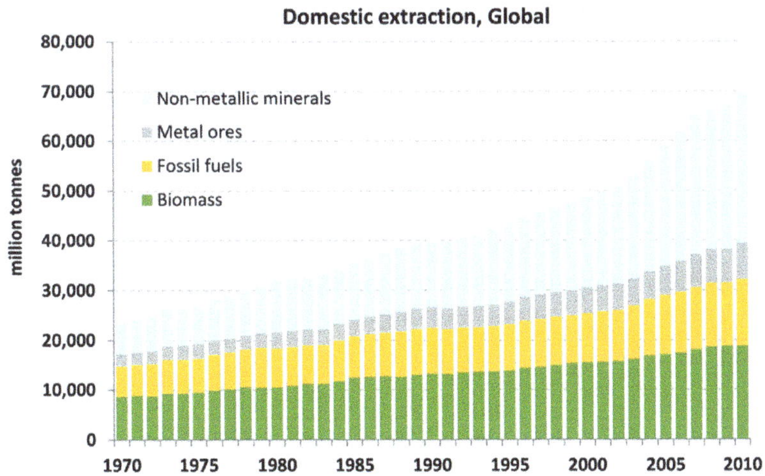

Figure 36.2. The global extraction of four major materials categories, 1970–2010 (Schandl *et al.*, 2018).

sharp increase in materials extraction over that period (Figure 36.2). Even more significant, the rate of increase grew over the 2000–2010 decade. Should that trend continue, it is likely to become increasingly challenging for supply to meet demand across the global economy. The sections that follow explore aspects of supply potential until mid-century for the major groups of materials. Fossil fuels are exempted from the prospective analyses in this book in view of the accelerating transition from fossil energy to renewable energy.

36.2.1 *Non-metallic minerals*

Non-metallic minerals are the largest category of domestic extraction (Figure 36.2), generally as a consequence of the use of sand and gravel for building construction and road building. In principle, at least, such uses are unlimited in the future, although energy for extraction and processing will be needed as will material transport from the source to the point of use. Of more interest in the present discussions are minerals whose properties are unique or unusual and which are significant for society. The obvious group of minerals in that context are those that enable productive agriculture: nitrogen, potassium, and phosphorous.

Nitrogen availability is not an issue, as the Haber–Bosch process extracts it from the atmosphere and converts it to ammonium nitrate. Potassium availability appears not to be a serious issue for the 21st century, as analyses indicate a reserves-to-production ration of about 500.

Phosphorus presents a more complex story. It is derived from phosphate rock that is processed to generate ammonium phosphate fertilizer, and current reserves of phosphate rock could satisfy present demand for several centuries. However, the phosphorus in fertilizer is dispersed in a number of ways, only about 15% being ultimately consumed in food. The rest is lost to inefficient farming practices, lack of recovery from livestock manure, and low food waste recovery efficiency. Furthermore, given increasing global populations and increasing per capita incomes, food demand is expected to significantly increase over time. Unless the capture and reuse of phosphorus in the food chain is sharply increased, phosphorus demand may exceed supply within several decades.

36.2.2 *Fibers*

Forest products have traditionally been used in the construction industry and for making a wide variety of paper products. In recent times, this product spectrum has been diversified; it now includes different types of processed construction products (cross-laminated timber, for example), as well as forest-based synthetic fibers. With cotton land being purchased for more lucrative uses, and the realization that synthetic fiber production and use is often environmentally harmful, wood-based fabrics and other new products are finding market opportunities.

Forest-based manufacturing depends, of course, on the continued existence and competent management of natural fiber sources. Surveys indicate that sufficient land is available and that annual growth in forests is adequate to support continued sustainable extraction and production, even as new uses proliferate.

36.2.3 *Metal ores*

The future prospects for metal ores have been researched much more extensively than has been the case for fibers and non-metallic minerals, and it is therefore appropriate to discuss metal ore status in some detail. We begin by providing the scenario results for prospective demand for seven major metals until 2050 (Figure 36.3). The details are discussed by Elshkaki *et al.* (2018), but the central message is easy to discern. The diagram shows derived scenario demands as colored wedges, with the vertical axis on the plot expressed in fractions of the "reserves to mid-century (RMC)" — the estimated geological supply available to meet the prospective scenario demands. RMC values are only roughly estimated, but the results suggest that for zinc and perhaps for copper, lead, and nickel, the geological resources may not be sufficient to meet demand.

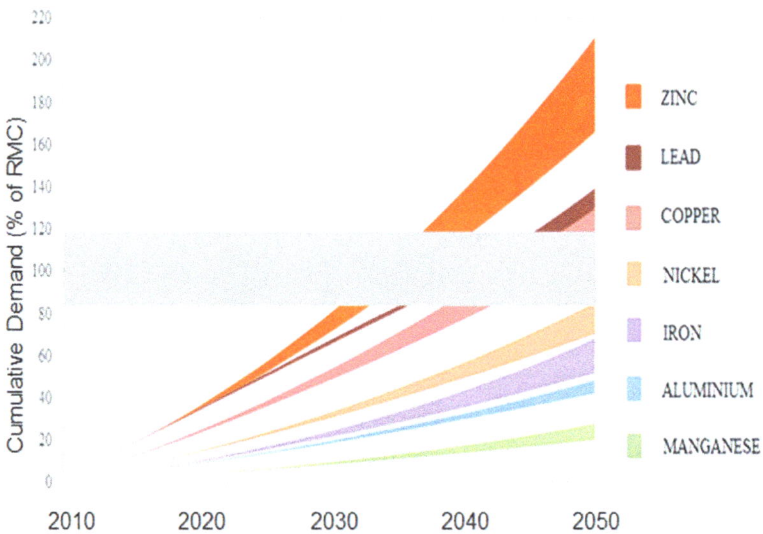

Figure 36.3. Cumulative global demand for seven major metals, 2020–2050, compared to Resources to Mid-Century (RMC), the very roughly estimated potential global resource production to 2050; the gray region suggests the uncertainty for the RMC estimates. The calculated demand for four different scenarios for a given metal is contained within its colored wedge (Elshkaki *et al.*, 2018).

Another mining result, for global and for the period 1950–2010, is relevant here. As seen in Figure 36.4, most new copper deposits discovered during the past 20 years or so are modest in size, much more so than in earlier epochs (diameter of circles), and they tend to be of lower ore grade. An explanation for the ore grade decreases over time is inspired by a diagram originally proposed by Brian Skinner, a Yale University geologist (Figure 36.5). Skinner's concept, supported by extensive historic data for moderate to high ore grade mining, was that metals are deposited in rock in two different ways: in a small quantity of high ore grade deposits and in a dispersed molecular fashion in common rocks. Between the two humps on the diagram is a barrier imposed by the processes of mineralogy that occasionally provide favorable circumstances that concentrate molecules into ores. The richest (right-hand) portion of that higher-grade peak has by now been largely mined out, and miners are now working on the Reserve — larger in quantity but poorer in grade. Still larger in quantity but even poorer in grade is the Reserve Base. Beyond that stage is the mineralogical barrier. Extensive evidence to validate Skinner's vision does not exist, as it would require extensive mining of extremely low-grade ores of negligible value. Nonetheless, Skinner's picture is

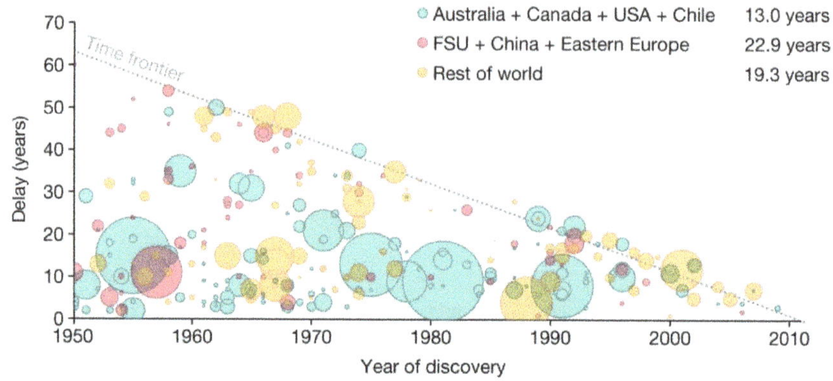

Figure 36.4. Mining delays in copper projects based on year of official discovery. The diameter of the symbols represents the projected size of the extractable deposit (Ali *et al.*, 2017).

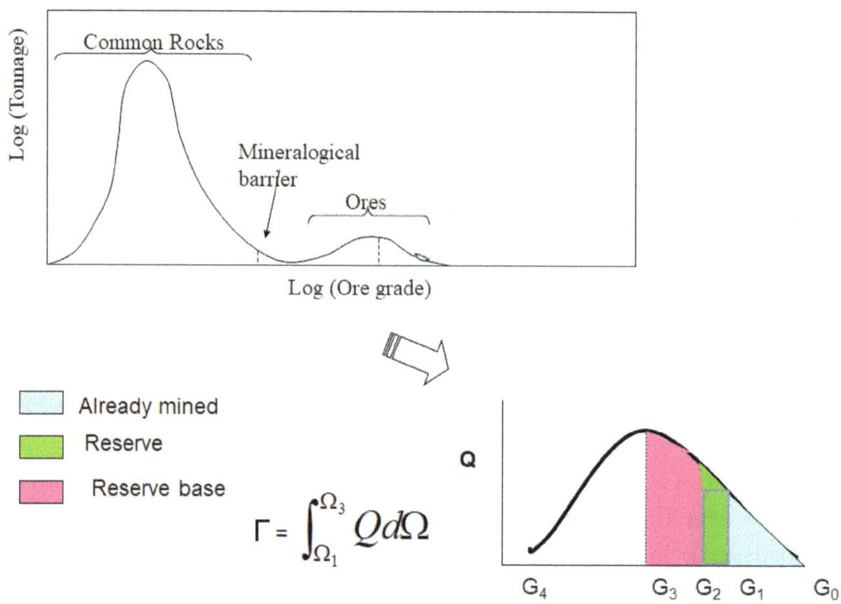

Figure 36.5. Skinner diagram for copper resources on Earth: already mined, reserves, reserve base, and in common rocks (adapted from Skinner, 1976).

generally accepted by the mineralogical community. If so, future mining is unlikely to locate an abundance of new rich ore deposits.

A final bit of interesting data appears in Figure 36.5 — that the time delay between the discovery of a new deposit and the point at which metal ore is being extracted is typically 15–20 years, a time period not obviously changed much over that period. The message is that planning for opening a new discovery, acquiring adequate financing, and meeting environmental and social requirements is such that the gap between discovery and metal availability remains long and is unlikely to decrease in the future.

36.2.4 *Energy considerations*

A final consideration in the future of metal ore availability is that mining and processing of metal ore are very energy intensive, and most mineral deposits are found in locales distant from convenient sources of energy. This tends to favor fossil fuel energy which is, of course, a major concern from a climate impact perspective. In fact, as ore grades decrease, the energy requirements form essentially a straight line on an energy use-ore grade plot. Metal supplies for future generations thus require (1) locating rich ore deposits, (2) receiving approval for mining and pre-processing, (3) acquiring sufficient non-polluting energy sources, and (4) final processing, purification, and marketing. This list presents a high bar for metal supply in a period in which every indication is that demand will continue to increase steadily over time.

36.3 Novel Sources of Supply

Should terrestrial repositories of resources prove inadequate to meet demands in the future, two major alternatives have been suggested: mining resources from the world's oceans, or from captured asteroids. These possibilities are explored in the following sections.

36.3.1 *Ocean mining: A new source?*

In the past several years, mining companies have begun to exploit a small number of mineral deposits that extend from land beneath shallow waters. This is the inspiration for proposals to provide much greater production of metals and minerals from deposits on or near the surfaces of the deep oceans. The concept is that the materials would be scooped from the seafloor, transferred to ships overhead, minimally processed on board the ships (the rejected material being returned to the ocean surface to sink back to depth), and the resulting material transported to onshore facilities for separation and processing. A vision of a part of the process is shown in Figure 36.6. As of this writing, some preliminary seafloor mining devices have been designed and some prototype machinery constructed. Although

Figure 36.6. Vision for a seafloor mining technology (MIT Mechanical Engineering, 2019).

Table 36.1. Estimated Ultimately Available Resources from Terrestrial and Seafloor Repositories (Hein *et al.*, 2013).

Element	Terrestrial resources (Tg)	Seafloor resources (Tg)
Lithium	1.6	1.4
Titanium	899	67
Vanadium	38	9.4
Manganese	5200	5990
Cobalt	13	44
Nickel	150	270
Copper	1000+	226
Arsenic	1.6	1.4
Yttrium	0.5	2
Niobium	3.0	0.46
Molybdenum	19	12
Tellurium	0.05	0.08
Tungsten	6.3	1.3

mineral explorations have not yet been extensive, preliminary estimates (Table 36.1) suggest that the seafloor resources could be useful supplements for terrestrial resources for some, but not all, minerals.

Although the engineering challenges of seafloor mining are agreed to be daunting, still more problematic is the paucity of information on marine life in the deep habitats where mining would occur. Many marine ecologists have pointed out that there has been very little study of deep ocean life, and these organisms could be severely disrupted by the mining that is contemplated. Even if this issue is resolved, the caution that will be needed to minimize biosystem damage, together with the engineering challenges that have yet to be worked out, make it unlikely that extensive seafloor mining will take place for several decades, at least.

In addition to mineral deposits on or under the seabed, the oceans hold vast quantities of dissolved minerals and metals. For some elements like magnesium, we already use seawater as a major source. For others like uranium, the oceans represent the largest known supply, but we have not yet found a way to extract them economically.

36.3.2 *Asteroid mining: A new source?*

Around the start of the 21st century, it was proposed that near-Earth asteroids might provide a source for metals and minerals if the asteroids were captured, brought to low Earth orbit, the ores mined and processed there, and the resulting material transported to Earth to supplement dwindling supplies from terrestrial resources. Because it would be too expensive to transport waste rock, the processing would have to take place on the asteroids themselves; recall that terrestrial metals are mined at metal concentrations of a few percent at best. A vision of how this might be done is shown in Figure 36.7. It is mind-boggling to imagine a sequence involving the manufacture of such equipment, transporting it to the

Figure 36.7. Vision for an asteroid mining technology (Barnatt, 2014).

target asteroid, mining the mineral deposit, separating the waste rock from the target material, and transporting the resulting impure metal to a processing facility on Earth (or perhaps the Moon). Zacny *et al.* (see Further Reading) (have examined this possibility; their conclusion is "We do not see an economic model for returning a whole asteroid to the Earth-Moon system". They did not explore the mining and processing aspects, but those prospects seem similarly discouraging. It does not appear that asteroids provide, in any way, a potentially attractive alternative source of mineral resources to significantly supplement those of Earth.

36.4 The Potential for Curbing Demand

As discussed in Chapter 2, all people wish to satisfy a reasonable set of basic needs: food, housing, safety, heat and light. After that, they value aspects of a more interesting, valuable, and rewarding life: transportation, adequate medical care, education, and so forth. Our interest is typically not in *how* but *whether* those needs and wants are met, because what we actually need are the services provided by technology, not the technology itself. And in many cases, we cannot see or do not know what materials are being used anyway. For example, building owners might be willing to have engineered wood used in place of concrete and steel in buildings, so long as the buildings serve their purposes for the occupants (see Chapter 20). Both materials provide structural stability, but engineered wood is far preferable environmentally. The functions that the materials provide can often be fulfilled by multiple options, leaving it to engineers to choose based on cost, performance, and hopefully environmental attributes (see Chapter 24), but in some cases, substitution is much more challenging. For a dozen different metals used in modern society, there are essentially no practical alternatives (see Graedel *et al.*, 2015).

The demand for materials is often influenced by tradition, an approach that can overuse materials to achieve a purpose that could be realized by more technologically sophisticated engineering approaches. One such approach is the use of tessellation in the manufacture of small parts and sections. Tessellation, a field of mathematics, is the covering of a plane using one or more geographic shapes, with no overlaps and no gaps. The simplest of these shapes are squares, equilateral triangles, and regular hexagons, the first two of which are shown in Figure 36.8(a). A related approach is castellation, in which a material form such as an I-beam is cut horizontally, one part moved over the other, and the parts welded together (Figure 36.8(b)). Not all of the metal of the I-beam is actually needed to handle

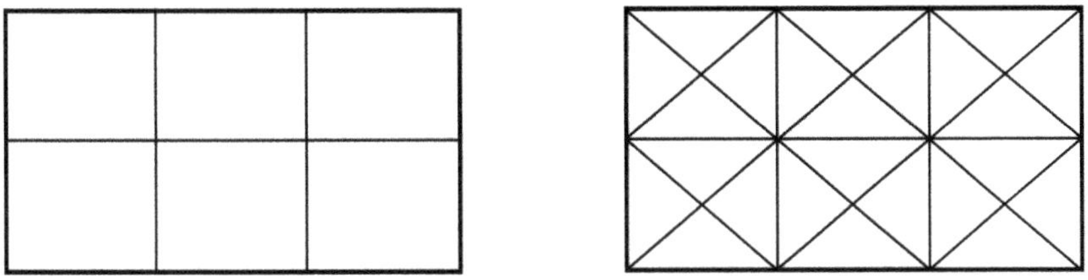

(a)

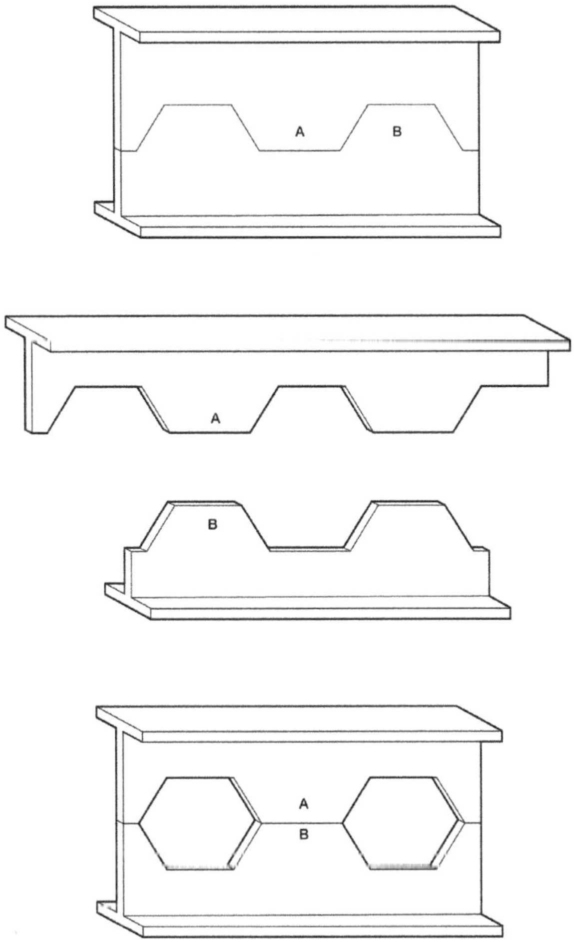

(b)

Figure 36.8. (a) Tessellated patterns on a plane. (b) A castellated beam formed from an I-beam. The continuous beam is first cut (top), then separated (center), and then welded together (bottom) to produce a beam of minimized weight and maximized strength.

loads, and so the resulting piece serves its function but with much less material and is lighter and easier to install.

These and other approaches for minimizing human impacts while maintaining or improving performance directed at human needs and wants are discussed in detail by Allwood and Cullen (2012), who calculate that such *material efficiency* approaches can minimize weight by as much as 30% and reduce scrap metal generation in manufacture by up to half. Initiatives such as these hold out the possibility of diminishing materials-related impacts even while populations are rising.

The challenge of creating a "modern life" for Earth's people is complicated further by the rising global population — projected at some nine billion people by mid-century and perhaps 10 billion by 2100. Will resource scarcity limit the human future?

36.5 Weighing the Options

There is a general agreement that a satisfactory lifestyle requires a few basic needs at minimum (see Chapter 2): food, drink, warmth, and rest. From a physical perspective, this list implies food production, clean water supplies, adequate housing, and personal safety. When these needs are satisfied, personal interactions are next on the list: companionship, mating, social groups, and the like; these are enabled by transportation, communications, and stable societies. From an industrial ecology perspective, most of these needs are met to greater or lesser degrees by the use of materials.

A growing global population implies a concomitant increase in food demand. The availability of productive agricultural land thus has a high priority, but growing more food is only part of the needed response — the food needs to be transported to the people who need it. In a teleconnected world, this realization implies large-scale food production and global and local shipping and thus trucks, merchant ships, distribution centers, and a variety of related infrastructure.

The need for shelter in turn requires materials for housing, procured by activities such as logging, mining, the production of industrial products such as beams and fasteners, intermediate product transport, and assembly equipment. Clean water implies water collection, treatment, transport, and delivery, and thus a variety of pumps, tubes, and pipes. Clearly, all of these activities and products require energy, lots of it. And, as global climate continues to evolve, the energy needed for increased cooling will also rise considerably.

Human companionship, urbanism, and well-being suggest substantial and increasing demand for connected communities, travel over moderate and longer ranges, and a host of life-improving technologies. Therefore, meeting human needs, even imperfectly, will require access to very large quantities of services if humanity is to possess even a modest level of its needs and wants in the future. How we supply these services in material terms is the big question. We are in the middle of a great transition, moving away from our existing fossil and non-renewable industrial system and toward renewable forms of energy, toward bio-based chemicals, toward greater circularity, and toward greater environmental consciousness. Whether the pace of environmental innovation and investment can counteract the increased demand from rising population and affluence is at the heart of sustainability in the 21st century.

Further Reading

Abrahamian, A. A. (2019). The asteroid bubble. *MIT Technology Review*, *122* (4), 62–65.

Ali, S. *et al.* (2017). Mineral supply for sustainable development requires resource governance. *Nature*, *543*, 367–372.

Allwood, J. M. and Cullen, J. M. (2012). *Sustainable Materials With Both Eyes Open.* Cambridge, UK: UIT Cambridge.

Barnatt, C. (2014). *Future Visions: Asteroid Mining*. Retrieved from: https://www.explainingthefuture.com/visions/vision_asteroid_mining.html.

Bhuwalka, K. *et al.* (2021). Characterizing the changes in material use due to vehicle electrification. *Environmental Science & Technology*, *55*, 10097–10107.

Elshkaki, A., Graedel, T. E., Ciacci, L., and Reck, B. K. (2018). Resource demand scenarios for the major metals. *Environmental Science & Technology*, *52*, 2491–2497.

Graedel, T. E., Harper, E. M., Nassar, N. T., and Reck, B. K. (2015). On the materials basis of modern society. *Proceedings of the National Academy of Sciences*, *112* (20), 6295–6300.

Hein, J. R., Mizell, K., Koschinsky, A., and Conrad, T. (2013). Deep ocean mineral deposits as a source of critical metals for high-end and green-technology applications: Comparison with land-based deposits. *Ore Geology Reviews*, *51*, 1–14.

Levin, L., Amon, D. J., and Lily, H. (2020). Challenges to the sustainability of deep-seabed mining. *Nature Sustainability*, *3*, 784–794.

Miller, K. A., Thompson, K. F., Johnston, P., and Santillo, D. (2018). An overview of sea bed mining including the current state of development, environmental impacts, and knowledge gaps. *Frontiers in Marine Science*, *4*, DOI: 10.3389/fmars.2017.00418.

MIT Mechanical Engineering (2019, December 11). Visualizing deep-sea mining [Video file]. Retrieved from: https://www.youtube.com/watch?v=Lwq1j3nOODA.

Schandl, H. *et al.* (2018). Global material flows and resource productivity: Forty years of evidence. *Journal of Industrial Ecology*, *22* (4), 827–838.

Skinner, B. J. (1976). A second iron age ahead? *American Scientist*, *64* (3), 258–269.

Vaccari, D. A., Powers, S. M., and Liu, X. (2019). Demand-driven model for global phosphate rock suggests paths for phosphorus sustainability. *Environmental Science & Technology*, *53*, 10417–10425.

Withers, P. J. A. *et al.* (2020). Towards resolving the phosphorus chaos created by food systems. *Ambio*, *49*, 1076–1089.

Zacny, K. *et al.* (2013). Asteroid mining. *American Institute for Aeronautics and Astronautics SPACE 2013 Conference and Exhibition*, San Diego, CA. https://doi.org/10.2514/6.2013-5304.

Chapter 37

The Circular Economy

Chapter Concepts

- A circular economy is one in which reuse, remanufacturing, and recycling are strongly favored over disposal
- Factors such as elemental complexity in products, designs that are impediments to efficient recycling, and basic thermodynamics make a comprehensive circular economy impossible
- Nonetheless, a semi-circular global economy has the potential to provide major benefits for Earth's people and for the planet itself

37.1 The Circular Economy Concept

The concept of the circular economy is of an economic system that aims to achieve the continuous use of material resources, made possible by activities such as remanufacturing and recycling. The Ellen MacArthur Foundation of the United Kingdom, an advocacy organization for the circular economy, promotes a three-pronged effort:

1. Keep products and materials in use.
2. Regenerate natural systems so as to maximize renewable resource use.
3. Design out waste and pollution.

A complete realization of that vision would eliminate waste and thereby reduce the demand for primary materials. The vision of a sustainable Earth system through a circular economy is closely aligned with earlier sustainability

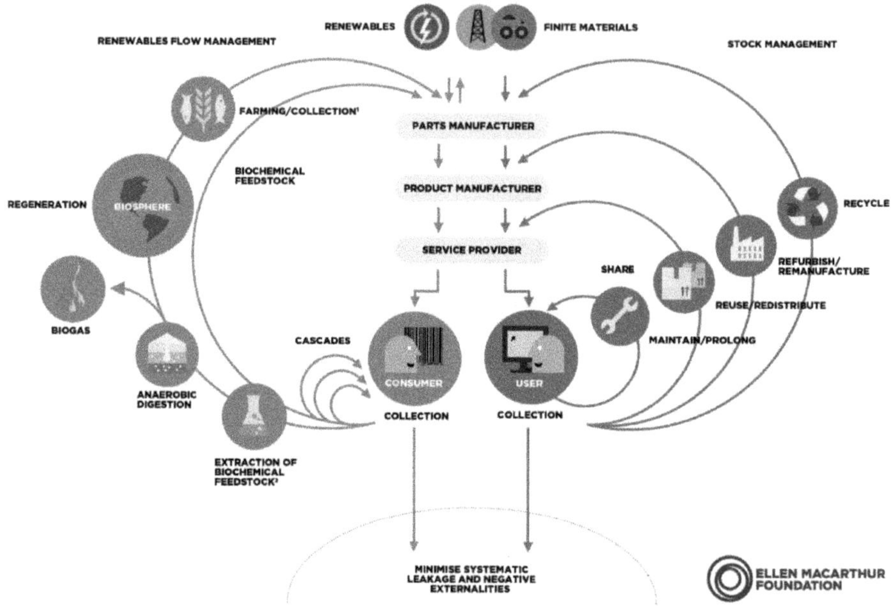

Figure 37.1. A conceptual picture of the circular economy (Ellen MacArthur Foundation).
Note: 1. Hunting and fishing. 2. Can take both post-harvest and post-consumer wast as an input.
Source: Eillen MacArthur Foundation Circular economy systems diagram (February 2019) www.ellenmacarthur-foundation.org. Drawing based on Braungart & McDonough, Cradle to Cradle (C2C).

frameworks such as that of Frosch and Gallopolous (see Chapter 5) and of McDonough and Braungart (2002); see Chapter 24 and References.

The basic idea of the circular economy is to transform our material society from the traditional linear material use approach ("dig it up, use it, dispose of it") to one resembling that of Figure 37.1, in which materials retained in the inner circles require less energy and fewer or no new resources to reuse than would be needed for alternative actions in the outer circles. The idea is inherently attractive and has led to circular economy goals and policies being taken up by numerous companies and by institutions such as the World Economic Forum (WEF) and the European Commission. The challenge, however, is to determine the degree to which such a transition from the present approach might be possible and, more importantly, might be desirable from technological, environmental, economic, social, and political perspectives. Clearly, if comprehensive recovery and reuse would require problematic environmental tradeoffs (massive energy use and greenhouse gas emissions, for example), the circular economy, while inherently attractive, should not be an end in itself.

An important caveat in all of this is that so long as material demand continues to grow, the processes of reuse, refurbishment, and recycling can offset only a portion of demand, as explained in Chapter 9. Constraining demand is thus a prerequisite for a totally circular economy, no matter how efficient.

37.2 Materials, Products, and the Circular Economy

Whether a circular economy is a desirable goal is a complex and far-reaching question, and one difficult to approach simplistically. To present the scope of the challenge of addressing circularity, it is instructive to discuss a few sample topics that illustrate some of the challenges and nuances of a circular economy that are likely to involve design engineers, consumers, policymakers, and other stakeholders. These and other topics are explored in more detail in references, for example, by Singh *et al.* (2021).

37.2.1 *Metals*

Unlike some other materials, metals are, in principle, capable of being reused indefinitely if they can be recovered with high efficiency and at sufficiently high purity. This is fairly close to the current situation with a number of widely used metals such as copper, in which much of the use is in pure form as an electrical conductor. In many cases, however, metals are used in an alloy form in which mixtures of a few to more than a dozen elements are employed. Many of the alloy metals are "companion metals" that are used largely in trace amounts in complex products. Recovery and reuse of these metals depends on the technological potential for separation during reprocessing, which generally involves the treatment of the alloy metals in one or more smelters. As shown in Chapter 9, the thermodynamics of the smelters and the mixtures of elements typically result in the recovery of some of the elements and the loss of others to slag (furnace waste) or vapor. Partly because recovery of minor alloy constituents has not traditionally been a focus of recycling technology, end-of-life recycling rates of most of these "companion metals" are quite low (see Figure 9.7(b)).

In any case, because a circular economy is meaningless if the environmental impacts of achieving it are disastrous, the energy required (and thus the climate impacts of energy use) would seem to discourage the recovery of small fractions of minor metals from alloys or other mixtures. However, recovering those metals from products is generally less energy intensive than mining them, partly because their concentrations in modern technological products are often higher than they are in mineral ores. Depending, therefore, on the degree to which these metals are being utilized in modern technology, the expenditure of significant amounts of energy to recover and reuse them may be a sensible choice, given the alternative of procuring them by mining and processing ores.

37.2.2 *Plastics*

The benefits of modern plastics are enormous and ubiquitous. They preserve food, provide inexpensive household and industrial products, and are major factors in modern medicine. However, at least 80% of all plastics that have ever been made have been discarded — a big loss of resources and a major environmental problem. A key reason for this poor performance is the failure to collect plastics following use, especially plastics in food packaging where contamination issues may be involved. In addition, the raw material from which plastics are made is, of course, fossil fuels, the use of which exacerbates climate change.

A circular economy for plastics would involve their separation and purification in a recycling facility, much the same goals as those of the metals industry. Plastics, however, are complex mixtures of polymers, stabilizers, colorants, biocides, antioxidants, ultraviolet absorbers, fillers, and other additives. These additives would all need to be removed or at least addressed in order to not inhibit reuse opportunities for recycled polymers.

A final challenge is that plastics entering recycling from consumers, health facilities, and other locales are often too contaminated to be reused directly. The removal of contaminants would be required during the recycling process, especially if the material is intended for food and medical applications; otherwise, downcycling into less valuable applications may be the only practical option.

Given the scope of harm caused by plastic pollution in streams, rivers, oceans, and elsewhere, enhanced efforts at collection seem almost certain to be a future focus. Extensive reuse, however, will require behavioral changes in society as well as massive technological development. Some of this will doubtlessly occur, but achieving a semi-circular plastics society appears challenging at best.

37.2.3 *Medical devices*

Medical devices present an interesting case because of the increasing prevalence of single-use disposable devices in the healthcare industry, an approach generally opposite in philosophy from that of the circular economy. In some cases, single-use devices are important for maintaining a sterile environment and reducing the risk of infections, but in many cases, devices can also be disinfected or sterilized and reused without risk to patients (see MacNeill *et al.*, 2020).

It is worth considering the scope of what a truly circular economy would demand of a technologically proficient multi-product medical device industry. As an example, recall the diversity of elements used by GE Healthcare, a large but not unique supplier of medical devices. Its "periodic table" of use includes (amazingly!) all but 11 of the elements below atomic number 90, including nearly a dozen that are considered critical. This incredible elemental diversity emphasizes the extremely broad scope of modern industrial use of almost all the elements in a variety of settings. Each element's use in medical devices or for medical examinations has a purpose, of course: better imaging of body organs, more precise detection of initial stages of disease, minimization of organism damage during testing, and more. A medical device maker adhering dogmatically to the circular economy vision would thus have to not only deal with contamination and sterilization issues, but also with the reuse, reprocessing, and remanufacturing of essentially the entire suite of the elements, or at least of many medical devices. This would be a major commitment for designers, product manufacturers, and supply chain managers, but would certainly be a part of a comprehensive transition to a circular economy.

37.3 Should Everything Be Reused or Recycled?

Reuse and recycling sound as if they are sensible approaches to deal with the accumulation of discarded products, and in general they are. However, there are instances where routine reuse and recycling may not be the ideal approach. One of the most obvious is where a discarded product contains a historically employed material that would not now be desired in the economy, particularly materials or assemblages not regarded as hazardous when first used but now of significant concern: toxic metals such as cadmium in aircraft landing gear, lead in paint, mercury in thermostats, or carcinogenic materials such as polychlorinated biphenyls in transformers. Kral *et al.* (2019) suggest that new product designs need to avoid such constituents and that selected older products leaving service or hazardous material dissipated during product use should eventually reach a *final sink*: a repository that either destroys an unwanted substance completely or retains it for a long period so that its reuse can be reconsidered in the future. The process is suggested schematically by Figure 37.2.

Examples of the establishment of final sinks are the deep repositories set up by some countries as repositories of waste material from nuclear power reactors. Because they store potentially hazardous materials, such repositories tend to be controversial, especially to those living nearby. Despite the challenges, however,

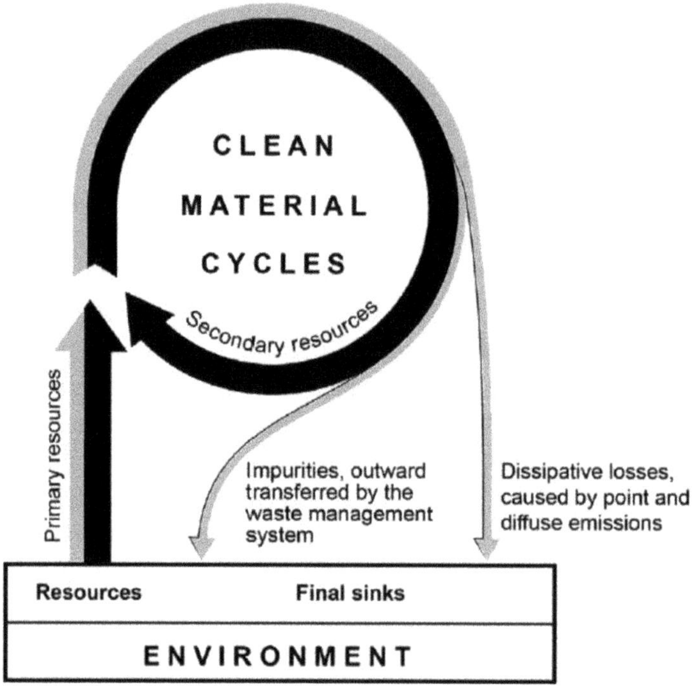

Figure 37.2. Material flows based on a clean cycle strategy (Kral *et al.*, 2019).

it is inappropriate to utilize materials known or suspected of toxicity and then to provide no way to deal with them when no longer desired. If these materials are deemed so beneficial to modern technology that society wishes to use them, the challenges to doing so need to be recognized, as well as provisions made for product employment that do not follow a circular economy approach.

37.4 The Spatial Challenge

One issue not commonly discussed by circular economy advocates is *where* the reuse, remanufacturing, and recycling can or should happen. For simpler materials such as deconstructed building components, the availability of local reuse markets will need to be relatively close by in order to avoid the expense and challenge of the stockpiling of recovered products. For more complex products, it will be necessary to ensure that they are transported to a facility fully capable of their remanufacture. In a technological world where intricately engineered products are often manufactured in a small number of specialized facilities, sold to users around the world, perhaps later resold or re-leased, and eventually discarded, product complexity and recycling technology cannot be assumed to exist around every corner. As Hagelüken *et al.* note (2016), the real challenge is to capture the end-of-life products once they are obsolete but before they become degraded and disassembled, and then to ensure that they are transported to a facility fully capable of their remanufacture. In the case of complex products, there will likely be few such facilities in the world, and the challenges of identification, transportation, and economics quickly become daunting.

The locational issues can be illustrated by a simple example, that of nickel in Australia, whose material cycle is shown in Figure 37.3(a). Australia has very large nickel deposits and a vigorous mining industry. As a result, nickel extraction and ore processing are substantial, but the resulting refined metal is largely exported. Much of this nickel goes to be utilized in stainless steel production elsewhere (Australia does not produce stainless steel, an alloy of nickel with about 75 parts iron, 15 parts chromium, and 10 parts nickel). However, imports of stainless steel to Australia (Figure 37.3(b)) might enable a circular economy of nickel if discarded stainless steel were captured for domestic use rather than being mostly exported. The message here is that, in a global economy, it is very unlikely that the facilities to enable a circular economy will be available everywhere; rather, extensive ocean shipping and international political and scientific coordination will almost certainly be required.

37.5 The Markov Chain Tyranny

As we have seen, there are many steps in the life cycles of materials, and losses are virtually certain to occur to some degree at every step. The result of such a sequence can be described mathematically by the Markov chain (MC) modeling. In such a system, a unit of material has a specific probability of successfully transitioning to the next stage in the sequence. MCs based on material flow analysis (MFA) use the distribution of physical flows of materials among economic sectors to populate a transition matrix. If the probabilities of each transition can be estimated, the loss of a material from the economic system over time can be computed. A sample result for nickel flow through 52 geographic regions is shown in Figure 37.4; it reveals that after 50 time steps, some 96% of the starting material has been lost, either to the environment or to absorption into iron alloys.

From a circularity perspective, the supremely recyclable nickel was found to enter new use over the world only three times. Other studies, for copper and paper, suggest even fewer lifetime uses. Inexorably, small losses at each life stage ultimately result in material loss rather than eternal reuse.

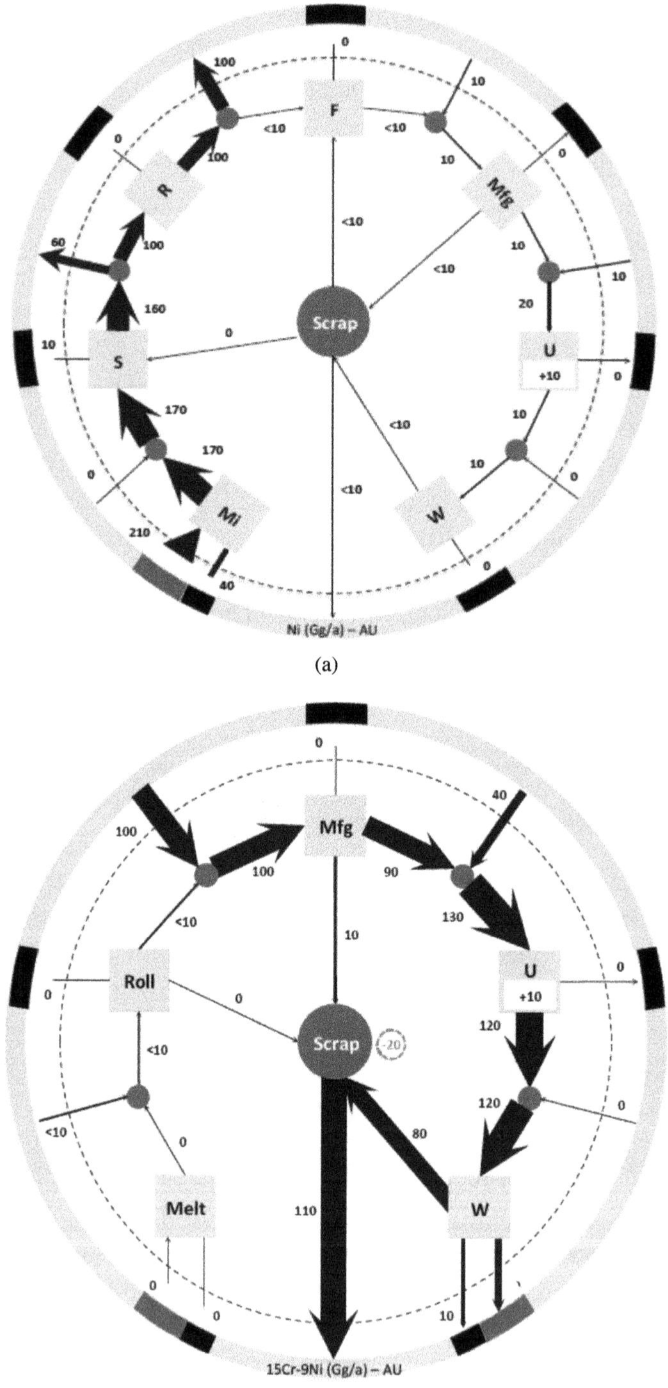

(b)

Figure 37.3. The Australian cycles of nickel (a) and stainless steel (b) in 2010. The units are Gigagrams (thousand metric tons) of metallic equivalent per year (Graedel *et al.*, 2019).

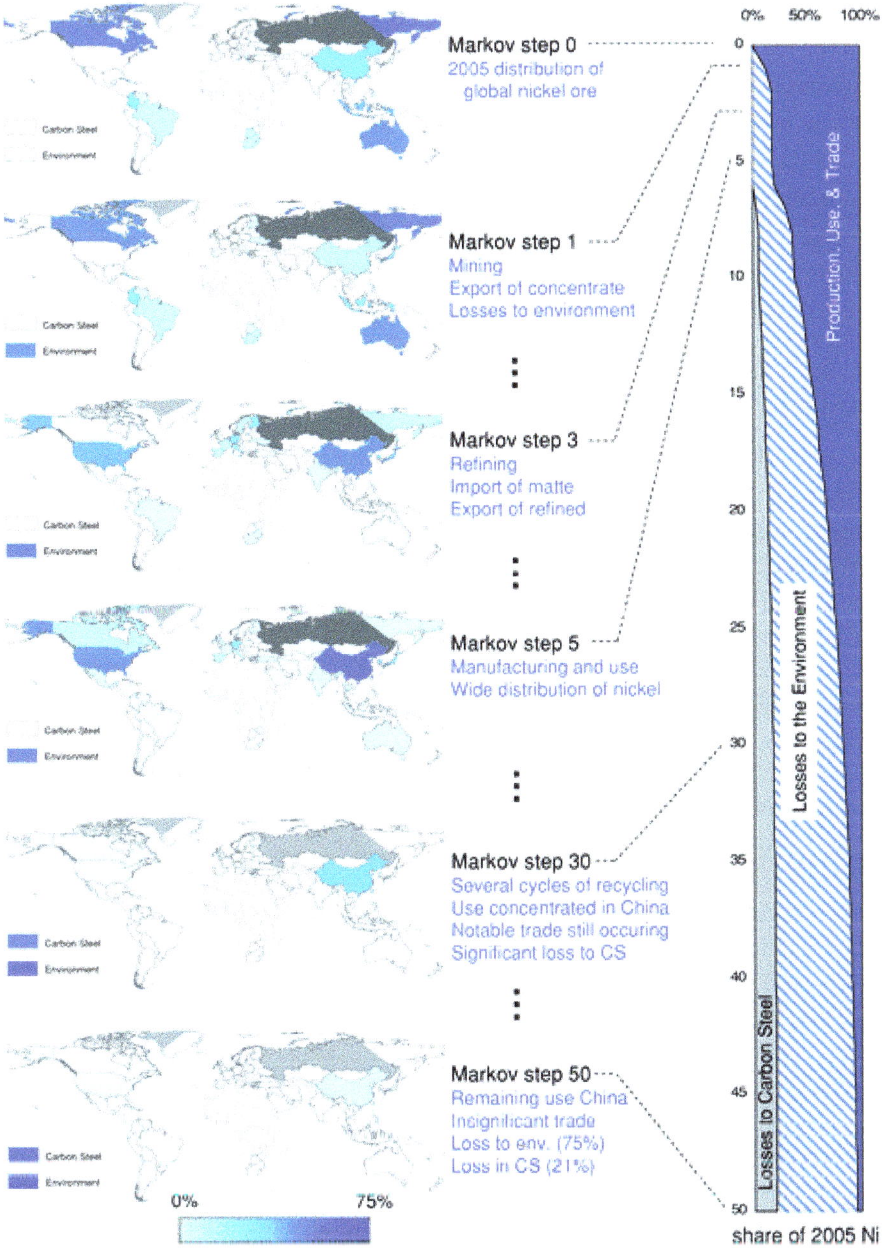

Figure 37.4. The stepwise global movement of nickel down 50 steps of the Markov chain, in percentage of total 2005 extraction for individual countries, regions, and global sinks (left) and globally aggregated (right) (Eckelman *et al.*, 2012).

The nickel recycling story is part of the broader story of thermodynamics, in which the use of energy at each stage of the circularity process will eventually lead to unsustainable levels of resource depletion and generation of waste, especially in a growing economy. This fact should not lead to despair, however. If indeed a completely circular economy is not possible, an extensive and well-functioning semi-circular global economy has the potential to provide major benefits for Earth's people and for the planet itself. We need not completely solve complex challenges to make major strides in making the situation better. From that perspective, the circular economy goal is one worth aiming for.

Further Reading

Bucknall, D. G. (2020). Plastics as a material system in a circular economy. *Philosophical Transactions of the Royal Society A*, *378*, 20190268.

Eckelman, M. J., Reck, B. K., and Graedel, T. E. (2012). Exploring the global journey of nickel with Markov models. *Journal of Industrial Ecology*, *16* (3), 334–342.

Graedel, T. E., Reck, B. K., Ciacci, L., and Passarini, F. (2019). On the spatial dimension of the circular economy. *Resources*, *8*, 32.

Guzzo, D., Calvalho, M. M., Balkenende, R., and Mascarenhas, J. (2020). Circular business models in the medical device industry: Paths toward sustainable healthcare. *Resources, Conservation and Recycling*, *160*, 104904.

Hagelüken, C., Lee-Shin, J. U., Carpentier, A., and Heron, C. (2016). The EU circular economy and its relevance to metal recycling. *Recycling*, *1*, 242–253.

Johnson, J. *et al.* (2008). The energy benefit of stainless steel recycling. *Energy Policy*, *36*, 181–192.

Korhonen, J., Honkasalo, A., and Seppälä, J. (2018). Circular economy: The concept and its limitations. *Ecological Economics*, *143*, 37–46.

Kral, U., Morf, L. S., Vyzinkarova, D., and Brunner, P. H. (2019). Cycles and sinks: Two key elements of a circular economy. *Journal of Material Cycles and Waste Management*, *21*, 1–9.

MacNeill, A. J. *et al.* (2020). Transforming the medical device industry: Road map to a circular economy. *Health Affairs*, *39* (12), 2088–2097.

McDonough, W. and Braungart, M. (2002). *Cradle-to-Cradle: Remaking the Way We Make Things*. Berkeley, CA: North Point Books.

Singh, S. *et al.* (2021). Thematic exploration of sectoral and cross-cutting challenges to circular economy implementation. *Clean Technologies and Environmental Policy*, *23* (3), 915–936.

Van den Bergh, J. C. J. M. (2020). Six policy perspectives on the future of a semi-circular economy. *Resources, Conservation and Recycling*, *160*, 104898.

Looking to the Future

Chapter 38

Sustainability

Chapter Concepts

- Humanity's current path of resource consumption and environmental degradation appears potentially unsustainable
- The United Nations Sustainable Development Goals provide a reasonable set of targets for achieving planetary-wide sustainability, equity, and lives worth living. Industrial ecology insights are crucial for achieving many of the goals
- Sustainability in the 21st century will require a future in which natural and human systems prosper together on an evolving planet

38.1 Is Humanity's Path Unsustainable?

Unpredicted societal collapse is an occasional feature of the history of humanity. The classic case is that of Easter Island in the southeastern Pacific Ocean. It is very remote, and was not settled until about 800 A.D. When the Polynesians arrived, they began to cut trees to create farmland and to make canoes. Soon, they began to erect the large statues for which the island is famous, and trees were used to transport the statues and erect them. Over time, the island's trees were all cut for these purposes. The lack of trees meant that Easter Island had no firewood, mulch, or canoes. Without the ability to catch dolphins from canoes, and with the depletion of nesting birds, the population came under severe food restrictions, and the island was too remote for help to come. There were no alternatives to a severe and ultimately permanent population collapse.

Easter Island is a special case, certainly, but it is not hard to find other cases in which the changing conditions or the misuse of technology has forever changed part of the planet (see Diamond, 2011). The discussion of collapse can be generalized by examining the alternative behavioral patterns for complex systems shown in Figure 38.1. The exponential path (a) traces the path of social progress for some 200 years. This pattern occurs when there are no constraints to growth or when innovation causes apparent limits to recede. Alternatively, the sigmoidal curve (b) is characteristic of the system with fixed constraints in which action is controlled by feedback based on a sense of the distance to the limits. To manage the smooth approach, the system must respond without significant lags and with accurate knowledge of the distance yet to go.

The curve showing oscillatory behavior (c) is typical of systems where feedback mechanisms are inaccurate and responses are slow. At the point that awareness of some limit becomes sufficient to produce action, it is too late to avoid

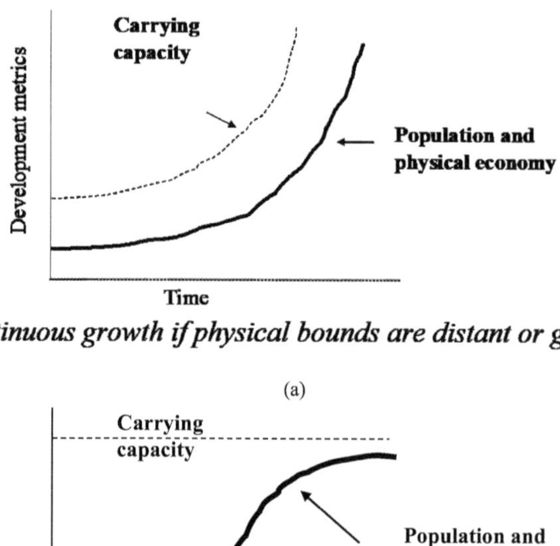

Continuous growth if physical bounds are distant or growing

(a)

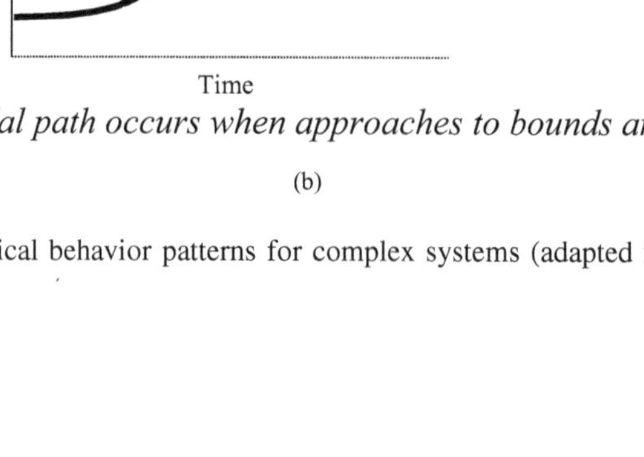

Sigmoidal path occurs when approaches to bounds are seen

(b)

Figure 38.1. Four typical behavior patterns for complex systems (adapted from Meadows *et al.*, 1992).

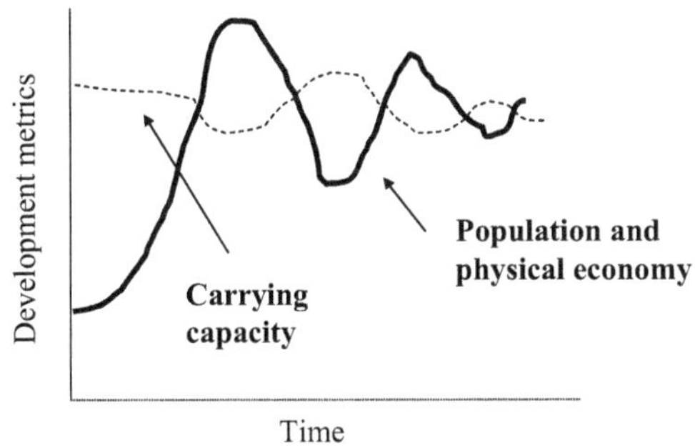

Oscillations occur with delayed signals but robust bounds

(c)

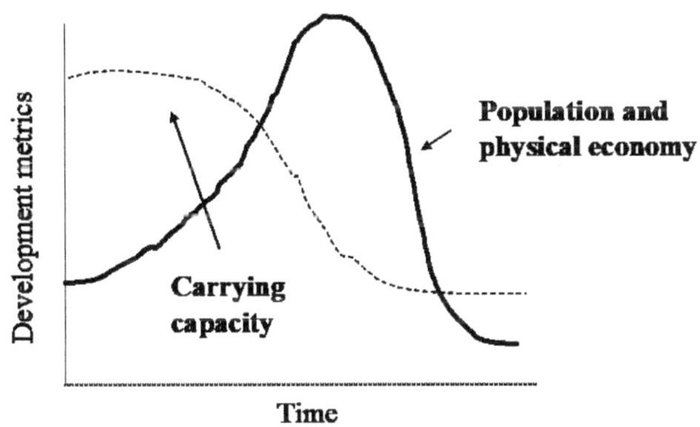

Collapse occurs with delays and erodable bounds

(d)

Figure 38.1. (*Continued*)

overstepping the limits and the system continues to move beyond what appears to be some long-run sustainable state. If the stress produced by the overshoot does not completely degrade the system, subsequent corrections can enable the system to oscillate about and approach the limit. Curve (d) depicts initial behavior somewhat like the third curve, but with a critical difference. Here, the system is insufficiently robust, corrections are insufficient, and collapse occurs. This is the Easter Island trajectory.

It is important to note that the initial stages of these curves are quite similar. We imagine that we are close to the origin, and further imagine intuitively that we are on the exponential growth pattern. If we are not, we must look sustainability in the face, and think hard about the robustness and stability of our technological society.

There are possible clues as to what may lie ahead if no significant changes to human resource supply and use are eventually agreed upon. An image may be useful in this regard: people having even a modest level of familiarity with bacteriology are familiar with the Petri dish — a round flat dish containing a nutrient growth medium onto which are deposited bacteria. As the bacteria grow, they occupy increasing fractions of the available space and consume more and more of the available nutrients. But when the bacteria eventually exhaust the food supply, they die. In similar fashion, our modern society consumes industrial food — metals, engineering plastics, fibers, and energy on demand. If or when industrial food supplies are no longer sufficient to meet demand, society must somehow reduce demand, switch to other industrial food sources, or disaster awaits.

The Petri dish is useful and defendable in bacteriology given its properties:

- a spatial boundary;
- a defined stock of nutrients;
- a growing population of consumers;
- stability of external conditions.

If our concern is the human population of the planet, however, the planetary version of the Petri dish has a set of properties that mimics the Petri dish in some aspects but not others:

- an increasing human population;
- an increasing rate of resource demand per individual;
- a potentially decreasing stock of available primary resources;
- a spatial boundary determined by regions suitable for human habitation;
- rapid changes in a variety of currently enabling conditions, including rising sea levels, increasing temperatures, and limits to agricultural production.

Most of these properties and their evolution have been discussed in earlier chapters, especially the rapid changes in planetary and societal regimes that have occurred in recent history. Suffice it to say that global change in the Anthropocene is pushing many Earth systems into unknown regimes, and those changes are likely to limit the potential for human societies to adapt given their instinctive tendencies (recall the pyramid of needs and wants in Chapter 2).

38.2 Defining a Sustainability Transition

In this last chapter of the book, it is useful at this point to reiterate what we mean by sustainability. Many have tried to formulate succinct definitions, but two stand out. John Ehrenfeld's is conceptual: "*Sustainability* is the possibility that human and other forms of life will flourish on the planet forever". The International Institute of Environment and Development defines *sustainable development* (often used as a synonym for sustainability) as "a development path that can be maintained indefinitely because it is socially desirable, economically viable, and ecologically sustainable". The words have resonance, but offer minimal guidance to engineers, scientists, political leaders, and citizens. Perhaps the most impactful guidance is a famous statement of Mahatma Gandhi: "There is enough on this planet for everyone's need but not for everyone's greed". If his philosophy is adopted as a guide for action, perhaps as much out of necessity as conviction, it suggests an uncommon but not undiscussed approach to the future: to move to the creation and maintenance of a modest lifestyle for all rather than a luxuriant lifestyle for a few. The future of the human species may, in fact, depend on whether this transition can be made.

38.3 Approaching Sustainability

As has often been said, "*sustainability*" is a term that everyone uses but no one defines. When the issue is actually addressed in a thoughtful manner, it is generally from the perspective of identifying aspects of behavior that are demonstrably *unstable*. This is a useful beginning, but only a beginning. A more useful activity is to measure the attributes of sustainability and then devise ways to move toward a more sustainable state. How is one to approach the challenge of providing sustainability guidance in such a way that it can be implemented? We explore here a few examples of how such guidance might begin to point the way so that the "journey toward sustainability" might continue.

In 2015, the United Nations (UN) published a detailed list of 17 "sustainable development goals [SDGs]" (briefly introduced in Chapter 3). The goals were accompanied by 169 targets that provided additional perspective on the actions necessary to achieve the goals. The aim of this initiative was to achieve (or make substantial progress toward) the goals by the year 2030.

As Figure 38.2 illustrates, the goals are quite diverse. Some are hopes for better lives around the world, such as SDG #1: No poverty. Some are environmental in nature, such as SDG #14: Life below water. Others clearly require major doses of resources and technology, such as SDG #7: Affordable and clean energy. However, the case could easily be made that achieving *each* of the 17 SDGs will require large amounts of resources, acquired sustainably, used responsibly, and reused extensively are brought to bear. A necessary condition, in fact, is for the world of materials to be linked in detail to the SDGs, for the suite of materials for each SDG to be identified, and for the amount of SDG-related employment of

Figure 38.2. The United Nations Sustainable Development Goals. The starred SDGs are those to which industrial ecology analyses and approaches are particularly relevant.

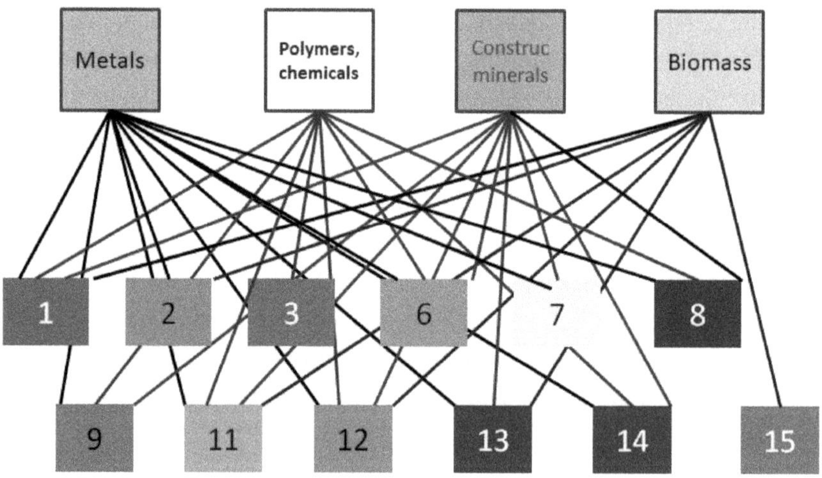

Figure 38.3. Resource group links to the SDGs.

those minerals to be quantified in a defendable manner. A starting map is shown in Figure 38.3, where each of the four principal resource groups is linked in use to several of the SDGs.

It is important in this regard not to become trapped into thinking that we must achieve a better world solely by drawing on the technological and social approaches that exist circa 2020. In fact, technology evolves in often meritorious ways and sometimes in transformative ones. Among the transformative examples of early 21st century technology are the following:

- Lighting by light-emitting diodes (LEDs): LEDs have transformed lighting by providing long-lasting, bright sources of light emanating from small amounts of materials and requiring very modest electricity to operate.
- Renewable electricity generation: Wind turbines and solar photovoltaic arrays have become common sights in many parts of the world, supplying carbon-free electricity in place of fossil fuels.
- Transport by electric vehicles: Electric vehicle transport is growing rapidly in both the commercial and consumer markets. These vehicles do not require fossil fuels, and their more limited number of moving parts contributes to enhanced reliability.
- Construction utilizing "engineered wood": Cross-laminated timber (CLT), glue-laminated lumber (glulam), and other evolutionary forest products have made wood, a widely available and low-density material, an increasingly common replacement for heavy traditional steel beams, studs, and other structural components, enabling significant reductions in greenhouse gas emissions while sharply decreasing construction times.

Game-changing technologies like these are generally recognized after the fact, but there is no reason to assume that innovation will not continue to recur. The most useful will be those that are capable of improving human lives worldwide, especially in the agricultural, medical, and educational sectors. New technologies that meet these criteria should be hunted out and promoted to the world to the greatest degree possible.

At the same time, technology is not a panacea for improving sustainability. If we wait for technological innovation to save the day, we may do severe damage in the meantime, and new technologies may have other environmental consequences that we don't foresee. Indeed, there is a school of thought that sees innovation as a driver of *increasing* resource consumption rather than decreasing, for as products and services become cheaper and more accessible, we tend to use more of them, a phenomenon known as the *rebound effect*.

As you have learned, industrial ecology is a systems approach, and its tools and methods explicitly link production and consumption through flows of materials and energy. Technology innovation tends to focus on the supply side of the equation, but making big improvements toward sustainability will require innovation and change on the demand side as well, shifting away from a culture of overconsumption and waste.

38.4 Global Ecology for the 21st Century

It is important in this final chapter to recall the framing of the first few chapters of this book: that many of the intellectual perspectives and methodologies of industrial ecology are adapted from those of biological ecology, and that sustainability on a planetary scale will require the efforts of both ecosystem ecologists and industrial ecologists. It is useful as well to recall the definition of industrial ecology presented in Chapter 4 (interestingly, differing very little from the definition presented in the 1995 textbook Industrial Ecology):

> Industrial ecology is the means by which humanity can deliberately approach and maintain sustainability, given continued economic, cultural, and technological evolution. The concept requires that an industrial system be viewed not in isolation from its surrounding systems, but in concert with them. It is a systems view in which one seeks to optimize the total materials cycle from virgin material, to finished material, to component, to product, to obsolete product, and to ultimate disposal.

This call to action retains its focus and relevance in the 21st century, providing insight into how far industrial ecology has come in some 25 years and how much its ideas have become widespread, but also how much is left to do.

Other organisms frequently modify their environments to suit their needs — beavers are an obvious example. Humans' ability to do so is unmatched, however. As human civilization has expanded through history, our modifications of the planet have been at increasing spatial scales and with increasing impacts. Industrial ecology's principal task is to continue to quantify those impacts and to work toward transformative approaches to the use of resources that minimize impacts on the planet and its peoples, while promoting better lives for humans and other organisms, large and small. One cannot imagine a bigger challenge, nor a more important one. We had all better get busy.

Further Reading

Board on Sustainable Development (1999). *Our Common Journey: A Transition Toward Sustainability*. Washington, DC: National Academy Press.

Diamond, J. (2011). *Collapse: How Societies Choose to Fail or Succeed* (Revised edition). New York: Penguin.

Ehrenfeld, J. R. (2007). Would industrial ecology exist without sustainability in the background? *Journal of Industrial Ecology*, *11* (1), 73–84.

Meadows, D. H., Meadows, D. L., and Randers, J. (1992). *Beyond the Limits*. White River Junction, VT: Chelsea Green.

Palmer, M.A. *et al.* (2005). Ecological science and sustainability for the 21st century. *Frontiers in Ecology and the Environment*, *3* (1), 4–11.

United Nations (2015). *Transforming Our World: The 2030 Agenda for Sustainable Development*. New York: United Nations.

United States Geological Survey (2022). *Mineral Commodity Summaries*. Reston, VA: United States Geological Survey.

Epilogue: The Universe in Your Hand

We close this book by recalling the connections between many of the topics that have been discussed — topics that link the universe and its generation of the elements with human acquisition of ores and fluids from Earth's crust, the processing of these starting materials into the forms used by industry, their employment in the manufacture of products, and the use of those products in our everyday lives.

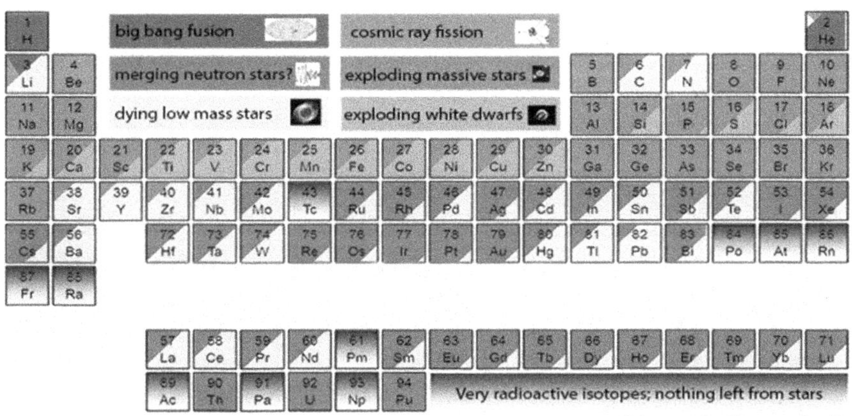

Graphic created by Jennifer Johnson
http://www.astronomy.ohio-state.edu/~jaj/nucleo/

Astronomical Image Credits:
ESA/NASA/AASNova

Our journey begins with an image from Chapter 5: the current astrophysical understanding of the origin of the elements. Those elements most familiar to us were created by exploding stars, while those less commonly known but widely used in technology were created by merging neutron stars or perhaps merging black holes.

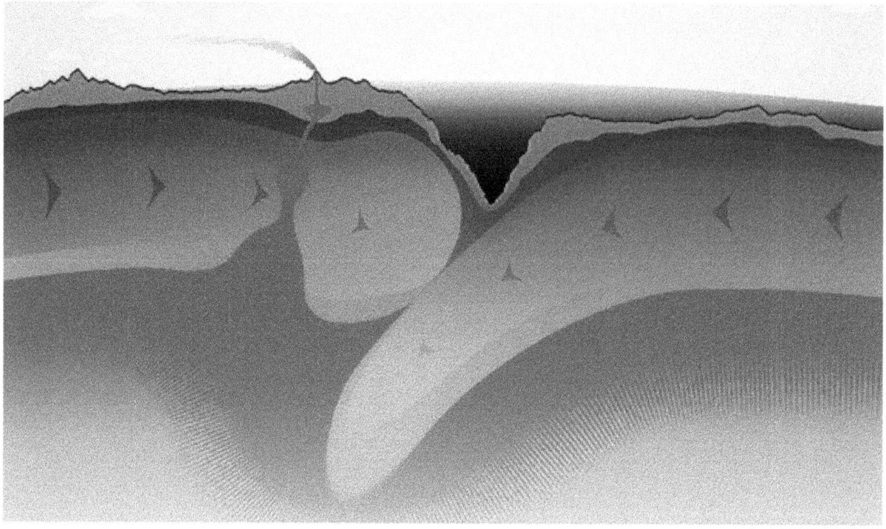

Earth formed from the detritus of these astrophysical processes. Over the eons of the planet's existence, some of this material was heated, melted, and transferred to the surface or nearly so by exploding volcanos. The volcanic ducts solidified in some cases, creating near-surface deposits of mineral ores. However, the deposits are often hard to find, small, or in remote areas.

At the surface, bogs and other habitats created by tectonic plate movements and the processes of nature provide the shelter and nutrients needed for the plants and animals of Earth.

Ore deposits are mined, and the rock is mechanically separated and then melted under high heat to recover the minerals and metals contained therein.

Purified elements recovered from the ore are processed into a very wide variety of products that make modern life what it is. Here, is a closeup of a computer chip (which would easily fit on your thumbnail). Of the fewer than 70 elements that could possibly be used in a chip, more than 60 are routinely employed for that purpose. This amazing elemental diversity and the challenge of chip recovery and reprocessing means that virtually all of these materials never see more than a single use.

Smartphones and other products are manufactured from computer chips and other parts and connections. The smallest wires along which information flows are fewer than 10 atoms wide. The components are housed in magnesium alloy bodies (rugged and lighter than other options).

The smartphones are placed in plastic or fiber containers for eventual delivery to customers — the final step in manufacturing.

Ocean shipping moves the smartphones from the manufacturing site to the world's customers.

Our journey ends with the culmination of the processes above: *You hold in your hand the products of merging black holes and exploding stars, created billions of years ago!*

LTE-A
LTE-A
LTE-A
LTE-A
LTE-A

Final Exercise for Students

Industrial ecology analytical tools apply to many stages of the transition from black hole to phone in your hand sequence. Identify the industrial ecology tools that can be used in as many of the steps on the sequence as you can.

Picture credits for this Epilogue

Origin of the Solar System Elements: Johnson, J., The origin of the elements, www.astronomy.ohio-state.edu/~jaj/nucleo. The diagram is made available through http://creativecommons.org/licensee/by_sa/4.0/.

Earth's tectonic plates: Shelton, J., Geophysicists find the crusty culprits behind sudden techtonic plate movements, Yale News, Jan. 9, 2015, https://news.yale.edu/2015/01/19/geophysicists-find-crusty-culprits-behind-sudden-tectonic-plate-movements.

Bog and surroundings: Image from Rob Routledge, Sault College, Bugwood.org, CC BY 3.0 US, accessed Jul. 22, 2021.

Smelting of ore: Hasanbeigi, A., 56 energy technologies for energy-efficiency and GHG emissions reduction in the iron and steel industry, Mar. 6, 2017, https://www.globalefficiencyintel.com/new-blog/2017/technologies-energy-emissions-steel-industry, accessed Jul. 22, 2021.

The computer chip: What is the difference between a wafer and a chip? Apr. 15, 2021, https://www.waferworld.com/difference-between-wafers-and-computer-chips/, accessed Jul. 22, 2021.

Assembling the smartphone: Can India boycott China and make phones locally? Yes, but it's a long, ardous journey, https://www.news18.com/news/tech/can-india-boycott-china-and-make-phones-locally-yes-but-its-a-long-arduous-journey-2677527.html, accessed Jul. 22, 2021.

Packaging the smartphone: Lin, D., VEGA Secret Note 2 Smartphone Packaging, https://www.packagingoftheworld.com/2015/01/vega-secret-note-2-smartphone-packaging/ Jan. 6, 2015, accessed Jul. 22, 2021.

Shipping the smartphone: Sahiner, D., Ocean freight explained, Sep. 1, 2017, https://www.morethanshipping.com/ocean-freight-explained, accessed Jul. 22, 2021.

The universe in your hand: Gartenberg, C., Sizing up the iPhone 12 mini and 12 Pro Max, Nov. 5, 2020, https://www.theverge.com/21551240/iphone-12-mini-vs-pro-max-hands-on-photos-comparison-apple-size, accessed Jul. 22, 2021.

Units of Measurement in Industrial Ecology

The basic unit of energy is the joule (= 1×10^7 erg). One will often see the use of the British Thermal Unit (Btu), which is 1.55×10^3 J. For very large energy use, a unit named the *quad* is common; it is a shorthand notation for one quadrillion British Thermal Units. Thus, 1 quad = 1×10^{15} Btu = 1.55×10^{18} J.

The units of mass in the environmental sciences and in this book are given in the metric system. Since many of the quantities are large, the prefixes given in Table A1 are common. Hence, we have such figures as 2 Pg = 2×10^{15} g. Where the word tonne is used, it refers to the metric ton = 1×10^6 g.

The most common way of expressing the abundance of a gas phase atmospheric species is as a fraction of the number of molecules in a sample of air. The units in common use are *parts per million* (ppm), *parts per billion* (thousand million; ppb), and *parts per trillion* (million million; ppt), all expressed as volume fractions and therefore abbreviated ppmv, ppbv, and pptv to make it clear that one is not speaking of fractions in mass. Any of these units may be called the *volume mixing ratio* or *mole fraction.* Mass mixing ratios can be used as well (hence, ppmm, ppbm, and pptm), a common example being that meteorologists use mass mixing ratios for water vapor. Since the pressure of the atmosphere changes with altitude and the partial pressures of all the gaseous constituents in a moving air parcel change in the same proportions, mixing ratios are preserved as long as mixing between air parcels can be neglected.

For constituents present in aqueous solution, as in seawater, the convention is to express concentration in volume units of moles per liter (designated M) or some derivative thereof [one mole (abbreviated mol) is 6.02×10^{23} molecules]. Common concentration expressions in environmental chemistry are millimoles per liter (mM), micromoles per liter (μM), and nanomoles per liter (nM), Sometimes one is concerned with the "combining concentration" of a species rather than the

Table A1. Prefixes for Large and Small Numbers.

Power of ten	Prefix	Symbol	Power of ten	Prefix	Symbol
+24	yotta	Y	–3	milli	m
+21	zetta	Z	–6	micro	mu
+18	exa	E	–9	nano	n
+15	peta	P	–12	pico	p
+12	tera	T	–15	femto	f
+9	giga	G	–18	atto	a
+6	mega	M	–21	zepto	z
+3	kilo	k	–24	yocto	y

absolute concentration. A combining concentration, termed an *equivalent*, is that concentration which will react with 8 grams of oxygen or its equivalent. For example, one mole of hydrogen ions is one equivalent of H^+, but one mole of calcium ions is two equivalents of Ca^{2+}. Combining concentrations have typical units of equivalents, milliequivalents, or microequivalents per liter, abbreviated eq/l, meq/l, and μeq/l. A third approach is to express concentration by weight, as mg/l or ppmw, for example. Concentration by weight can be converted to concentration by volume using molecular weight as a conversion factor.

Acidity in solution is expressed in pH units, pH being defined as the negative of the logarithm of the hydrogen ion concentration in moles per liter. In aqueous solutions, an acidity of pH = 7 is neutral at 25°C; lower pH values are characteristic of acidic solutions, and higher values are characteristic of basic solutions.

Diagrams and Displays Used in Industrial Ecology (and Elsewhere)

Visual displays of information (often in color) are common tools for understanding the implications of data and for presenting scientific and technological results to others. They are very widely used in industrial ecology and sustainability science. Some combinations of graphics, statistics, color theory, map theory, and the like are too complex to discuss in this book. However, for those new to industrial ecology or to industrial ecology presentations, we here name, define, and describe some two score approaches used herein. Some of the display techniques go by more than one name; we have tried to select the names in most common use.

A diagram is a symbolic representation of quantitative information. It employs visualization techniques for enhanced comprehension. Much industrial ecology and sustainability information is complex and is best communicated by one or more graphical display approaches. A number of these are defined and illustrated in this appendix, with reference to their utilization for industrial ecology purposes in specific figures in this book. The words "chart", plot", and "display" are used more or less interchangeably in the world of visualization diagrams. This appendix is not comprehensive but covers most of the common approaches and some of the uncommon.

This book also includes depictions, which are drawings or diagrams of qualitative information.

Age structure diagram A diagram that visualizes the demographic structure of a population.

Example in this book: Figure 35.2

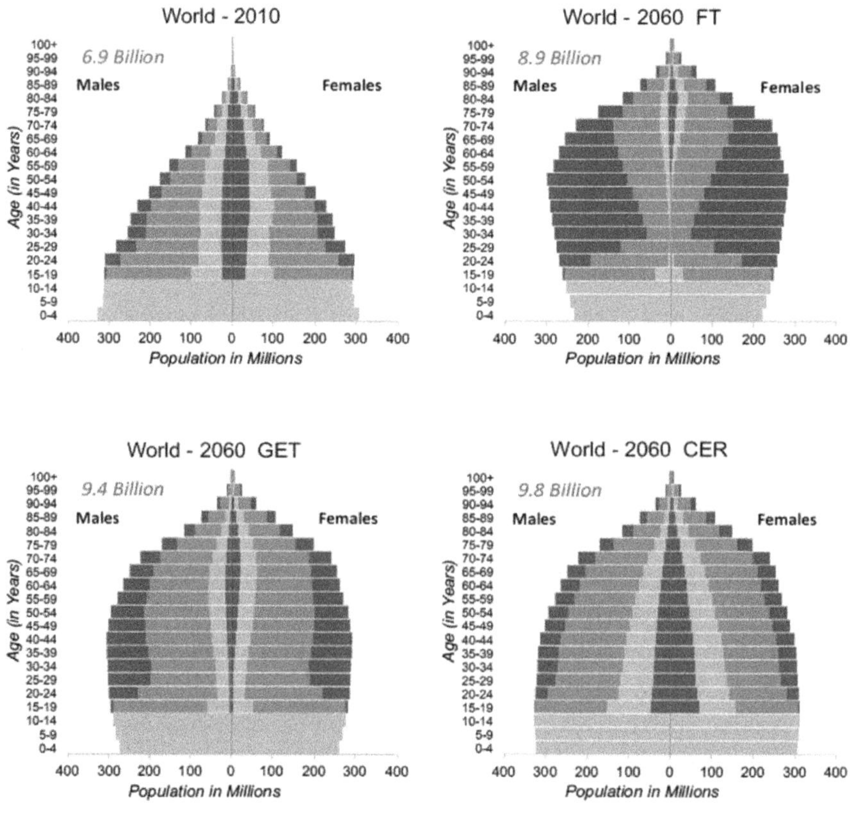

Alternate state diagram A visual approach that illustrates how different selections of a similar variable can be visually compared.

Example in this book: Figures 6.3(b) and 6.3(c)

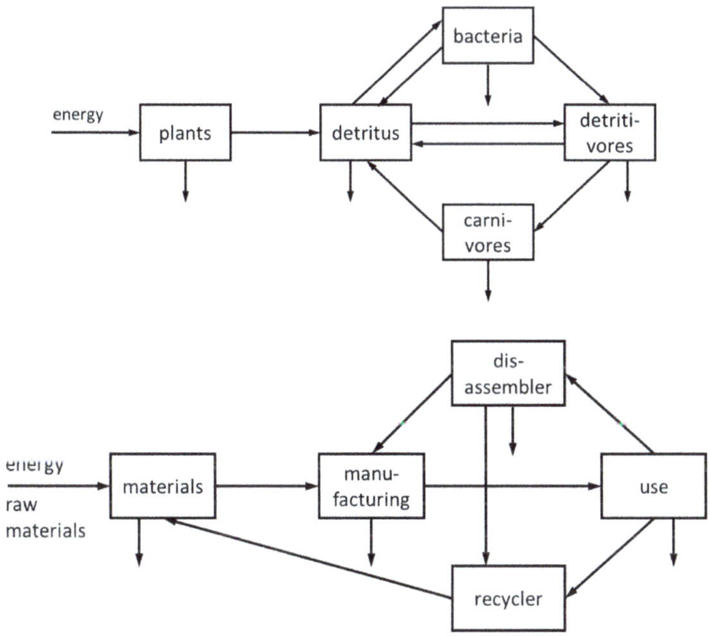

Area chart A stacked series of line charts that clearly illustrate relative differences among items.

Example in this book: Figure 1.2

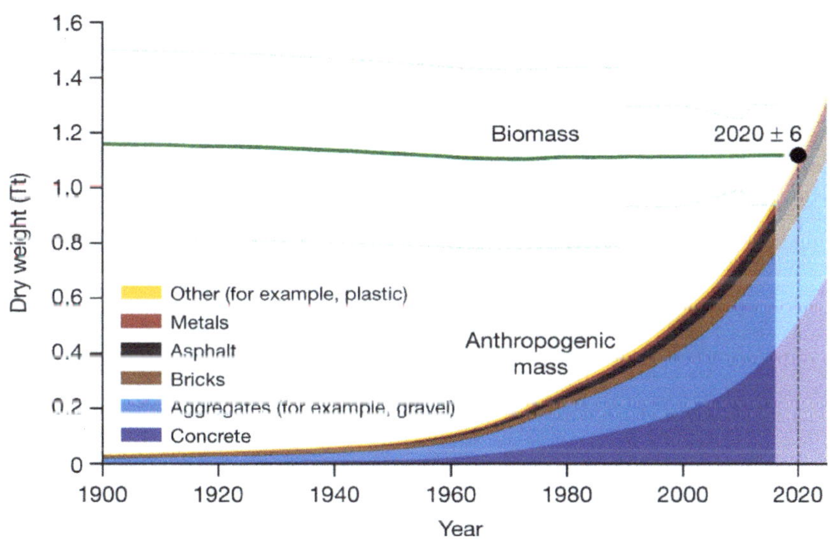

Ashby diagram A depiction of the two-dimensional domain occupied by different material groups.

Example in this book: Figure 24.1

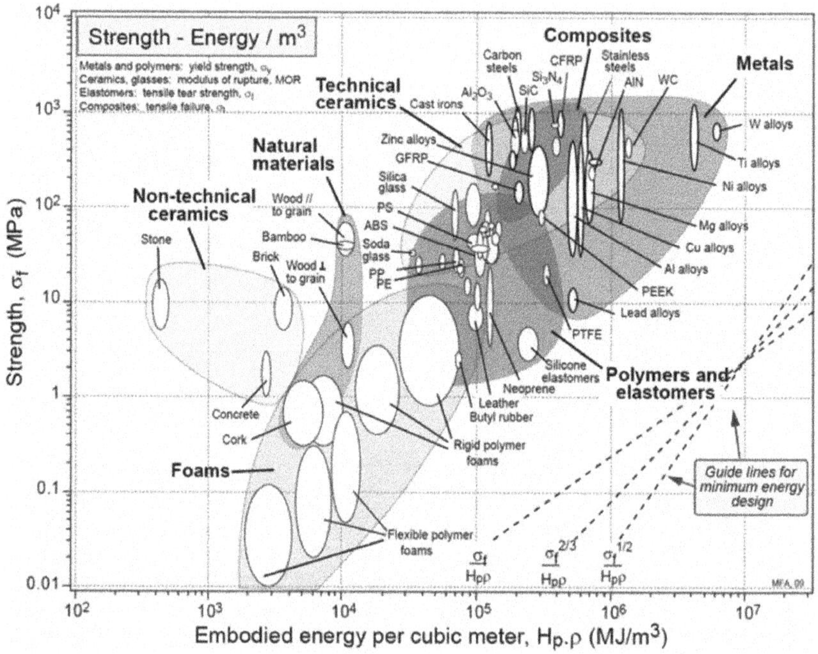

Bar chart A display that presents categorial data using bars with heights that are proportional to the values to which they relate.

Example in this book: Figure 33.3(b)

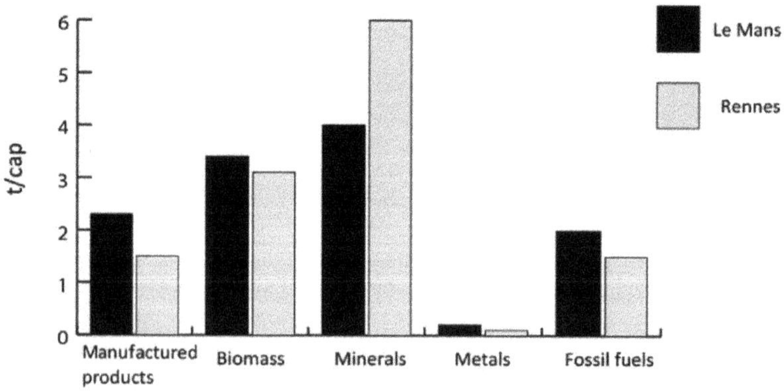

Bubble chart A display in which the points locate a variable in three dimensions: the *x* and *y* coordinate values and a third value expressed as the diameter of the bubble.

Example in this book: Figure 36.4

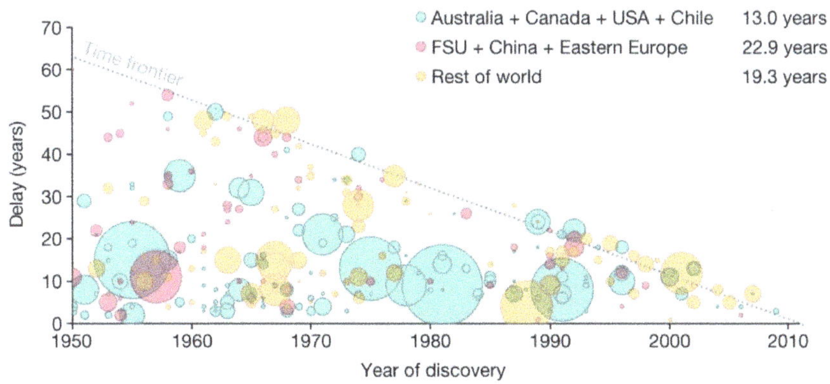

Cartogram A map in which distortions convey information on an alternative variable.

Example in this book: Figure 16.1

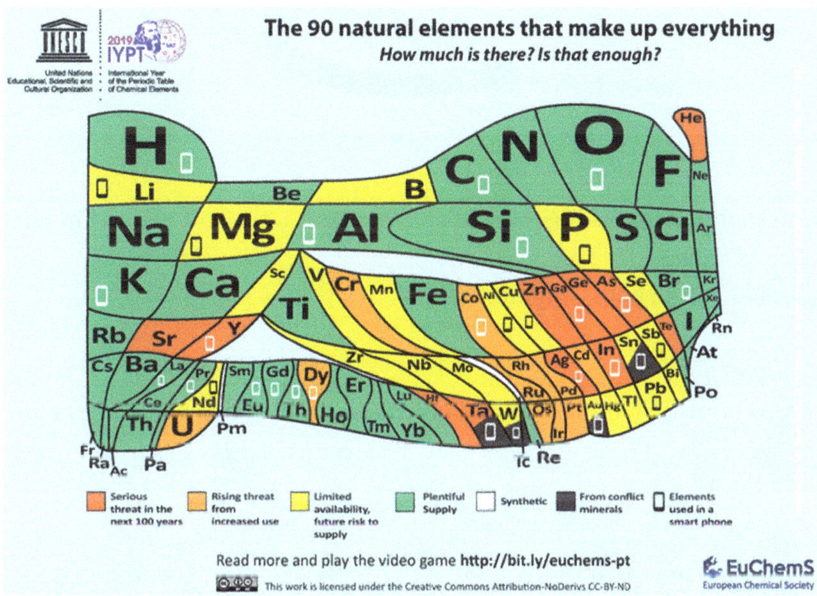

Chord diagram A diagram that uses bundling of adjacent edges that connect entities organized in a hierarchy; this approach decreases clutter and improves the perception of significant relationships.

Example in this book: Figure 21.4

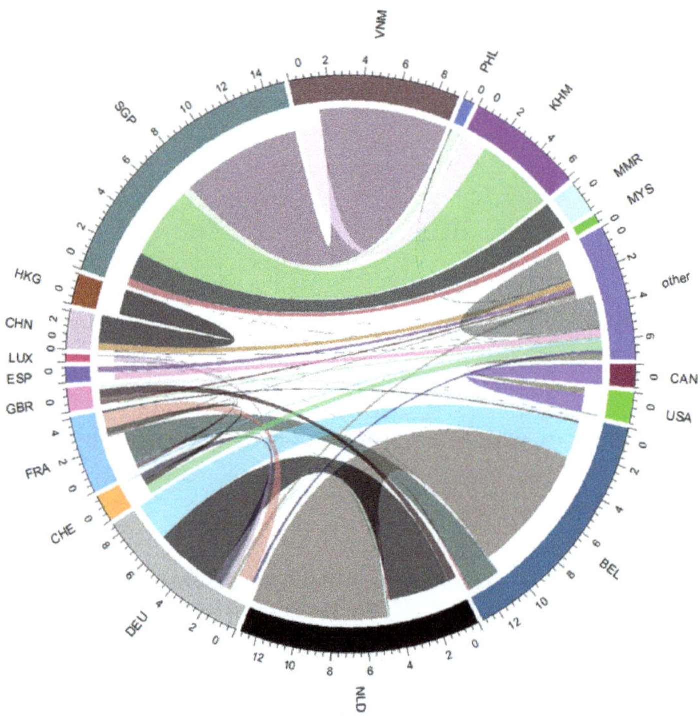

Choropleth diagram A map that displays regimes that are colored in relation to a numeric variable.

Example in this book: Figure 17.1(a)

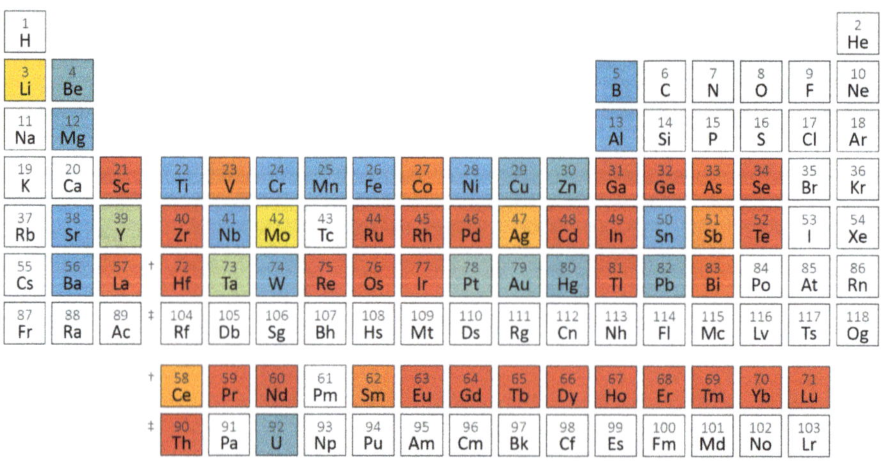

Data overlay map A map upon which data are superimposed.

Example in this book: Figure 27.7

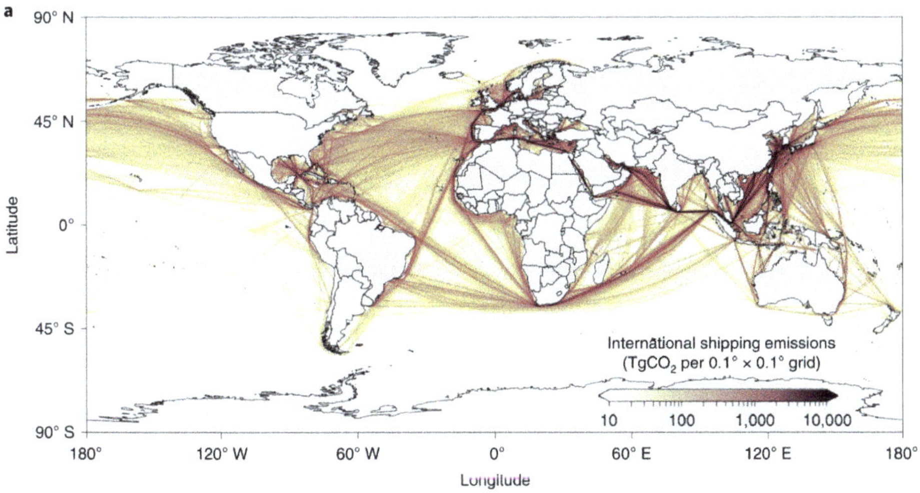

Diagram map A representation of an event distribution on a surface by means of symbols representing individual nodes or values.

Example in this book: Figure 33.5

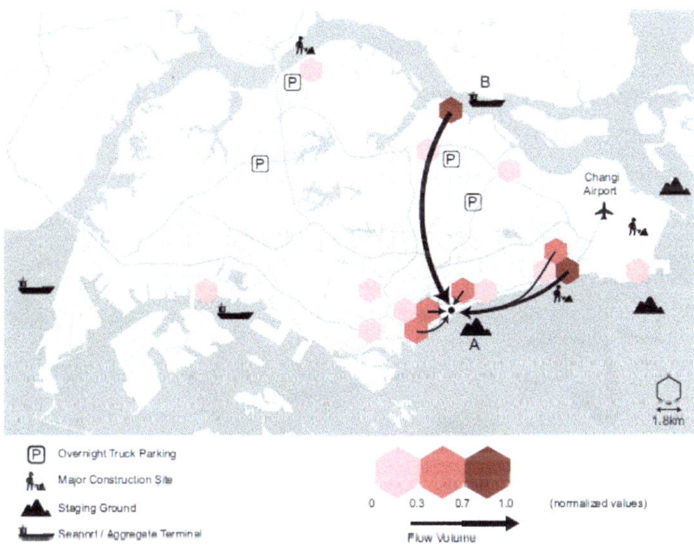

Exploded diagram A display that shows a component or components of an object or diagram suspended in surrounding space. This approach allows simultaneous viewing of the entire data set plus a selected portion of it or an entire product design.

Examples in this book: Figure 9.1(b)

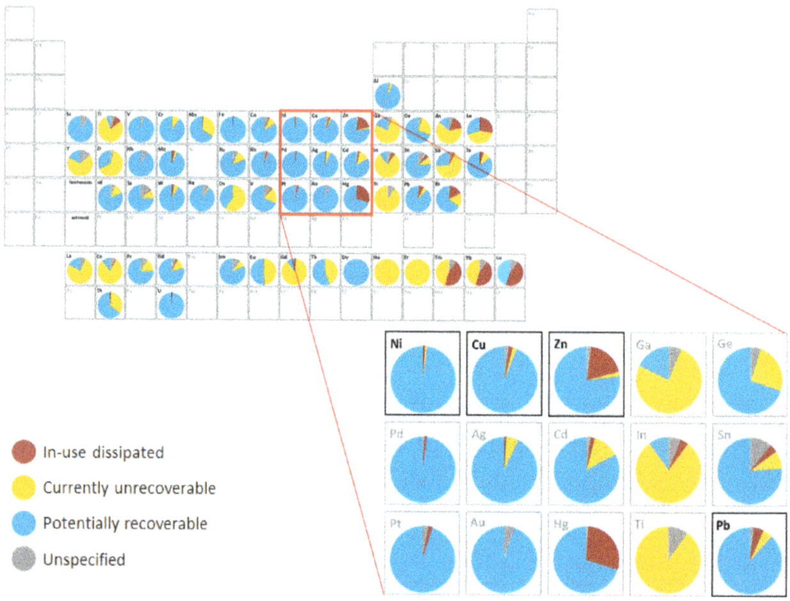

Flow map A map that illustrates the flows of a variable from one or more locations to target destinations.

Example in this book: Figure 27.3

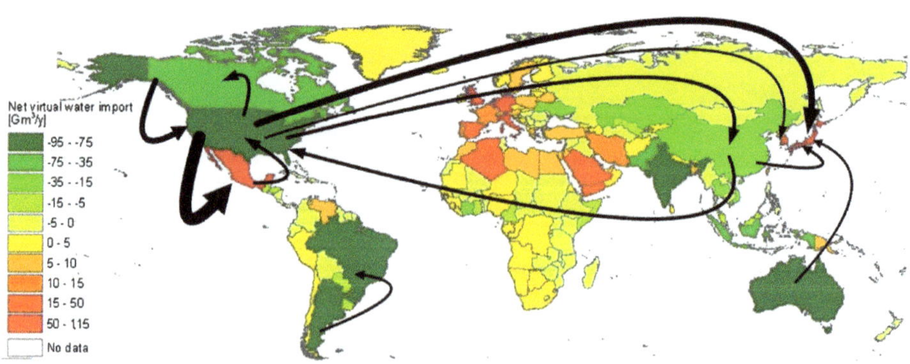

Heat map This display illustrates the state or magnitude of a phenomenon in two or more dimensions, using color hue or intensity.

Example in this book: Figure 17.1(b)

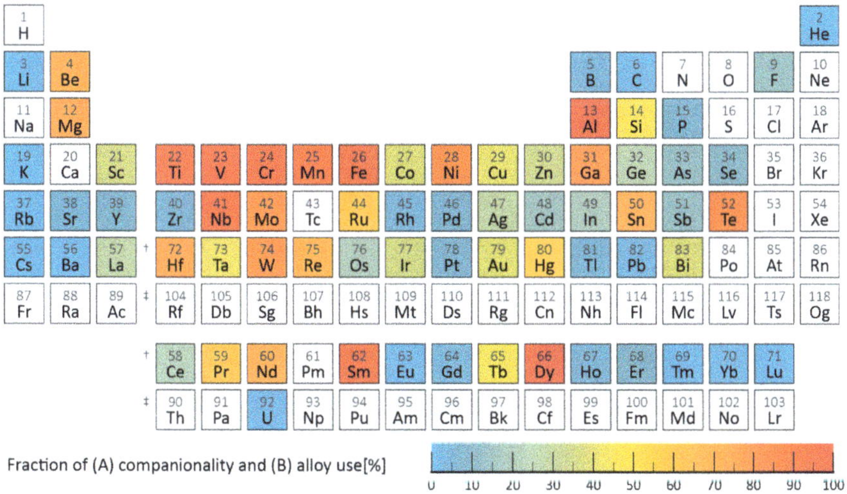

Impact trajectory diagram A depiction of the results of a future scenario calculation, often preceded by a line chart of historic data.

Example in this book: Figure 34.5

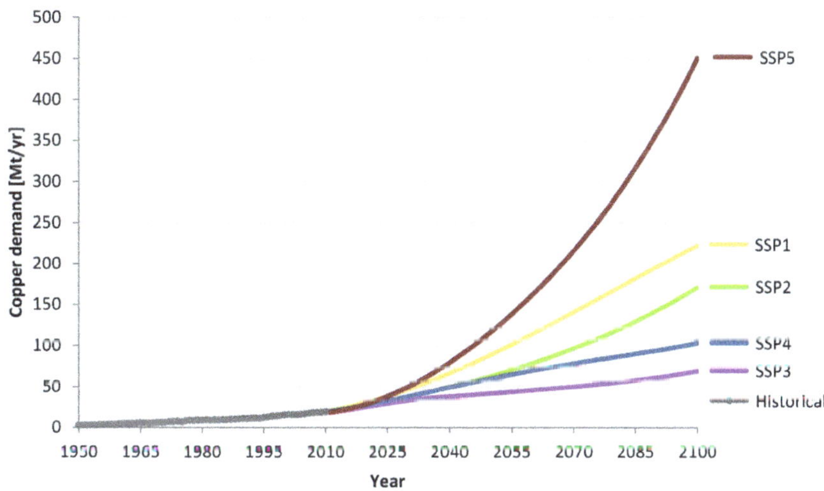

Line chart A diagram that displays information as a series of data points connected by line segments.

Example in this book: Figure 7.6(b)

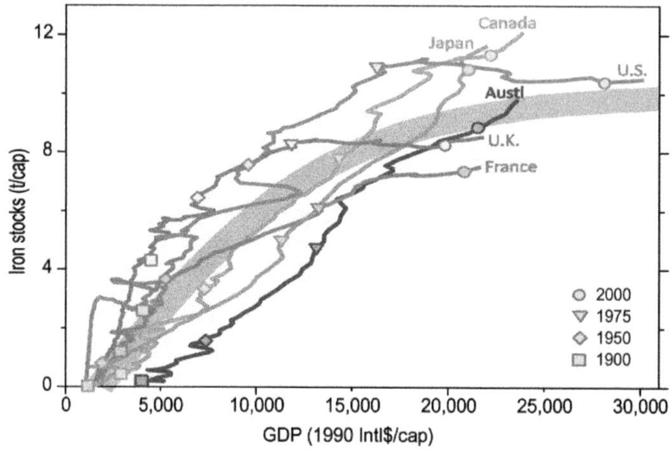

Network diagram A visual representation of a network — telecom, computer, chemical reactions, etc. The diagram may or may not imply control of one network element over others.

Example in this book: Figure 9.5

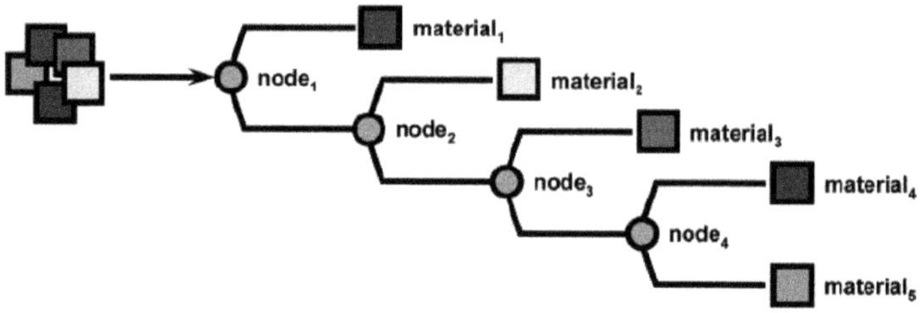

Normalized area chart An area chart that has been normalized to as to illustrate relative fractional contributions by different sectors.

Example in this book: Figure 7.4(b)

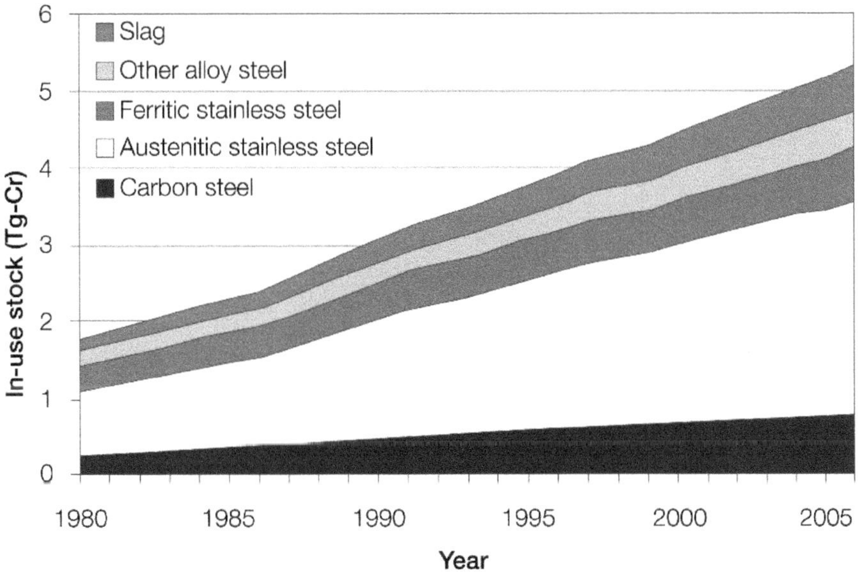

Pie chart A graph in which a circle is divided into sectors that represent portions of the whole.

Example in this book: Figure 21.2

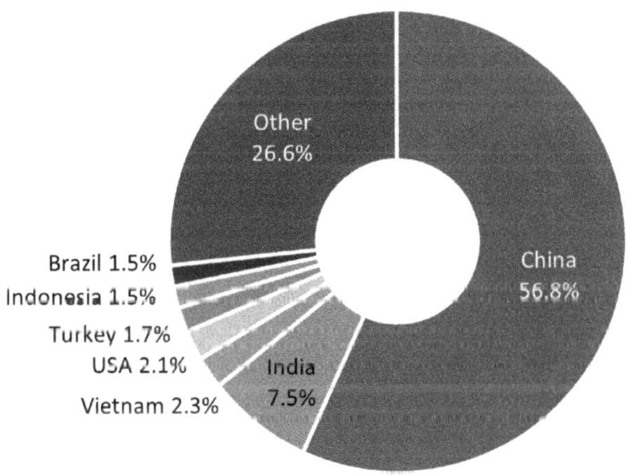

Pyramid diagram A chart used to visualize hierarchical structures.

Example in this book: Figure 29.3

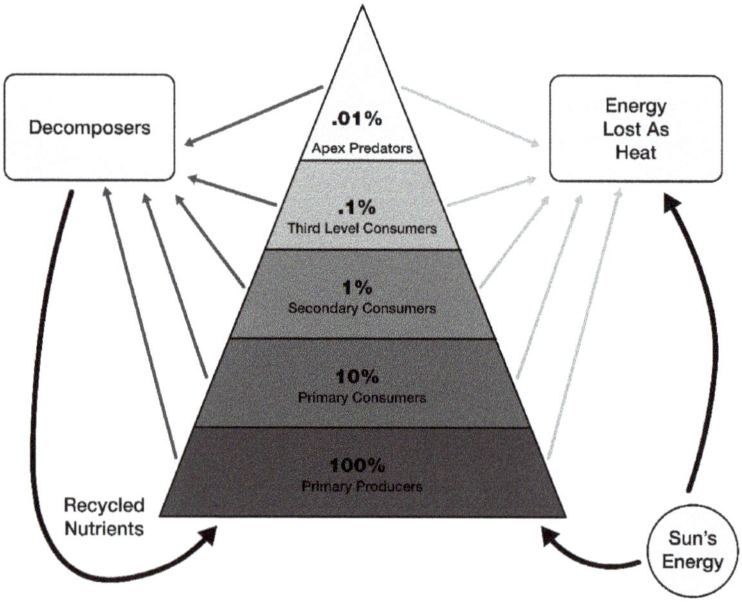

Reck diagram A display that combines a network diagram and a Sankey diagram with the concept of a circular economy.

Example in this book: Figure 8.4

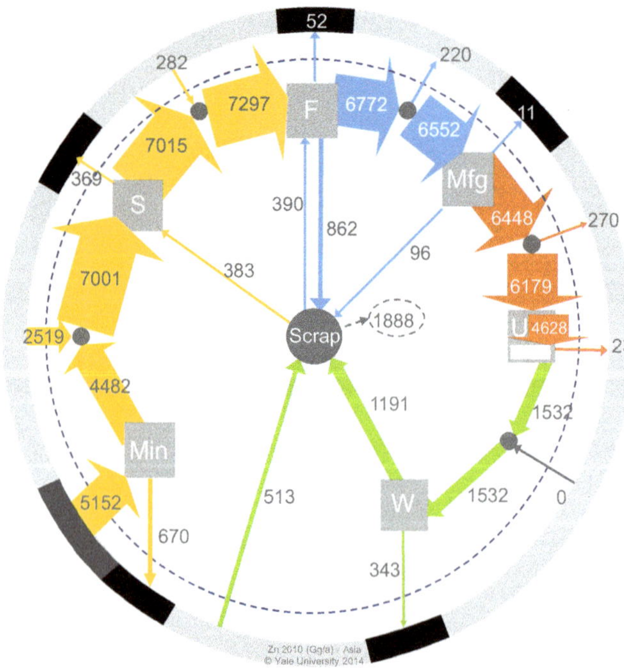

Ring chart A multilevel pie chart that uses concentric circles to visualize relational data.

Example in this book: Figure 18.7

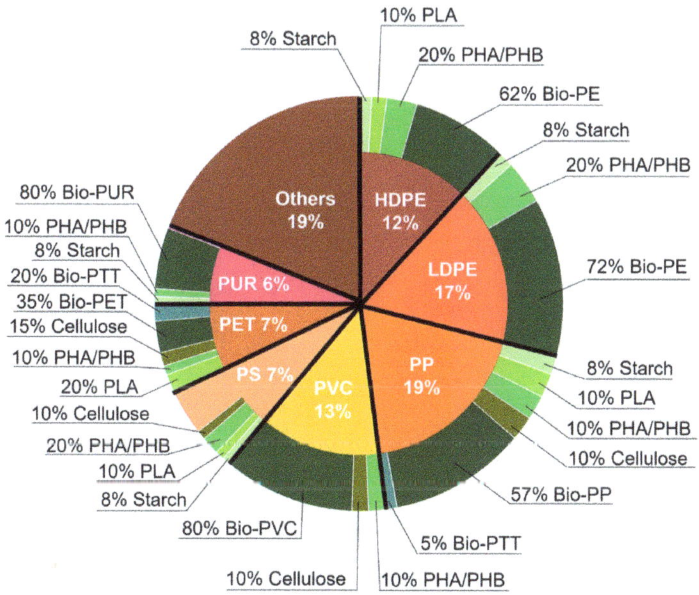

Sankey diagram A network diagram in which flow is depicted by the width of the link from one stage to the next. Different colors are generally used to distinguish different types of flows.

Example in this book: Figure 8.7

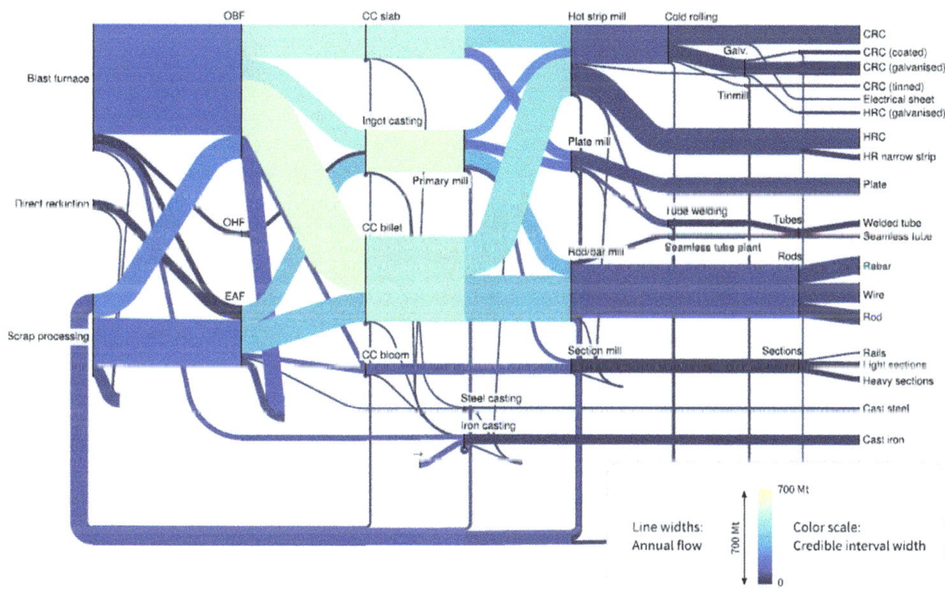

Scatterplot A widely used diagram that illustrates the relationship between data for two numerical variables, often with a line or lines that provide statistical fits to the data.

Example in this book: Figure 7.2

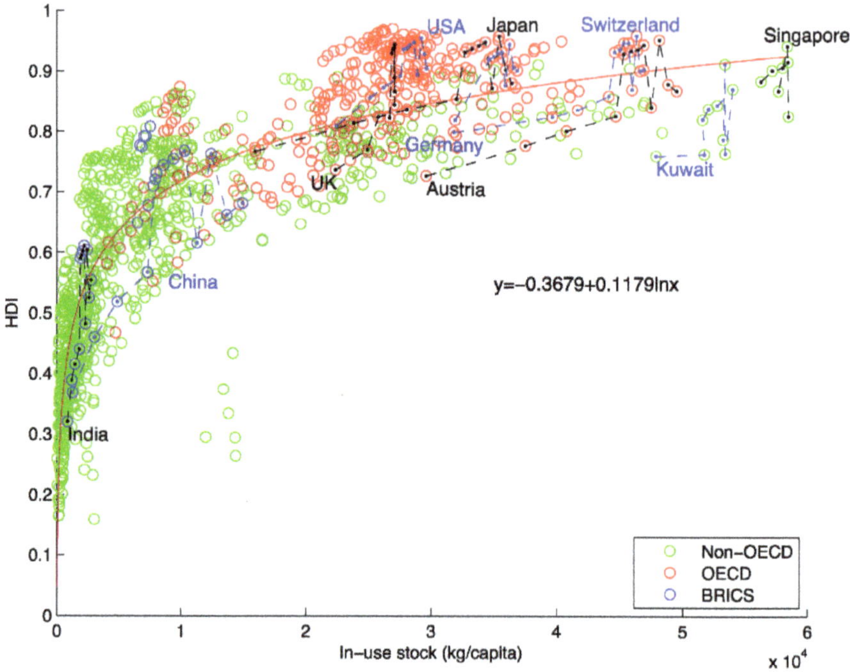

Segmented bar graph A stacked bar chart in which each bar will show 100% of the described value.

Example in this book: Figure 32.1(b)

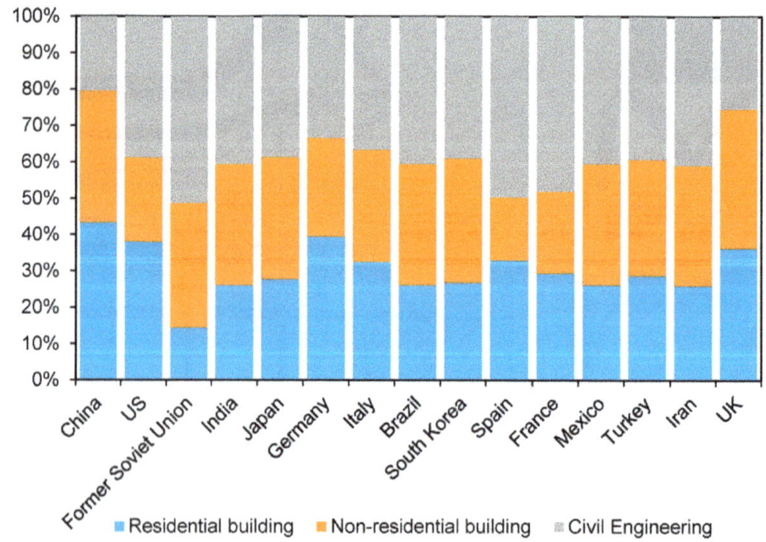

Sequence diagram A type of network diagram that shows object or activity interactions arranged in a temporal or structural sequence.

Example in this book: Figure 26.1

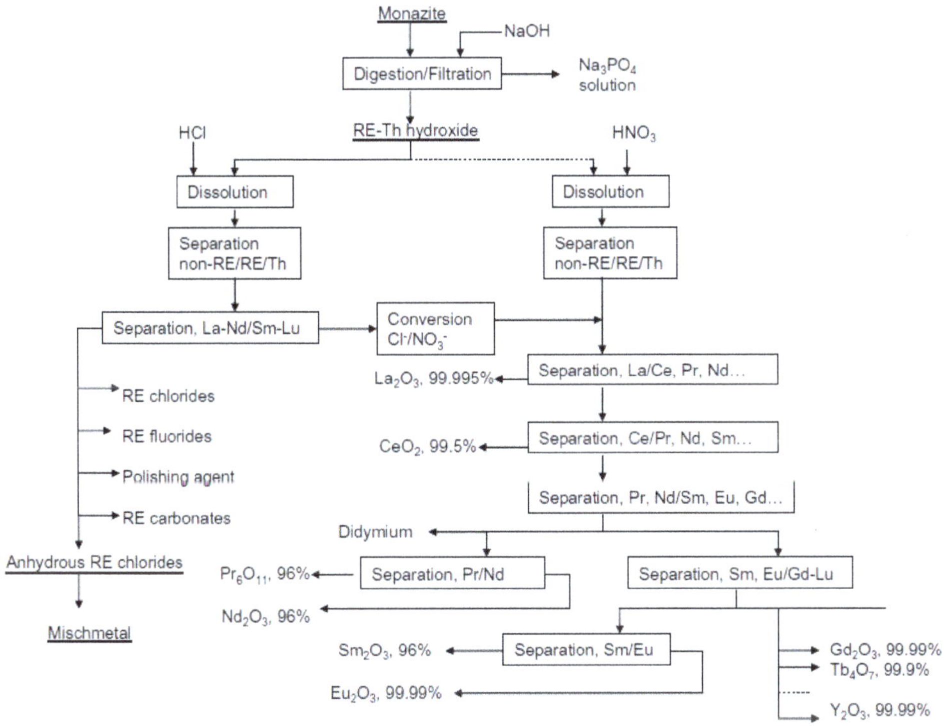

Stacked area chart A stacked bar chart that illustrates how a larger category is divided into smaller categories.

Example in this book: Figure 19.4

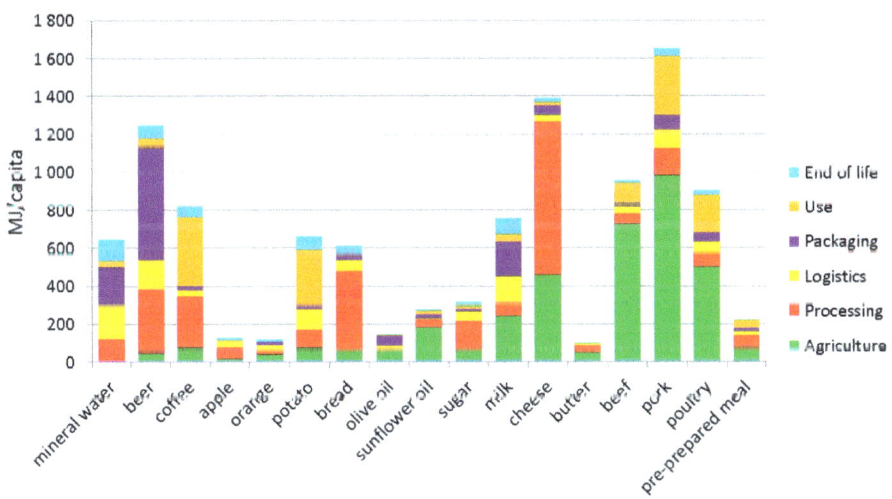

Tree chart A quantitative version of a flow diagram that depicts a central item that branches until the extent of information associated with the central item is exhausted. This chart is a quantitative version of an exploded diagram.

Example in this book: Figures 7.1(a) and 7.1(b)

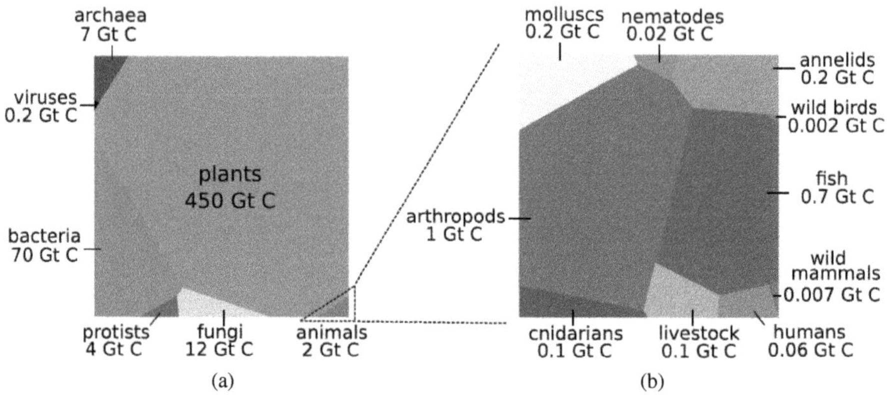

Venn diagram A display used to picture sets and the relationships between them. Circles and their interiors usually represent the sets, but almost any shape can be used.

Example in this book: Figure 29.5

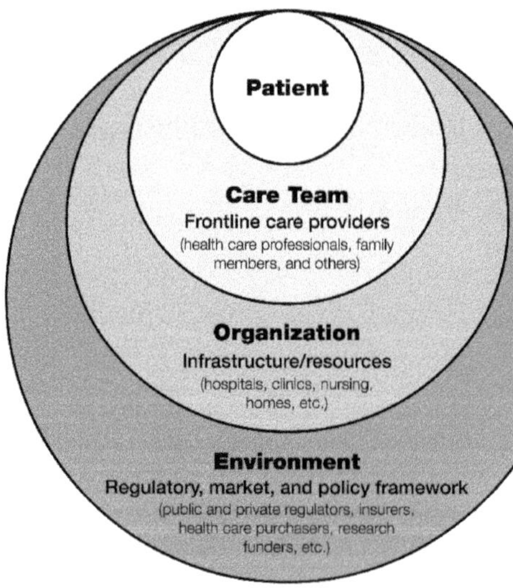

Depiction A drawing intended to convey qualitative information.
Example in this book: Figure 34.2

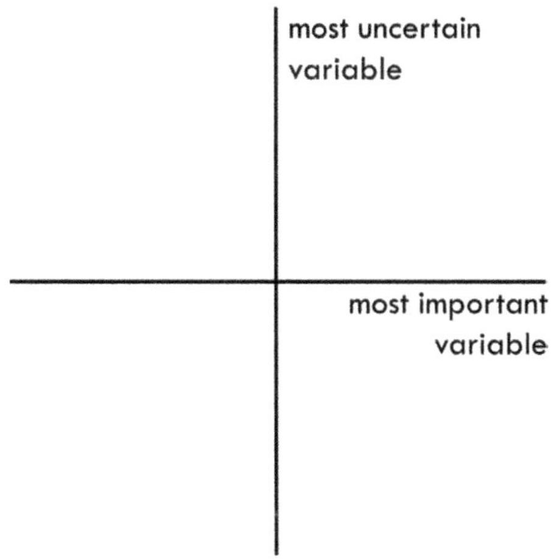

Glossary

A

Adaptive management — Management of complex socio-technical systems that recognizes the importance of technological information, the inadequacy of predictive activity, and the need to act nonetheless to guide systems to preferred outcomes over time.
Aggregate — Crushed stones, typically 1 cm to 3 cm in diameter, employed in concrete and for roadbed construction.
Alloy — A material made of more than one metallic element.
Anthropogenic — Derived from human activities.
Asphalt — A combination of small stones, mineral dust, and a few percent of bitumen (the lowest boiling fraction resulting from the refining of crude oil).
Atom economy — A metric in the field of green chemistry used to measure the percent of atoms that are input into a chemical synthesis that actually end up in the final product.

B

Bioaccumulation — The concentration of a substance by an organism above the levels at which that substance is present in the ambient environment. Examples include some forms of heavy metals and chlorinated pesticides such as dichlorodiphenyltrichloroethane (DDT).
Biomagnification — The increasing concentration of a substance as it passes into higher trophic levels of a food web. Many substances which are bioaccumulated are also biomagnified.
Blue water — A metric used in water footprinting to measure water used for irrigation.

Budget — A balance sheet of the magnitudes of all of the sources and sinks for a particular species or group of species in a single reservoir.
Burden shifting — A phenomenon in environmental management where, in the process of reducing one type of environmental burden (such as emissions or waste), another type of environmental burden is exacerbated. Burden shifting can be direct (e.g., air pollution is captured and becomes a solid waste to be disposed) or indirect (e.g., use phase emissions are decreased at the expense of upstream manufacturing emissions).
Byproduct — A useful product that is not the primary product being produced. In life cycle analysis (LCA), byproducts are treated as co-products.

C

Carcinogen — A substance that causes cancer.
Cement — A binder used to adhere materials together. Portland cement is the most common type used in the production of concrete and is made from limestone, clay minerals, and gypsum.
Circular economy — A set of principles of closed-loop design of products and the general economy, in order to minimize the generation of waste and the need for new materials.
Characterization — The process of quantitatively determining the impact resulting from the stress indicated by LCA inventory values.
Classification — The process of assigning raw LCA data on flows of materials and energy to particular environmental concerns.
Closed-loop recycling — A recycling system in which a particular mass of material is remanufactured into the same product (e.g., glass bottle into glass bottle). Also known as "horizontal recycling".
Commensalism — A type of symbiosis in which one organism benefits from, and is often completely dependent on, the other for food, shelter, or locomotion.
Commons — A geographical region or entity under common ownership or without ownership (e.g., a village green, the oceans, or the atmosphere).
Companionality — A metric used in resource management and geology that measures the degree to which a resource naturally occurs with another resource, such as multiple metallic elements within a single deposit.
Composite material — A material made of more than one substance.
Concrete — A composite material made of cement, fine aggregate such as sand, and coarse aggregate such as gravel.
Consumption — The organism-induced transformation of materials and energy.

Coproduct — A marketable byproduct from a process. This includes materials that may be traditionally defined as waste such as industrial scrap that is subsequently used as a raw material in a different manufacturing process.
Core — A term used in remanufacturing denoting a previously sold, worn, or non-functional product or part, intended for the remanufacturing process.
Cultural construct — A social idea or approach that has become absolute and unquestioned.
Cycle — A system consisting of two or more connected reservoirs, where a large part of the material of interest is transferred through the system in a cyclic manner.

D

Decoupling — Breaking the link between economic growth and the concomitant growth of environmental degradation and/or use of resources.
Dematerialization — An absolute or relative reduction in the use of materials per unit of value added or output.
Depletion time — The time required to exhaust a resource if the present rate of use remains unchanged.
Design for environment — An engineering perspective in which the environmentally related characteristics of a product, process, or facility design are optimized.
Directed network — A network in which flows occur in one direction.
Disposal — Discarding of materials or products at the end of their useful life without making provision for recycling or reuse.
Dissipative product — A product that is irretrievably dispersed when it is used (e.g., paint, fertilizer).
Downstream — In life cycle assessment, processes or emissions that occur after the life cycle stage under consideration.

E

Eco-industrial parks — An area of concentrated industrial activity in which different industries share resources including material byproducts, energy, water, services, or infrastructure.
Ecology (biological) — The study of the distribution and abundance of organisms and their interactions with the physical world.
Economic flow — In life cycle assessment, a flow into or out of a unit process that is connected to another unit process.

Edges — Components of networks or graphs that connect two nodes. In industrial ecology, edges are typically used to represent flows in network models.
Elemental flow analysis — An analysis of the flows of a specific element within and across the boundaries of a particular geographical region.
Embodied energy — The energy employed to bring a particular material or product from its initial physical reservoir or reservoirs to a specific physical state.
Embodied material emissions — The carbon dioxide and greenhouse gas equivalent emissions involved in bringing a particular material or product from its initial physical reservoir or reservoirs to a specific physical state.
Emergent behavior — The behavior of a system at a particular holonic level which is impossible to predict from detailed knowledge of adjacent holonic levels.
Emissions — Flows of chemical substances to an environmental compartment (e.g., air, water, soil) from natural or human activities.
Emissions inventory — A quantitative assessment of the types, magnitudes, and receptors of losses from a manufacturing process, a facility, a corporation, or a geographical entity.
Endpoint — In life cycle assessment, an indicator of environmental impacts that is measured in terms of damages or costs.
Energy audit — An accounting of input flows, output flows, and losses of energy within an industrial process, a facility, a corporation, or a geographical entity.
Enterprise resource planning (ERP) — The utilization of software to link the activities of a corporation with the associated requirements for labor, materials, financial systems, management, reporting, etc.
Environmental flow — In life cycle assessment, a flow into or out of a unit process that is not connected to another unit process, and instead comes from or goes directly to nature.
Enzyme — A molecule that acts as a catalyst in a biological chemical reaction.
Evolvability — The rate of evolution in a biological or industrial organism.
Exergy — The available energy in a system. Physically speaking, the work that could be extracted by taking a system with the same temperature and pressure as the environment reversibly to the same chemical composition as in the environment.
Exposure — Contact between a hazard and the target of concern, which may be an organ, an individual, a population, a biological community, or some other system. The confluence of exposure and hazard gives rise to risk.
Externality — A cost that is not captured within the economic system and thus is not reflected in prices.

F

Facultative — A type of symbiosis in which the two species engage in a symbiotic partnership through choice but can survive individually.
Final sink — In environmental chemistry, a permanent location or environmental compartment for substances leaving a reservoir or for material flows leaving the economy.
Flux — The rate of emission, absorption, or deposition of a substance from one reservoir to another. Often expressed as the rate per unit area of surface.
Food chain — A sequence in which resources flow in a linear fashion from one trophic level to the next.
Food web — A pattern in which resources flow largely from one trophic level to the next, but may also flow across trophic levels in a non-linear fashion.
Fossil fuel — A general term for combustible geological deposits of carbon in reduced (organic) form and of biological origin, including coal, oil, natural gas, oil shales, and tar sands.
Fugitive emissions — Emissions from valves or leaks in process equipment or material storage areas that are difficult to measure and do not flow through pollution control devices.
Full-cost accounting — An accounting system in which environmental costs are built directly into the prices of products and services.
Functional unit — The quantified function that a product or service will deliver, and that will serve as a basis for comparing products, services, or product or service scenarios.

G

Gangue — The non-metalliferous or worthless materials contained in ore.
Global warming — The theory that elevated concentrations of certain anthropogenic atmospheric constituents are causing or will cause an increase in Earth's average temperature.
Goal and scope — The first phase of life cycle assessment, in which the objectives and stakeholders are clearly identified, and the scope and methodological approach are defined.
Gray water — A metric used in water footprinting to measure the theoretical freshwater required for dilution of polluted water in order to meet water quality regulations.

Green accounting — An informal term referring to management accounting systems that specifically delineate the environmental costs of business activities rather than include those costs in overhead accounts.
Green chemistry — Employing chemical techniques and methodologies that reduce or eliminate the use or generation of feedstocks, products, byproducts, solvents, and reagents that are hazardous to the environment.
Green engineering — The design and implementation of engineering solutions that take environmental issues into account throughout the life cycle of the design.
Green water — A metric used in water footprinting to measure rainwater used.
Greenhouse effect — The trapping by atmospheric gases of outgoing infrared energy emitted by Earth. Part of the radiation absorbed by the atmosphere is returned to Earth's surface, causing it to warm.
Greenhouse gas — A gas with absorption bands in the infrared portion of the spectrum that leads to the greenhouse effect. The principal anthropogenic greenhouse gases are carbon dioxide (CO_2), methane (CH_4), nitrous oxide (N_2O), chlorofluorocarbons (CFCs), hydrofluorocarbons (HFCs), and perfluorocarbons such as sulfur hexafluoride (SF_6). Water vapor (H_2O) and ozone (O_3) are also greenhouse gases.

H

Heuristics — The intuitive rules used by human experts in arriving at decisions. In expert systems, representations of heuristics are combined with algorithms (precise definitions and procedures) to reach conclusions concerning problems whose characteristics cannot be completely specified.
Hidden flow — The indirect flows of materials such as resources, pollution, or waste that occur upstream in a production process but that are not physically embodied in the product itself.
Hitchhiker resource — A resource acquired principally as a byproduct of the mining of other resources.
Holarchy — A network of holons.
Holon — An individual entity in a system of systems.
Home scrap — The waste produced within a fabricating plant, such as rejected material, trimmings, and shavings. Home scrap is recirculated within the fabricating plant and does not become external waste.
Horizontal recycling — See **Closed-loop recycling**.

I

Impact assessment — The third phase of life cycle assessment, in which the flows described by the life cycle inventory are linked to various categories of environmental impacts. An impact assessment method is a formal collection of models that cover a specified set of impact categories.
Impact category — In life cycle assessment, a type of impact that is included in a life cycle impact assessment, for example, eutrophication or stratospheric ozone depletion.
Interpretation — The fourth stage of life cycle assessment, in which results are discussed in order to make recommendations for improving the environmental plan of a process, product, or facility.
Indicator — A non-quantitative measure of the status of a chosen parameter, environmental or otherwise.
Industrial ecology — An approach to the design of industrial products and processes that evaluates such activities through the dual perspectives of product competitiveness and environmental interactions.
Industrial enzyme — An industrial process or piece of equipment that results in a transformation of materials or energy.
Industrial symbiosis — See **Symbiosis**.
Infrastructure — The basic facilities, equipment, and installations needed for the functioning of a community or industrial operation, often characterized by the service it provides (e.g., water infrastructure and energy infrastructure).
In-use product stock — The physical quantity of products in 'active use', including those being repaired for return to the original user.
Inventory analysis — The second stage of life cycle assessment, in which the system-wide inputs and outputs of materials and energy are determined for a process, product, or facility.
Inviolate's list — A list of design decisions never allowed to be taken by product or process designers.

L

Leachate — The solution that is produced by the action of percolating water through a permeable solid, as in a landfill.
Level — A unit of analysis located in a particular position on a scale.
Life cycle — The stages of a product, process, or package's life, beginning with raw materials acquisition, continuing through processing, materials manufacture,

product fabrication, and use, and concluding with any of a variety of waste management options.
Life cycle assessment — A concept and a methodology to evaluate the environmental effects of a product or activity holistically, by analyzing the entire life cycle of a particular material, process, product, technology, service, or activity.
Lock-in — In which a particular process or product standard becomes embedded in product design, manufacturing, or consumption patterns, thus stifling innovation. A common example is the notably inefficient QWERTY keyboard.

M

Markov chain — A stochastic model describing a sequence of possible transitions in which the probability of each transition occurring depends only on the state attained in the previous transition.
Mass balance — A fundamental principle in industrial ecology and all of science and engineering that uses the physical conservation of mass as a means for quantifying inputs, outputs, and changes to stocks in a system.
Material efficiency — A set of product engineering and design approaches that aims to reduce the amount of material needed in products without compromising their function.
Material flow analysis — An analysis of the flows of materials within and across the boundaries of a particular geographical region.
Material flux (material exchange rate) — The physical quantity of material flowing between nodes of a system.
Material requirements planning (MRP) — The utilization of software to link industrial production schedules with the associated requirements for parts, components, and materials, as well as the status of current inventories.
Metabolic analysis — The analysis of the aggregate of physical and chemical processes taking place in an organism, biological or industrial.
Metabolism — The physical and chemical processes taking place in an organism, biological or industrial.
Metabolite — An intermediate product of chemical transformations within an organism.
Metric — A quantitative measure of the status of a chosen parameter, environmental or otherwise.
Midpoint — In life cycle assessment, an indicator of environmental impacts that is measured in terms of physical changes in the environment.

Mineral — A distinguishable solid phase that has a specific chemical composition, e.g., quartz (SiO_2) or magnetite (Fe_3O_4).
Model system — A system used for research purposes that has been extensively characterized and/or can be somehow controlled, for the purpose of learning about specific phenomena.
Module — A self-contained unit or item, such as an assembly or segment of a larger product, which itself performs a defined task and can be linked with other such units to form a larger system.
Mutualism — A type of symbiosis in which the two species engage in a mutually beneficial symbiotic partnership.

N

Negative feedback — A feature of a dynamic system that decreases the rate of change of the system.
Network analysis — A field of study that applies graph mathematics techniques to determine various features of a network, or to characterize the relationships among discrete nodes of a network.
New scrap — See **Prompt scrap**.
Nodes — Components of networks or graphs that are connected by edges. In industrial ecology, nodes can have a variety of representations, including individuals, processes, facilities, sectors, or countries.
Non-point source — See **Source**.
Normalization — In life cycle assessment, the process of relating environmental impact values derived at the Characterization step to reference values in order to arrive at common indicator values.

O

Obligate — A type of symbiosis in which the two species engage in a symbiotic partnership without which neither could survive.
Old scrap — See **Postconsumer solid waste**.
Omnivory — The acquisition of resources from organisms at several different trophic levels.
Open-loop recycling — A recycling system in which a product from one type of material is recycled into a different type of product (e.g., plastic bottles into fence posts). The product receiving recycled material itself may or may not be recycled. Also known as "cascade recycling".

Ore — A natural rock assemblage containing an economically valuable resource.
Organism — An entity internally organized to maintain vital activities.
Original equipment manufacturer (OEM) — The manufacturer of original parts or equipment, including items manufactured, assembled, and installed during construction of a new product.
Overburden — The material to be removed or displaced that is overlying the ore or material to be mined.

P

Packaging, primary — The level of packaging that is in contact with the product. For certain beverages, an example is the aluminum can.
Packaging, secondary — The second level of packaging for a product that contains one or more primary packages. An example is the plastic rings that hold several beverage cans together.
Packaging, tertiary — The third level of packaging for a product that contains one or more secondary packages. An example is the stretch wrap over the pallet used to transport packs of beverage cans.
Pathway — The sequence of chemical reactions that connects a particular starting material with the final material that is produced.
Plastic — A category of materials primarily derived from polymers that can be molded, extruded, or pressed into a variety of shapes.
Plating — The act of coating a surface with a thin layer of metal.
Point source — See **Source**.
Pollution prevention — The design or operation of a process or item of equipment so as to minimize environmental impacts.
Positive feedback — A feature of a dynamic system that increases the rate of change of the system.
Postconsumer scrap — See **Postconsumer solid waste**.
Postconsumer solid waste — A material that has served its intended use and has become a part of the waste stream; also called **Old scrap** or **Postconsumer scrap**.
Process mass intensity (PMI) — A metric in the field of green chemistry used to measure the mass of inputs required to synthesize a unit mass of final product.
Prompt scrap — Waste produced by users of semifinished products (turnings, trimmings, etc.), also called **New scrap**. This scrap must generally be returned to the materials processor if it is to be recycled.

R

Resource-resource mutualisms — A type of mutualism in symbiosis in which the two species exchange resources such as nutrients.
Rebound effect — A consumption decision that appears beneficial, but which in practice stimulates behavior that nullifies the benefits of the original decision.
Recycling — The reclamation and reuse of output or discard material streams useful for application in products.
Refurbishment — Modification of an object to increase or restore performance and/or functionality or to meet applicable technical standards or regulatory requirements, with the goal of arriving at a fully functional product.
Remanufacture — The process of bringing large amounts of similar products together for purposes of disassembly, evaluation, renovation, and reuse.
Repair — Addressing a specified fault or faults and/or replacing defective components in order to enable the object to be used for its originally intended purpose.
Reserves — The total known amount of a resource that can be mined with today's technology at today's market prices.
Reserve base — The total known amount of a resource that can be mined, without regard for technology or market prices.
Reservoir — A receptacle defined by characteristic physical, chemical, or biological properties that are relatively uniformly distributed.
Residence time — The average time spent in a reservoir by a specific material.
Resolution — The precision used in measurement.
Resource productivity — The efficiency with which an economy or a production process uses material resources. It is often expressed as a ratio of materials used per monetary unit of production.
Reuse — Re-employing materials and products in the same use without the necessity for recycling or remanufacture.
Reverse logistics — Activities involved in recapturing the value of products, parts, and materials once they have reached end-of-use or end-of-life.
Risk — The combination of exposure and hazard; a statistical concept reflecting the probability that an undesirable outcome will result from specified conditions (such as exposure to a certain substance for a certain time at a certain concentration).
Risk assessment — An evaluation of potential consequences to humans, wildlife, or the environment caused by a process, product, or activity, including both the likelihood and the effects of an event.

S

Scale — A spatial, temporal, quantitative, or analytical dimension used to measure and study an object or process of interest.
Scenario — An alternative vision of how the future might unfold.
Self-organizing holarchic open (SOHO) system — A conceptually or causally-linked system of objects or processes grouped along an analytical scale, but with a structure that is self-imposed.
Service-resource mutualisms — A type of mutualism in symbiosis in which one species provides resources to the other and receives some type of service in return.
Servicizing — Satisfying a need by selling a service rather than by selling physical products that meet that need.
Sink — In environmental chemistry, the process or origin from which a substance is lost from a reservoir.
Slag — The fused residue that results from the separation of metals from their ores during pyrometallurgy.
Social ecology — The study of societies and their evolutionary behavior from the perspective of societal use of energy and materials.
Solute — The substance dissolved in a solvent.
Solution — A mixture in which the components are uniformly distributed on an atomic or molecular scale. Although liquid, solid, and gaseous solutions exist, common nomenclature implies the liquid phase unless otherwise specified.
Solvent — A medium, usually liquid, in which other substances can be dissolved.
Source — In environmental chemistry, the process or origin from which a substance is injected into a reservoir. Point sources are those where an identifiable source, such as a smokestack, can be identified. Non-point sources are those resulting from diffuse emissions over a large geographical area, such as pesticides entering a river as runoff from a general agricultural area.
Standing stock — In ecology, the amount (by number, mass, or some other convenient metric) of a particular species or species group in a particular area.
Stainless steel — One of a number of steel alloys that is resistant to corrosion.
Stochastic processes — A process in which there is an element of randomness.
Stock — The contents of a reservoir.
Streamlined Life Cycle Assessment — A simplified methodology to evaluate the environmental impacts of a product or process by analyzing the most significant flows or environmental impacts in the life cycle.
Stressor — A set of conditions that may lead to an undesirable environmental impact.

Substance flow analysis — An analysis of the flows of a specific chemically identifiable substance within and across the boundaries of a particular geographical region. The word "substance" typically incorporates atoms and molecules of the entity of interest without regard for chemical form.
Succession — In ecology, a change or evolution of a complex system from one state to another.
Sustainable development goals (SDGs) — A set of 17 worldwide goals set forth by the United Nations in 2015 to guide global development toward sustainability, with a target of 2030.
Sustainable engineering — The design and implementation of engineering solutions that take environmental and sustainability issues into account throughout the life cycle of the design.
Sustainability — In the context of industrial ecology, sustainability is the state in which humans living on Earth are able to meet their needs over time while nurturing planetary life-support systems.
Symbiosis — A relationship within which at least two willing participants exchange materials, energy, or information in a mutually beneficial manner.

T

Tailings — The residue remaining after ore has been ground and the target metallic minerals separated and retained.
Telecoupling — The connection of human–nature interactions over a distance.
Textile — Fibrous material suitable for being spun or woven into fabric.
Trophic level — A group of organisms that perform similar resource exchanges as part of natural food chains or food webs.
Troposphere — The lowest layer of the atmosphere, ranging from the ground to the base of the stratosphere at 10 km to 15 km altitude, depending on latitude and weather conditions. About 85% of the mass of the atmosphere is in the troposphere, where most weather features occur. Because its temperature decreases with altitude, the troposphere is dynamically unstable.

U

Upgrade — The act of raising a product to a higher standard, with the objective of improving performance, efficiency, and/or functionality by adding or replacing components, including electronics and/or software.
Upstream — In life cycle assessment, processes or emissions that occur before the life cycle stage under consideration.

Urban metabolism — A sub-field of industrial ecology that applies material flow analysis techniques to understand resource flows within cities.
Urban mining — The recovery and recycling of resources from urban waste streams.

V

Valuation — In life cycle assessment, the process of assigning economic value or damages to impact assessment results.
Virtual water — In life cycle assessment and water footprinting, the amount of water required by a product or process.

W

Waste — Material thought to be of no practical value. One of the goals of industrial ecology is the reuse of resources, and hence the minimization of material regarded as waste.
Waste audit — An accounting of output flows and losses of wastes within an industrial process, a facility, a corporation, or a geographical entity.
Weighted network — A network in which the edges have values associated with them.
Weighting — In life cycle assessment (LCA), the process of assigning factors to different impact categories based on their perceived relative importance, for the purpose of simplifying decision-making.

Index

I

J

K

L

M

www.ingramcontent.com/pod-product-compliance
Ingram Content Group UK Ltd.
Pitfield, Milton Keynes, MK11 3LW, UK
UKHW052113170926
8624IPUK00026B/374

9 789811 277603